Lecture Notes in Mathematics 1563

Editors:
A. Dold, Heidelberg
B. Eckmann, Zürich
F. Takens, Groningen

Subseries: Fondazione C.I.M.E., Firenze
Adviser: Roberto Conti

E. Fabes M. Fukushima L. Gross
C. Kenig M. Röckner D.W. Stroock

Dirichlet Forms

Lectures given at the 1st Session of the Centro Internazionale Matematico Estivo (C.I.M.E.) held in Varenna, Italy, June 8-19, 1992

Editors: G. Dell'Antonio, U. Mosco

Springer-Verlag
Berlin Heidelberg New York
London Paris Tokyo
Hong Kong Barcelona
Budapest

Authors

Eugene Fabes
Department of Mathematics, University of Minnesota
Minneapolis, MN 55455, USA

Masatoshi Fukushima
Department of Mathematical Science, Faculty of Engineering Science
Osaka University, Toyonaka, Osaka, Japan

Leonard Gross
Department of Mathematics, Cornell University
Ithaca, NY 14853, USA

Carlos Kenig
Department of Mathematics, University of Chicago
Chicago, IL 60637, USA

Michael Röckner
Institut für Angewandte Mathematik, Universität Bonn
Wegelerstrasse 6, D-53115 Bonn, Germany

Daniel W. Stroock
M.I.T., Rm 2-272
Cambridge, MA 02139, USA

Editors

Gianfausto Dell'Antonio
Umberto Mosco
Dipartimento di Matematica, Università "La Sapienza"
Piazzale Aldo Moro, 5, I-00185 Roma, Italy

Mathematics Subject Classification (1991): Primary: 46-xx
Secondary: 31-xx, 35-xx, 60-xx

ISBN 3-540-57421-2 Springer-Verlag Berlin Heidelberg New York
ISBN 0-387-57421-2 Springer-Verlag New York Berlin Heidelberg

Typesetting: Camera-ready by author/editor
46/3140-543210 - Printed on acid-free paper

Preface

In recent years the theory of Dirichlet forms has witnessed some very important developments both in its theoretical foundations and in its applications, which have ranged from the theory of Stochastic Processes to Quantum Field Theory to the theory of highly inhomogeneous materials.

It was therefore felt that it was due time to have on this subject a CIME school, in which leading experts in the field would present both the basic theoretical aspects and some of the recent applications, pointing to areas of active research.

The school was organized in six courses, covering

- Foundations of the theory and connection with Potential Theory and Stochastic Processes (M. Fukushima, M. Roeckner)
- Regularity results and a-priori estimates for solutions of elliptic equations in general domains (E. Fabes, C. Kenig)
- Hypercontractivity of semigroups and relation with spectral properties (L. Gross)
- Logarithmic Sobolev inequalities and connections with Statistical Mechanics (D. Stroock)

In the afternoons of the last three days a workshop was held, with financial support from the CNR project "Irregular Variational Problems", on further applications of the theory of Dirichlet Forms, also in the noncommutative setting.

The seminars were given by

. L. Accardi (Roma II) Noncommutative Stochastic Processes and applications
. S. Albeverio (Bochum) Construction of infinite-dimensional processes
. M. Biroli (Pol. Milano) Asymptotic Dirichlet Forms
. F. Guerra (Roma I) Annealing and Cavitation in Spin Glass Models
. M. Lindsay (Nottingham) Noncommutative Semigroups
. S.R.S. Varadhan (Courant Institute) Hydrodynamic Limit and Entropy Estimates I, II
. Ma Zhiming (Academia Sinica) Nonsymmetric Dirichlet Forms

The lectures of the School were followed with constant and active participation by many young researchers, both from Italy and from abroad.

We believe that the School was successful in reaching its aims, and we wish to express here our appreciation to the speakers for the high quality of their lectures and for their availability for discussions during the School.

We would also like to thank prof. R. Conti and the CIME Scientific Committee for the invitation to organize the School, and prof. P. Zecca and the staff of the Centro Volta in Como for their very effective help.

Gianfausto Dell'Antonio
Umberto Mosco

TABLE OF CONTENTS

Gaussian Upper Bounds on Fundamental Solutions of Parabolic Equations; the Method of Nash

by

E.B. Fabes

Table of Contents

Introduction

In the past decade tremendous progress has been made in understanding the pointwise behavior of fundamental solutions of parabolic operators in divergence form. One can see this progress in reading the monograph of E.B. Davies [3]. The interest and progress continue to this day. (See, for example, the related recent work of Saloff-Coste [9],[10].)

Historically the first paper establishing Gaussian upper and lower bounds for the fundamental solution $\Gamma(x,t;y,s)$ (x and $y \in R^n$, $t > s > 0$) was D.G. Aronson's paper [1]. Here, in the nondegenerate case, Aronson proved there exists a constant $C > 0$ such that

$$C^{-1}(t-s)^{-n/2}e^{-C|x-y|^2/(t-s)} \le \Gamma(x,t;y,s) \le C(t-s)^{-n/2}e^{-C^{-1}|x-y|^2/(t-s)}$$

for all x, y, t, s with $t > s$. Ideas in this paper continue to influence recent work on estimating Γ in the case of degenerate parabolic operators with nonsmooth coefficients. (See [5]). Almost 10 years before Aronson's work the celebrated paper of John Nash appeared ([8]). In this work Nash established the Hölder continuity of weak solutions of (nondegenerate) parabolic equations in divergence form. Nash's method in fact concentrated on the fundamental solution of the parabolic operator, on establishing estimates for the fundamental solution that did not depend on the smoothness of the coefficients. From these apriori estimates came the Hölder continuity of solutions. Later in 1964 and 1967 Moser proved a Harnack inequality for nonnegative solutions and from the Harnack inequality obtained Nash's result on the Hölder continuity of weak solutions ([6],[7]). It was with the help of the Harnack inequality of Moser that Aronson established the Gaussian lower bound for the fundamental solution ([1]).

In 1985 D.W. Stroock and I returned to Nash's paper and found that his original ideas could be used to obtain a Gaussian lower bound on the fundamental solution ([4). Our proof relied on some apriori spatial decay at infinity of the fundamental solution. A Gaussian upper bound certainly sufficed and such decay had been established by D.G. Aronson in [1] and by E.B. Davies in [3, p. 89] without the need of Harnack's inequality. Hence a proof of upper and lower bounds for the fundamental solution in terms of Gaussians could be achieved without Harnack's inequality and, indeed, such estimates on the fundamental solution could be used to prove Moser's Harnack inequality. (See [4].)

The methods to obtain the upper bound estimates for the fundamental solution developed by Aronson and Davies are powerful and can be used to obtain upper bounds of fundamental solutions associated with certain degenerate parabolic operators. (See [3],[5].) The direct relationship of these methods with those of Nash is not clear. Certainly the initial estimate of Nash on the fundamental solution Γ, namely,

$$\Gamma(x,t;y,s) \leq C(t-s)^{-n/2}$$

plays an important role in both methods; but the direct connection stops here and the methods diverge into interesting new directions. Of particular note is that neither method pays attention to what Nash calls "The moment bound" and refers to it as "essential to all subsequent parts of this paper."

In Part I, will review Nash's proof of the moment bound for the fundamental solution and then show that his ideas can be extended to estimate all moments of the fundamental solution. Moreover the estimates for the moments will be such that they will imply the Gaussian decay. In Part II, the ideas of Nash will be applied to obtain a Gaussian upper bound on the fundamental solution of the heat equation for a class of complete Riemannian manifolds which includes those with a lower bound hypothesis on the Ricci curvature. In this regard the reader should also see the recent work of L. Saloff-Coste ([9],[10]).

Before beginning the main body of this paper, I would like to express my thanks to Noah Herschel Moss Fabes for his encouragement and active help in preparing this manuscript.

Part I. The Nondegenerate Case The notation we will use for Sections 1,2, and 3 of Part I is rather standard. The letter x, y, z, and ξ will denote points in the Euclidean space R^n, $n \geq 3$. The letters r, s, and t will denote real numbers.

We consider parabolic operators of the form

$$Lu(x,t) = \sum_{i,j=1}^{n} D_{x_i}(a_{ij}(x,t)D_{x_j}u(x,t)) - D_t u(x,t)$$

where the matrix $a(x,t) \equiv (a_{ij}(x,t))$ is symmetric, measurable, and uniformly nondegenerate, i.e. There exists $\lambda > 0$ such that for all $(x,t) \in R^{n+1}, \xi \in R^n$,

$$\lambda|\xi|^2 \leq \sum_{i,j=1}^{n} a_{ij}(x,t)\xi_i\xi_j \leq \frac{1}{\lambda}|\xi|^2.$$

We let $\Gamma \equiv \Gamma(x,t;y,s)$ denote the fundamental solution corresponding to the operator L. At times it will be necessary to show the correspondence of the fundamental solution with the operator L through the matrix $a(x,t)$. At such times we will use the notation $\Gamma_a(x,t;y,s)$.

The letter C will stand for a constant which depends only on the ellipticity parameter λ and the spatial dimension Γ. The constant need not be the same at each occurrence.

We will often write

$$\sum_{i,j=1}^{n} D_{x_i}(a_{ij}(x,t)D_{x_j}u(x,t)) = \text{div } a(\nabla u)$$

where we understand that div and ∇ denote only the spatial divergence and spatial gradient.

Section 1. The Fist Moment Bound.

In this section we will review Nash's proof of the following moment bound:

Theorem 1.

There exists a constant C depending only on λ and n, such that for $s < t$ and all x,

$$\int_{R^n} |x-y|\Gamma(x,t;y,s)dy \le C(t-s)^{1/2}.$$

We begin the proof with the basic pointwise bound which serves as a good estimate along the diagonal.

Lemma 1.1. $\Gamma(x,t;y,s) \le C(t-s)^{-n/2}$, $s < t$.

Proof

For $t > 0$ set $p(x,t) = \Gamma(x,t;0,0)$ and $u(t) = \int p^2(x,t)dx$. Then $u'(t) = 2\int pp'dx = 2\int p \text{ div } (a\nabla p)dx$ and so

$$u'(t) \le -2\lambda \int |\nabla p|^2 dx.$$

Applying the interpolation inequality

$$(\int p^2)^{1+2/n} \le C_n(\int |\nabla p|^2)(\int p)^{4/n}$$

and remembering that $\int pdx = 1$, we have

$$u'(t) \le -\frac{1}{C}u(t)^{1+2/n}, \qquad i.e. \qquad \frac{d}{dt}(u^{-2n}) \ge \frac{1}{C}\,.$$

We integrate this last inequality from $\frac{t}{2}$ to t and obtain

$$u(t) \equiv \int \Gamma(x,t;0,0)^2 dx \le Ct^{-n/2}. \tag{1.2}$$

Now for $s < t$

$$\Gamma_a(x,t;y,s) = \Gamma_a(x+w,t-u;x+z,t-v)|_{\substack{w=0,u=0\\ z=y-x,v=t-s}}$$

and $\Gamma_a(x+w,t-u;x+z,t-v) = \Gamma_{a(x+\cdot,t-\cdot)}(z,v;w,u)$. As a consequence of (1.2) we have

$$\int \Gamma_a(x,t;y,s)^2 dy \le C(t-s)^{-n/2}. \tag{1.3}$$

Similarly

$$\Gamma_a(x,t;y,s) = \Gamma_a(x,t-u;y,t-v)|_{\substack{u=0\\ v=t-s}}$$
$$= \Gamma_{a(\cdot,t-\cdot)}(y,t-s;x,0)$$

and from (1.3) it also follows

$$\int \Gamma_a(x,t;y,s)^2 dx \le C(t-s)^{-n/2}. \tag{1.4}$$

Finally the reproducing property of Γ_a implies (for $s < t$)

$$\Gamma_a(x,t;y,s) = \int \Gamma_a(x,t;z,\frac{t+s}{2})\Gamma_a(z,\frac{t+s}{2};y,s)dt.$$

Schwartz's inequality together with (1.3) and (1.4) give the conclusion of Lemma (1.1).

Lemma (1.5): (The first moment bound)

There exists a constant C depending only on λ and n such that

$$\int |x-y|\Gamma(x,t;y,s)dx \le C(t-s)^{1/2}.$$

Proof. As in the proof of Lemma (1.1) we may take $y = 0$ and $s = 0$.

Now set

$$M_1(t) = \int |x|p(x,t)dx$$

where again $p(x,t) = \Gamma(x,t;0,0)$.

$$M_1'(t) = \int |x|\text{div } a(\nabla p)dx = -\int \frac{x}{|x|}\cdot a(\nabla p)dx$$
$$\le C(\int \frac{|\nabla p(x,t)|^2}{p(x,t)}dx)^{1/2}.$$

To estimate the interesting quantity on the right-hand side of the above inequality Nash introduces the function

$$Q(t) = -\int p(x,t)\log p(x,t)dx.$$

This function is introduced since

$$Q'(t) = \int \frac{a(\nabla p)\cdot\nabla p}{p}dx, \ i.e.$$

$Q'(t)$ is equivalent to $\int \frac{|\nabla p|^2}{p}dx$. Hence

$$M_1'(t) \le C(Q'(t))^{1/2} \le Ct^{1/2}(Q'(t)+\frac{1}{t}).$$

Integrating this last inequality from ϵ to t we obtain

$$M_1(t) - M_1(\epsilon) \leq 2C(t^{1/2} - \epsilon^{1/2}) + C(Q(t)t^{1/2} - Q(\epsilon)\epsilon^{1/2}) - \frac{C}{2}\int_\epsilon^t \frac{Q(s)}{s^{1/2}}ds$$

with $C > 0$. Set $Q^- = \max(0, -Q)$. Then

$$M_1(t) - M_1(\epsilon) \leq 2Ct^{1/2} + CQ(t)t^{1/2} + CQ^-(\epsilon)\epsilon^{1/2} + \frac{c}{2}\int_\epsilon^t \frac{Q^-(s)}{s^{1/2}}ds. \tag{1.6}$$

From Lemma 1.1 $-\log p(x,t) \geq -C + \frac{n}{2}\log t$ and since $\int p(x,t)dx = 1$ we conclude

$$Q(t) \geq -C + \frac{n}{2}\log t. \tag{1.7}$$

In particular, $Q^-(t) \leq C + \frac{n}{2}|\log t|$ and therefore after letting $\epsilon \to 0+$ we have from 1.6,

$$M_1(t) \leq 2Ct^{1/2} + CQ(t)t^{1/2} + \frac{c}{2}\int_0^t \frac{|\log s|}{s^{1/2}}dx. \tag{1.8}$$

In order to estimate $M_1(t)$ from below Nash observes that for any fixed σ,

$$\min_p(p\log p + \sigma p) = -e^{-\sigma-1}.$$

Hence for any constants a and b,

$$\int(p\log p + (a|x| + b)p)dx \geq -e^{-b-1}\int e^{-a|x|}dx$$

that is

$$-Q(t) + aM_1(t) + b \geq -C_n e^{-b}a^{-n}, \quad C_n = e^{-1}\int e^{-|x|}dx.$$

Now set $a = \frac{1}{M_1(t)}$ and $e^{-b} = a^n$. Then

$$Q(t) \leq C_n + 1 + b = C_n + 1 + n\log M_1(t)$$

and therefore

$$e^{\frac{1}{n}Q(t)} \leq CM_1(t). \tag{1.9}$$

Inequalities (1.7),(1.8), and (1.9) imply several bounds: there exists $C > 0$, depending only on λ and n such that

i) $|Q(1)| \leq C$

ii) $\frac{1}{C} \leq M(1) \leq C$.

Finally $\Gamma_a(x,t;0,0) = \Gamma_a(\sqrt{t}\frac{x}{\sqrt{t}}, t\cdot 1; 0,0) = t^{-n/2}\Gamma_{a(\sqrt{t}\cdot, t\cdot)}(\frac{x}{\sqrt{t}}, 1;0,0)$ and so

$$\frac{1}{C}t^{1/2} \leq M_1(t) \leq Ct^{1/2}.$$

Remark. Lemma (1.5) is the same as Theorem 1 since, once again,

$$\begin{aligned}\Gamma_a(x,t;y,s) &= \Gamma_a(x+w, t-u; x+z, t-v)\Big|_{\substack{w=0,u=0\\ z=y-x,v=t-s}}\\ &= \Gamma_{a(x+\cdot,t-\cdot)}(y-x, t-s; 0,0)\end{aligned}$$

2. Iteration of Moment Bounds

In this section we will give an inductive procedure controlling higher order moments of the fundamental solution in terms of lower order ones. For j a nonnegative integer set

$$M_j(t) = \int |x|^j \Gamma(x,t;0,0)dv.$$

We will prove

Theorem 2. For each $j \geq 0$ there exists $C_j > 0$ depending only on j, λ, and n such that

$$M_j(t) \leq C_j t^{j/2}.$$

Proof.

Take $j \geq 2$ and again set $p(x,t) = \Gamma(x,t;0,0)$. Then

$$\begin{aligned}M_j'(t) &\leq Cj\int |x|^{j-1}|\nabla p(x,t)|dx\\ &\leq CjM_j(t)^{1/2}\Big(\int |x|^{j-2}\frac{|\nabla p|^2}{p}dx\Big)^{12},\end{aligned}$$

and so

$$M_j(t)^{1/2} \leq Cj\int_0^t \Big(\int_{R^n} |x|^{j-2}\frac{|\nabla p|^2}{p}dx\Big)^{1/2}ds. \tag{2.1}$$

Now set

$$Q_j(t) = -\int |x|^j p\log p\,dx.$$

It is easily seen that

$$\begin{aligned}Q_j'(t) = j\int |x|^{j-2}x\cdot a(\nabla p)(1+\log p)dx\\ + \int |x|^j\frac{a(\nabla p)\cdot\nabla p}{p}dx.\end{aligned}$$

Therefore

$$\int |x|^j \frac{|\nabla p|^2}{p} dx \leq \frac{1}{\lambda} Q_j'(t) + \frac{C}{\lambda} j \int |x|^{j-1} |\nabla p|(1 + |\log p|) dx$$

$$\leq \frac{1}{\lambda} Q_j'(t) + \frac{C}{\lambda} j (M_{j-1}(t) + \int |x|^{j-1} p (\log p)^2 dx)^{1/2} (\int |x|^{j-1} \frac{|\nabla p|^2}{p} dx)^{1/2},$$

and so

$$\int |x|^j \frac{|\nabla p|^2}{p} dx \leq \frac{1}{\lambda} Q_j' + \frac{C_j}{\lambda} [(M_{j-1}(t) + \int |x|^{j-1} p (\log p)^2 dx) + \int |x|^{j-1} \frac{|\nabla p|^2}{p} dx].$$

Iterating the last inequality we obtain

$$\begin{aligned} \int |x|^j \frac{|\nabla p|^2}{p} dx &\leq \sum_{k \leq j} C_{k_j} Q_k'(t) + \sum_{k<j} \tilde{C}_{k_j} M_k(t) \\ &\quad + \sum_{k<j} \tilde{C}_{k_j} \int |x|^k p (\log p)^2 dx \end{aligned} \tag{2.2}$$

where $C_{k_j} > 0$ and $\tilde{C}_{k_j} \geq 0$ depend only on k, j, λ and n.

Putting (2.2) into (2.1) we have for $j \geq 2$

$$\begin{aligned} M_j(t)^{1/2} &\leq \int_0^t (\sum_{k \leq j-2} C_{k_j} Q_k'(s) + \sum_{k<j-2} \tilde{C}_{k_j} M_k(s) \\ &\quad + \sum_{k<j-2} \tilde{C}_{k_j} \int |x|^k p (\log p)^2 dx)^{1/2} ds \end{aligned} \tag{2.3}$$

(Note if $j = 2$ only the term involving $C_o Q'$ appears on the right hand side of 2.3.)

We continue the proof of Theorem 2 with a lemma which helps control the right hand side of (2.3).

Lemma 2.4. Set $P_k(t) = \int |x|^k \Gamma(x, t; 0, 0) \log^2 \Gamma(x, t; 0, 0) dx$. There exists a positive constant C_k depending only on k, λ and n such that

$$P_k(t) \leq C_k (1 + \log^2 t)(M_k(t) + \frac{1}{\sqrt{t}} M_{k+1}(t) + t^{k/2})$$

Proof

The idea of the proof is taken from Nash's proof of the upper bound estimate for the first moment, $M_1(t)$, which we presented in Section 1. (See the proof of inequality 1.8.).

Since $\Gamma_a(x, t; 0, 0) = \Gamma_{a(\sqrt{t}\cdot, t\cdot)}(\frac{x}{\sqrt{t}}, 1; 0, 0) t^{-n/2}$, we may assume $t = 1$. Now set $p(x) = \Gamma(x, 1; 0, 0)$. We want to show

$$\int |x|^k p(x) (\log p(x))^2 dx \leq C_k (M_k(1) + M_{k+1}(1) + 1).$$

Now $\int_{p>1/e}|x|^k p(\log p)^2 dx \leq C\int |x|^k p^2 \leq \tilde{C}M_k(1)$ since p is bounded (Lemma 1.1). For each real positive number r,

$$\max_{0\leq p\leq 1/e} p[(\log p)^2 - r] = e^{-(1+r)^{1/2}-1}(2(1+r)^{1/2}+2).$$

Hence

$$\begin{aligned}\int_{0<p<\frac{1}{e}} |x|^k p(\log p)^2 dx &= \int_{0<p<1/e} |x|^k p[(\log p)^2 - |x|]dx + M_{k+1}(1)\\ &\leq C\int_{0<p<1/e} |x|^k e^{-(1+|x|)^{1/2}/2}dx + M_{k+1}(1)\\ &\leq C_k + M_{k+1}(1).\end{aligned}$$

Let's return to the proof of Theorem 2. Recall inequality (2.3); with the above rotation

$$M_j(t)^{1/2} \leq \int_0^t \left\{ \sum_{k\leq j-2} C_{k_j}Q_k'(s) + \sum_{0\leq k<j-2} \tilde{C}_{k_j}(M_k(s)+P_k(s))\right\}^{1/2} ds \tag{2.5}$$

where $C_{k,j}>0$ and $\tilde{C}_{k,j}\geq 0$ depend only on k,j,λ, and n. We can now proceed by induction on the statement of Theorem 2.

$M_o(t)\equiv 1$ and $M_1(t)\leq C_1 t^{1/2}$ by Lemma 1.5. Now take $j\geq 2$ and assume $M_k(t)\leq C_k t^{k/2}$ with C_k depending only on λ, n, and k for all $0\leq k<j$. Using (2.5), Lemma 2.4, and the induction assumption,

$$M_j(t)^{1/2} \leq \int_0^t (\sum_{k\leq j-2} C_{k_j}Q_k'(s) + C_j \sum_{o\leq k<j-2} s^{k/2}(1+(\log s)^2))^{1/2}ds,$$

where $C_j\geq 0$ depends only on j,λ, and n ($C_j=0$ if $j=2$). Now take $t=1$. Then

$$M_j(1)^{1/2} \leq \int_0^1 (\sum_{k\leq j-2} C_{k_j}Q_k'(s) + C_j s^{-1})^{1/2}ds.$$

As Nash observed if a and $(a+b)$ are positive then $(a+b)^{1/2}\leq a^{1/2}+\frac{b}{2a^{1/2}}$. Hence

$$\begin{aligned}M_j(1)^{1/2} &\leq C_j^{1/2}\int_0^1 \frac{1}{s^{1/2}}ds + \frac{1}{2C_j^{1/2}}\int_0^1 (\sum_{k\leq j-2} C_{k_j}Q_k(s))' s^{1/2}ds\\ &\leq 2C_j^{1/2} + \frac{1}{2C_j^{1/2}}\sum_{k\leq j-2} C_{k_j}Q_k(1) - \frac{1}{4C_j^{1/2}}\int_0^1 (\sum_{k\leq j-2} C_{k_j}Q_k(s))\frac{1}{s^{1/2}}ds.\end{aligned}$$

Once again using induction and Lemma 2.4 (note that $|Q_k(t)|\leq P_k(t)^{1/2}M_k(t)^{1/2}$) there exists $\tilde{C}_j>0$ depending only on λ, j, and n such that

$$M_j(1)\leq \tilde{C}_j.$$

A dilation argument concludes the proof of Theorem 2.

Corollary (2.6). For each $j \geq 0$ there exists $C_j \geq 0$ depending only on j, n, and λ such that

$$\int |x-y|^j \Gamma(x,t;y,s)dy \leq C_j(t-s)^{j/2}.$$

Proof

We need only recall that

$$\Gamma_a(x,t;y,s) = \Gamma_{a(x+\cdot,t-\cdot)}(y-x,t-s;0,0).$$

Section 3. The Gaussian Upper Bound.

In this section we will prove

Theorem 3. There exists a constant $C > 0$ depending only on λ and n such that

$$\Gamma(x,t;y,s) \leq C(t-s)^{-n/2}e^{-|x-y|^2/C(t-s)}.$$

We begin the proof with

Lemma (3.1). Set $\phi_j(t) = \int |x|^{2j}\Gamma(x,t;0,0)^2 dx$. There exists a positive constant C depending only on n and λ such that

$$\phi_j(t) \leq C^j j! t^{-n/2+j}.$$

Proof.

From Lemma (1.1) and Theorem 2,

$$\phi_j(t) \leq Ct^{-n/2}\int |x|^{2j}\Gamma(x,t;0,0)dx \leq CC_j t^{-n/2+j}$$

with C depending only on λ and n and C_j depending only on λ, n, and j. Hence we can find B depending only on λ and n such that

$$\phi_j(t) \leq B^j j!\, t^{-n/2+j} \quad \text{for} \quad 0 \leq j \leq [\frac{n}{2}]+1$$

where $[\frac{n}{2}]$ denotes the largest integer $\leq \frac{n}{2}$.

Now take $j > [\frac{n}{2}]+1$ and set $p(x,t) = \Gamma(x,t;0,0)$.

$$\phi_j'(t) \leq \tilde{C}j\int |x|^{2j-1}p|\nabla p|dx - \lambda\int |x|^{2j}|\nabla p|^2 dx$$

with $\tilde{C}$ depending only on λ and n. Hence for such a constant $\tilde{C}$ we have

$$\phi_j'(t) \leq \tilde{C}j^2\phi_{j-1}(t).$$

Since $j > [\frac{n}{2}] + 1$, $\phi_j(0) = 0$, and therefore

$$\phi_j(t) \le \tilde{C}j^2 \int_0^t \phi_{j-1}(s)ds.$$

Assume now that $\phi_{j-1}(s) \le C^{j-1}(j-1)!s^{-n/2+(j-1)}$. Then

$$\begin{aligned}\phi_j(t) &\le C^{(j-1)}(j-1)^2\tilde{C}j^2 \int_0^t s^{-n/2+j-1}ds \\ &\le C^{(j-1)}j!\tilde{C}j(j-n/2)^{-1}t^{-n/2+j}.\end{aligned}$$

For $j > [\frac{n}{2}] + 1$, $j(j-\frac{n}{2})^{-1} \le D_n$ with $D_n = \frac{[\frac{n}{2}]+1}{1+\frac{n}{2}-[\frac{n}{2}]}$. So if we choose

$$C \ge \max(B, \tilde{C}D_n)$$

we can successfully apply induction and obtain the conclusion of Lemma (3.1).

Corollary (3.2). There exists a constant C, depending only on λ and n, such that for all $j \ge 0$,

$$|x|^j\Gamma(x,t;0,0) \le C^j(j!)^{1/2}t^{-n/2+j/2}.$$

Proof.

Using the reproducing property of Γ we can write

$$\Gamma(x,t;0,0) = \int \Gamma(x,t;y,t/2)\Gamma(y,t/2;0,0)dy$$

So

$$\begin{aligned}|x|^j\Gamma(x,t;0,0) \le 2^j &\int |x-y|^j\Gamma(x,t;y,t/2)\Gamma(y,t/2;0,0)dy \\ &+ 2^j \int \Gamma(x,t;y,t/2)|y|^j\Gamma(y,t/2;0,0)dy.\end{aligned}$$

The proof of Corollary (3.2) is completed by applying Schwartz's inequality to each integral and using the estimate of Lemma (3.1).

Proof of Theorem 3.

As we have seen, we may take $y = 0$ and $s = 0$.

$$e^{\delta|x|^2/t}\Gamma(x,t;0,0) = \sum_{j=0}^{\infty} \frac{\delta^j t^{-j}}{j!}|x|^{2j}\Gamma(x,t;0,0).$$

Applying Corollary (3.2),

$$t^{n/2}e^{\delta|x|^2/t}\Gamma(x,t;0,0) \le \sum_{j=0}^{\infty} \delta^j C^j \frac{[(2j)!]^{1/2}}{j!}.$$

Choosing $\delta = \frac{1}{2C}$ and using Sterling's formula we see that the series on the right side of the above inequality is bounded by a constant depending only on λ and n.

Part II. The Nash Method Applied to Heat Kernels on Riemannian Manifolds.

In this part of the paper we will show that the ideas of Nash described in the previous sections carry over to obtain Gaussian-type upper bound estimates on the heat kernel or heat flow on a class of complete Riemannian manifolds (M, g) containing those with a global lower bound hypothesis on the Ricci curvature of g. Having announced this, let me confess that my understanding of Riemannian geometry is woefully lacking. Because of this, I prefer to consider the more specialized setting of R^n endowed with a Riemannian metric, $g(x) \equiv (g_{ij}(x))$, i.e. a smooth choice of a positive definite symmetric matrix $g(x)$. Setting

$$(g^{ij}(x)) = (g_{ij}(x))^{-1}$$

we define the inner product

$$< \xi, \eta >_x = \sum_{i,j} g^{ij}(x)\xi_i\eta_j$$

and the modulus of a vector ξ at x by

$$|\xi|_x^2 = \sum_{i,j=1}^{n} g^{ij}(x)\xi_i\xi_j.$$

The notation $d(x, y)$ will stand for the distance function induced by the Riemannian metric, and $B_r(x)$ will denote the ball defined through d of radius $r > 0$ and center x.

The volume element on the manifold is given by

$$m(dx) = w(x)dx$$

where $w(x) = \sqrt{det(g_{ij}(x))}$. We set $|E| = m(E)$, the volume of E.

For f a smooth function on R^n we define its intrinsic gradient $\nabla_g f(x)$ by

$$(\nabla_g f(x))_i = \sum_{j=1}^{n} g^{ij}(x)\frac{\partial f}{\partial x_j}(x), \; i = 1, \ldots, n.$$

We will use the notation $|\nabla_g f|$ to denote the function

$$|\nabla_g f(x)|_x =< \nabla f(x), \nabla f(x) >_x^{1/2}$$

where ∇ denotes the standard gradient.

The intrinsic Laplacian is defined by

$$\Delta_g f(x) = \frac{1}{w(x)} \sum_{i,j=1}^{n} \frac{\partial}{\partial x_i}(w(x)g^{ij}\frac{\partial f}{\partial x_j}(x)).$$

In this way we have

$$\int \Delta_g h(x) f(x) m(dx) = \int h(x) \Delta_g f(x) m(dx)$$
$$= -\int < \nabla h(x), \nabla f(x) >_x m(dx).$$

$\Gamma(x,y,t)$ will denote the heat kernel or fundamental solution for the operator $\frac{\partial}{\partial t} - \Delta_g$.

When we have need to make explicit the dependence of the above defined quantities on the Riemannian metric we will write

$$\Gamma \equiv \Gamma_g, \; d \equiv d_g, \; B_r \equiv B_r^{d_g}, \; m_g(dy), \text{ etc.}$$

In Part II we consider (R^n, g) and Γ_g with the following properties:

$II(i)$
$$\Gamma_g(x,y,t) \leq \frac{C}{|B_{\sqrt{t}}(x)|}$$

for all $x \in R^n, y \in R^n$, $\underline{0 < t \leq 1}$. Here C is an absolute constant independent of x and t.

II(ii) There exists a constant $C > 0$ such that for all $x \in R^n$ and for all $s > 0$

$$|B_s(x)| \leq C s^n$$

II(iii) $m(dx)$ satisfies the local doubling condition, i.e. there exists a constant $C > 0$ such that for all $x \in R^n$ and all $\underline{0 < s \leq 1}$,

$$|B_{2s}(x)| \leq C |B_s(x)|$$

II(iv) There exist constants $C > 0$, $\tau > 0$, and $K > 1$ such that for all $s > 0$, $R \geq 1$, and $x \in R^n$

$$|B_{Rs}(x)| \leq C R^\tau e^{(K-1)Rs} |B_s(x)|.$$

Observation. It is important to observe if (R^n, g) and Γ_g satisfy II(i)-(II(iv) then for $\underline{0 < t \leq 1}$ the space $(R^n, g(\sqrt{t}\cdot))$ and the fundamental solution $\Gamma_{g(\sqrt{t}\cdot)}$ satisfy II(i)-II(iv) with no change in the structural constants. This is so since

$$\Gamma_{g(\sqrt{t}\cdot)}(x,y,s) = t^{n/2} \Gamma_g(\sqrt{t}x, \sqrt{t}y, ts)$$

and

$$|B^{d_g}_{\sqrt{t}s}(\sqrt{t}x)| = t^{n/2} |B_s^{d_{g(\sqrt{t}\cdot)}}(x)|.$$

In Part II we will prove the

Theorem. Assume (R^n, g) and Γ satisfy II(i)- II(iv). Then there exist constants C, C_1, C_2, and L depending only on the constants appearing in II(i)-II(iv) such that for $0 < t \leq 1$

$$\Gamma(x,y,t) \leq \frac{(C_1 + C_2 \log \frac{t^{n/2}}{|B_{\sqrt{t}}(x)|})^L + (C_1 + C_2 \log \frac{t^{n/2}}{|B_{\sqrt{t}}(y)|})^L}{|B_{\sqrt{t}}(x)|^{1/2} |B_{\sqrt{t}}(y)|^{1/2}} e^{-d^2(x,y)/Ct}.$$

Section 4. Moment Bounds

In this section we will prove

Theorem 4. Assume (R^n, g) and Γ satisfy II(i)- II(iv). For each nonnegative integer j, there exist positive constants C_j^1 and C_j^2 depending only on j, n and the structural constants in II(i)-II(iv) such that for $0 < t \leq 1$,

$$\int d(x,y)^j \Gamma(x,y,t) m(dy) \leq t^{j/2} (C_j^1 + C_j^2 \log \frac{t^{n/2}}{|B_{\sqrt{t}}(x)|})^j.$$

Proof.

Our proof will follow very closely the ideas of Sections 1-3. For this reason we will abbreviate the proof emphasizing only those details involving some additional explanation due to the Riemannian structure we have put on R^n.

Lemma 4.1. (The first moment bound).

Assume (R^n, g) and Γ satisfy II(i)-II(iv). Then there exist constants $C_1 > 0$ and $C_2 > 0$ such that for $0 < t \leq 1$

$$\int d(x,y) \Gamma(x,y,t) m(dy) \leq \sqrt{t} (C_1 + C_2 \log \frac{t^{n/2}}{|B_{\sqrt{t}}(x)|})$$

Proof.

With x fixed set $p = p(y,t) = \Gamma(x,y,t)$, $d(y) = d(x,y)$, and $M_1(t) = \int d(y) p(y,t) m(dy)$. Since

$$|\nabla_g d(y)| \leq 1 \text{ } (a.e. m(dy)) \text{ and } \int p(y,t) m(dy) = 1,$$

$$\begin{aligned} M_1'(t) &\leq \int |\nabla_g d(y)| |\nabla_g p(y,t)| dm(y) \\ &\leq (\int \frac{|\nabla_g p|^2}{p} m(dy))^{1/2}. \end{aligned}$$

Set

$$Q(t) = -\int p \log p \, m(dy).$$

Then

$$Q'(t) = \int \frac{|\nabla_g p|^2}{p} m(dy)$$

and therefore

$$M_1(t) \le \int_0^t Q'(s)^{1/2} ds.$$

Using II(i) and II(iii)

$$\text{(4.2)} \qquad Q(t) \ge -C + \log|B_{\sqrt{t}}(x)| \ge -C - C\log\frac{1}{t} - \log\frac{1}{|B_1(x)|},$$

and proceeding as in the proof of Lemma 1.5, there are positive constants C_1, C_2, C_3, and C_4 such that for $0 < t \le 1$,

$$\text{(4.3)} \qquad M_1(t) \le C_1 t^{1/2} + C_2 Q(t) t^{1/2} + C_3 \int_0^t \frac{\log 1/s}{s^{1/2}} ds + C_4\sqrt{t}\log\frac{1}{|B_1(x)|}.$$

For the estimate of $M_1(t)$ from below in terms of $Q(t)$ we recall the ideas of Lemma 1.5. For positive numbers a and b

$$-Q(t) + aM_1(t) + b \ge -e^{-b-1}\int e^{-ad(x,y)} m(dy).$$

Therefore

$$\begin{aligned} Q(t) &\le aM_1(t) + b + e^{-b-1}\sum_{j=1}^{\infty}\int_{(\frac{j-1}{a})<d<(\frac{j}{a})} e^{-ad(x,y)} m(dy). \\ &\le aM_1(t) + b + e^{-b}\sum_{j=1}^{\infty} e^{-j}|B_{j/a}| \\ &\le aM_1(t) + b + \frac{e^{-b}}{a^n}\sum_{j=1}^{\infty} j^n e^{-j}. \end{aligned}$$

Hence

$$Q(t) \le aM_1(t) + b + Ce^{-b}a^{-n}.$$

Exactly as in Lemma 1.5 we conclude

$$\text{(4.4)} \qquad e^{\frac{Q(t)}{n}} \le CM_1(t).$$

Inequalities (4.2),(4.3), and (4.4) imply

$$e^{Q(1)} \le C_1 + C_2 Q(1) + C_3 + C_4 \log\frac{1}{|B_1(x)|}.$$

Hence if $Q(1) \ge 0$ we obtain

$$0 \le Q(1) \le \log(C_5 + C_4\log\frac{1}{|B_1(x)|})$$

(Remember $|B_1(x)| \leq C$ for all x from II(ii)). From (4.3),

$$M_1(1) \leq \tilde{C}_1 + \tilde{C}_3 \log \frac{1}{|B_1(x)|}. \tag{4.5}$$

In general, for $0 < t \leq 1$,

$$\begin{aligned} M_1(t) &= \int d_g(x,y)\Gamma_g(x,y,t)m_g(dy) \\ &= t^{n/2}\int d_g(x,\sqrt{t}y)\Gamma_g(x,\sqrt{t}y,t)m_{g(\sqrt{t}\cdot)}(dy). \end{aligned}$$

We observe

$$\Gamma_g(x,\sqrt{t}y,t\cdot 1) = t^{-n/2}\Gamma_{g(\sqrt{t}\cdot)}(x/\sqrt{t},y,1) \tag{4.6}$$

and

$$dg(x,\sqrt{t}y) = \sqrt{t}d_{g(\sqrt{t}\cdot)}(x/\sqrt{t},y). \tag{4.7}$$

Hence

$$M_1(t) = \sqrt{t}\int d_{g(\sqrt{t}\cdot)}(x/\sqrt{t},y)\Gamma_{g(\sqrt{t}\cdot)}(x/\sqrt{t},y,1)m_{g(\sqrt{t}\cdot)}(dy). \tag{4.8}$$

As previously observed, the space $(R^n, g(\sqrt{t}\cdot))$ and $\Gamma_{g(\sqrt{t}\cdot)}$ satisfy conditions II(i)-II(iv) with the same structure constants provided $0 < t \leq 1$. Recall also that

$$|B_s^{d_{g(\sqrt{t}\cdot)}}(x/\sqrt{t})| = t^{-n/2}|B^{dg}_{\sqrt{t}s}(x)|.$$

Therefore there exist constants $C_1 > 0$ and $C_2 > 0$ such that

$$M_1(t) \leq \sqrt{t}(C_1 + C_2 \log \frac{t^{n/2}}{|B_{\sqrt{t}}(x)|}).$$

<u>Lemma (4.9)</u>. Set $M_j(t) = \int d(x,y)^j\Gamma(x,y,t)m(dy)$,

$$Q_k(t) = -\int d(x,y)^k\Gamma(x,y,t)\log\Gamma(x,y,t)m(dy),$$

and $P_k(t) = \int d(x,y)^k\Gamma(x,y,t)\log^2\Gamma(x,y,t)m(dy)$. For $j \geq 2$,

$$M_j(t)^{1/2} \leq \int_0^t \{\sum_{k\leq j-2} C_{k_j}Q_k'(s) + \sum_{k<j-2}\tilde{C}_{k_j}(M_k(s)+P_k(s))\}^{1/2}ds$$

where all constants depend only on n, k, j and the structural constants in II(i)-II(iv).

<u>Proof</u>.

The proof is a verbatim repetition of the proof of the corresponding inequality (2.3).

Lemma (4.10). There exists a constant C_k depending on k and the structural constants in II(i)-II(iv) such that for $0 < t \leq 1$,

$$P_k(t) \leq C_k[t^{k/2} + (1 + \log^2 \frac{1}{t} + \log^2 \frac{t^{n/2}}{|B_{\sqrt{t}}(x)|})M_k(t) + \frac{1}{t}M_{k+2}(t)]$$

Proof.

First, we consider the inequality for $t = 1$. Recall $\Gamma(x, y, 1) = \frac{C}{|B_1(x)|}$. Set $|B| = |B_1(x)|$

$$\begin{aligned}P_k(1) &\leq C_1 \int d^k \Gamma \log^2 |B|\Gamma \; m(dy) + C_1(\log^2 \frac{1}{|B|})M_k(1) \\ &\leq \frac{C_1}{|B|}(\int_{|B|\Gamma < \frac{1}{e}} d^k |B|\Gamma \log^2 |B|\Gamma \; m(dy)) + C_2 M_k(1)(1 + \log^2 \frac{1}{|B|}).\end{aligned}$$

$$\begin{aligned}\int_{|B|\Gamma < \frac{1}{e}} d^k |B|\Gamma \log^2 |B|\Gamma m(dy) = &\int d^k |B|\Gamma(\log^2 |B|\Gamma - Kd^2)m(dy) \\ &+ K|B|M_{k+2}(1).\end{aligned}$$

Recall from Lemma 2.4; for each $r > 0$,

$$\max_{0 \leq p \leq \frac{1}{e}} p[(\log p)^2 - r] = e^{-(1+r)^{1/2} - 1}(2 + 2(1 + r)^{1/2}).$$

Hence

$$\begin{aligned}P_k(1) \leq \frac{C_1}{|B|} &\int d^k e^{-dK} m(dy) + C_1 M_{k+2}(1) + \\ &+ C_2 M_k(1)(1 + \log^2 \frac{1}{|B|}).\end{aligned}$$

Now

$$\begin{aligned}\int d^k e^{-Kd} m(dy) &= \sum_{\ell=1}^{\infty} \int_{(\ell-1) < d \leq \ell} d^k e^{-dK} m(dy) \\ &\leq \sum_{\ell=1}^{\infty} \ell^k e^{-(\ell-1)K} |B_\ell|.\end{aligned}$$

Using property II(iv) of the volume measure, $\int d^k e^{-dK} m(dy) \leq C_k|B|$. (Remember $|B| = |B_1|$.) Finally

$$P_k(1) \leq C_k + C_1 M_{k+2}(1) + C_2 M_k(1)(1 + \log^2 \frac{1}{|B_1(x)|}),$$

and from this inequality the usual dilation argument gives the conclusion of Lemma (4.10).

Lemma (4.11). There exist positive constants C and C_k such that

$$Q_k(t) \geq -(C + \log \frac{1}{|B_t(x)|}) M_k(t)$$

and

$$Q_k(t) \leq \frac{M_{k+1}(t)}{\sqrt{t}} + M_k(t) + C_k t^{k/2}.$$

The constant C depends only on the structure constants of II(i)-II(iv) while C_k in addition depends on k.

Proof.

The first inequality is a consequence of II(i). For the second inequality we recall that for positive numbers a and b

$$\int d^k(x,y)\Gamma(x,y,t)(\log \Gamma(x,y,t) + ad(x,y) + b)m(dy) \geq -e^{-b-1} \int d^k e^{-ad} m(dy),$$

that is

$$-Q_k(t) + aM_{k+1}(t) + bM_k(t) \geq -e^{-b-1} \int d^k e^{-ad} m(dy).$$

As in the proof of Lemma (4.1),

$$\int d^k e^{-ad} m(dy) \leq C_k a^{-n-k}.$$

If we set $a = \frac{1}{\sqrt{t}}$ and $b = 1$ we have

$$Q_k(t) \leq \frac{1}{\sqrt{t}} M_{k+1}(t) + M_k(t) + C_k t^{n/2} t^{k/2}.$$

With the above lemmas we begin the final part of the proof of Theorem 4. We are to show there exist positive constants depending only on j, n, and the structural constants in II(i)-II(iv) such that for $0 < t \leq 1$

$$M_j(t) \leq t^{j/2} (C_j^{(1)} + C_j^{(2)} \log \frac{t^{n/2}}{|B_{\sqrt{t}}(x)|})^j . \tag{4.12}$$

We proceed by induction noting that $M_o(t) \equiv 1$ and that (4.12) is valid for $M_1(t)$ by Lemma (4.1). We now fix $j \geq 2$ and assume (4.12) for nonnegative integers $k < j$. Using the inequality for $M_j(t)^{1/2}$ gives by Lemma (4.9), Lemmas (4.10) and (4.11), induction, and the conclusion of the proof of Theorem 2, we find that (4.12) holds for $M_j(1)$. The usual dilation argument gives (4.12) for $0 < t \leq 1$.

Section 5. The Gaussian Upper Bound

In this section we will prove

Theorem 5. There exist positive constants C, C_1, C_2, and L depending only on the structural constant of II(i)-II(iv) such that for $0 < t \leq 1$,

$$\Gamma(x,y,t) \leq \frac{(C_1 + C_2 \log \frac{t^{n/2}}{|B_{\sqrt{t}}(x)|})^L + (C_1 + C_2 \log \frac{t^{n/2}}{|B_{\sqrt{t}}(x)|})^L}{|B_{\sqrt{t}}(x)|^{1/2}|B_{\sqrt{t}}(y)|^{1/2}} e^{-d(x,y)^2/Ct}.$$

Proof. We begin the proof, as in Section 3, with

Lemma (5.1). Set $\psi_j(t) = \int d(x,y)^{2j}\Gamma(x,y,t)^2 m(dy)$. There exist positive constants C, C_1, C_2, and L depending only on the structural constants of II(i)-II(iv) such that for all j and all $0 < t \leq 1$,

$$\psi_j(t) \leq C^j j! \frac{t^j}{|B_{\sqrt{t}}(x)|}(C_1 + C_2 \log \frac{t^{1/2}}{|B_{\sqrt{t}}(x)|})^L$$

Proof.

From II(i) and Theorem 4, there exist positive constants A, C_1, C_2 and L such that the above conclusion holds for $j = 0, 1, \ldots L$ with $C = A$. We may also take L large enough so that $\psi_j(0) = 0$ for $j \geq L - 1$. (Condition II(iii), the local doubling condition, implies $|B_{\sqrt{t}}(x)| \geq \delta t^\nu |B_1(x)|$ for some (large) constant ν and (small) constant δ.) Exactly as in the proof of Lemma (3.1),

$$\psi_j(t) \leq \tilde{C} j^2 \int_0^t \psi_{j-1}(s) ds.$$

Induction now gives positive constants A, C_0, C_1, C_2, and ν such that

$$\psi_j(1) \leq C_0 A^j j! \frac{j}{j-\nu}(C_1 + C_2 \log \frac{1}{|B_1(x)|})^L \frac{1}{|B_1(x)|}$$

Dilation finishes the proof of Lemma (5.1).

Corollary (5.2) There exist positive constants C, C_1, C_2, and L depending only on the structure constants in II(i)-II(iv) such that

$$d^j(x,y)\Gamma(x,y,t) \leq t^{j/2}(j!)^{1/2} C^j \frac{(C_1 + C_2 \log \frac{t^{n/2}}{|B_{\sqrt{t}}(x)|})^L + (C_1 + C_2 \log \frac{t^{n/2}}{|B_{\sqrt{t}}(y)|})^L}{|B_{\sqrt{t}}(x)|^{1/2}|B_{\sqrt{t}}(y)|^{1/2}}.$$

Proof.

We proceed as in Corollary (3.2).

$$\Gamma(x,y,t) = \int \Gamma(x,z,t)\Gamma(z,y,t/2) m(dz).$$

Hence

$$
\begin{aligned}
d(x,y)^j\Gamma(x,y,t) \le 2^j &\int d(x,z)^j\Gamma(x,z,t)\Gamma(z,y,t/2)m(dz)+ \\
&+2^j\int \Gamma(x,z,t)d(z,y)^j\Gamma(z,y,t/2)m(dz) \\
&\le 2^j\psi_j(x,t)^{1/2}\Big(\int \Gamma^2(z,y,t/2)m(dz)\Big)^{1/2} \\
&+2^j\psi_j(y,\frac{t}{2})^{1/2}\Big(\int \Gamma^2(x,z,t)m(dz)\Big)^{1/2} \\
&\le 2^jC\psi_j(x,t)^{1/2}\frac{1}{B_{\sqrt{t}}(y)^{1/2}}+2^jC\psi_j(y,\frac{t}{2})\frac{1}{|B_{\sqrt{t}}(x)|^{1/2}}.
\end{aligned}
$$

An application of Lemma (5.1) gives the conclusion of Corollary (5.2).

Exactly as in Section 3 Corollary (5.2) gives Theorem 5.

References

1. Aronson, D.G., Bounds for the fundamental solution of a parabolic equation, Bulletin of the AMS 73 (1967), 890-896.
2. Aronson, D.G., Non-negative solutions of linear parabolic equations, Ann. Sci. Norm. Sup. Pisa, 22 (1968), 607-94.
3. Davies, E.B., Heat Kernels and Spectral Theory, Cambridge Tracts in Mathematics, 92, Cambridge University Press, 1989.
4. Fabes, E.B., Stroock, D.W., A new proof of Moser's parabolic Harnack inequality via the old ideas of Nash, Arch. Rat. Mech. Anal. 96 (1986), 327-338.
5. Gutierrez, C.E., Wheeden, R.L., Bounds for the fundamental solution of degenerate parabolic equations, to appear in the Communications of P.D.E.
6. Moser, J. A Harnack inequality for parabolic differential equations, Comm. Pure Appl. Math. 17 (1964), 101-134.
7. Moser, J. Correction to 'A Harnack inequality for parabolic differential equations', Comm. Pure Appl. Math. 20 (1967), 232-236.
8. Nash, J., Continuity of solutions of parabolic and elliptic equations, Amer. J. Math. 80 (1958), 931-954.
9. Saloff-Coste, Laurent, Uniformly elliptic operators on Riemannian manifolds, Journal Differential Geometry.
10. Saloff-Coste, Laurent, A note on Poincaré, Sobolev, and Harnack inequalities, International Mathematics Research Notices, Duke Journal, No. 2 (1992), 27-37.

Two topics related to Dirichlet forms: quasi everywhere convergences and additive functionals

Masatoshi Fukushima

Department of Mathematical Science
Faculty of Engineering Science
Osaka University, Toyonaka, Osaka, Japan

Chapter 1 (r,p)–capacities and quasi everywhere convergences

1 (r,p)-capacities and capacitary maximal inequalities

1.1 (r,p)-capacities associated with Markovian semigroups

Let X be a Lusinian separable metric space and m be a positive σ- finite Borel measure on X with $\mathrm{supp}[m] = X$. For $p > 1$, we denote by L^p or $L^p(X;m)$ the real L^p-space with norm $\| \ \|_p$. Let $\{T_t, t > 0\}$ be a family of strongly continuous contraction semigroup of linear operators which are Markovian: $f \in L^p, 0 \le f \le 1 \Rightarrow 0 \le T_t f \le 1$. We introduce for $r > 0$ the Gamma transform of $\{T_t, t > 0\}$:

$$V_r f = \frac{1}{\Gamma(r/2)} \int_0^\infty t^{\frac{r}{2}-1} e^{-t} T_t f dt, f \in L^p. \tag{1}$$

V_r is Markovian. It is contractive:

$$\|V_r f\|_p \le \|f\|_p, \ f \in L^p. \tag{2}$$

Further

$$V_r V_{r'} = V_{r+r'}, \ r, r' > 0. \tag{3}$$

In fact

$$\Gamma(\frac{r}{2})(\frac{r'}{2}) V_r V_{r'} f = \int\int s^{\frac{r}{2}-1} t^{\frac{r'}{2}-1} e^{-(t+s)} T_{t+s} f dt ds = \int \xi^{\frac{r+r'}{2}-1} e^{-\xi} T_\xi f d\xi \cdot B(\frac{r}{2}, \frac{r'}{2}),$$

where we have set $t = \xi \cos^2 \theta, \ s = \xi \sin^2 \theta$.

Suppose $V_r f = 0$ for some $r > 0$ and $f \in L^p$, then $V_{r'} f = 0$ for all $r' \ge r$ by (3) and consequently $T_t f = 0, t > 0$, which implies $f = 0$. Thus each V_r is injective on L^p and accordingly the following definition makes sense and gives us a Banach space $(\mathcal{F}_{r,p}, \| \ \|_{r,p})$:

$$\begin{aligned} \mathcal{F}_{r,p} &= V_r(L^p) \\ \|u\|_{r,p} &= \|f\|_p \ \ u = V_r f, \ f \in L^p. \end{aligned} \tag{4}$$

The associated set function $C_{r,p}$ is then defined, for open set G, by

$$C_{r,p}(G) = \inf\{\|u\|_{r,p}^p : u \in \mathcal{F}_{r,p}, u \ge 1 \ m - a.e. \text{on } G\} \tag{5}$$

and, for any set E,

$$C_{r,p}(E) = \inf\{C_{r,p}(G) : E \subset G, G \text{ open}\}. \tag{6}$$

We call $C_{r,p}$ the (r,p)-capacity for $\{T_t, t > 0\}$.

Lemma 1.1 (i) $m(E) \le C_{r,p}(E)$

(ii) $E \subset F \Rightarrow C_{r,p}(E) \le C_{r,p}(F)$

(iii) $r < r' \Rightarrow C_{r,p}(E) \le C_{r',p}(E)$

(iv) $p < p' \Rightarrow C_{r,p}(E) \le C_{r,p'}(E)$

(v) $C_{r,p}(\cup_n E_n) \le \sum_n C_{r,p}(E_n)$

(ii) is trivial. By (2) and (4), we have $\|u\|_p^p \le \|u\|_{r,p}^p$, from which (i) follows. By (2) and (3), we have, for $r < r', \mathcal{F}_{r,p} \supset \mathcal{F}_{r',p}$ and $\|u\|_{r,p} \le \|u\|_{r',p}, u \in \mathcal{F}_{r',p}$,which lead us to (iii).

We prove (iv) and (v) by using the next lemma.

Lemma 1.2 *For any open set G with finite (r,p)-capacity, there exists a unique element u_G in $\mathcal{F}_{r,p}$ such that*

$$u_G \ge 1 \ m-a.e. \text{ on } G \text{ and } \|u_G\|_{r,p}^p = C_{r,p}(G). \tag{7}$$

u_G admits an expression $u_G = V_r f$ for some non-negative $f \in L^p$.

Proof $(\mathcal{F}_{r,p}, \| \ \|_{r,p})$ is (by V_r^{-1}) isometric to $(L^p, \| \ \|_p)$ and hence uniformly convex (J.A.Clarkson[5]); for any $\epsilon > 0$ and $M > 0$, there is a $\delta > 0$ such that $\|u\|_{r,p} \le M, \|v\|_{r,p} \le M, \|u - v\|_{r,p} > \epsilon$ implies $\|u + v\|_{r,p} \le 2M - \delta$. (When $p = 2$, this is clear from $\|u+v\|_2^2 + \|u-v\|_2^2 = 2(\|u\|_2^2 + \|v\|_2^2)$). Suppose $u_G, \tilde{u}_G \in \mathcal{F}_{r,p}$ both satisfy (7) and $\|u_G - \tilde{u}_G\| > \epsilon$ for some $\epsilon > 0$. Then $w = (u_G + \tilde{u}_G)/2$ satisfies $\|w\|_{r,p} \le C_{r,p}^{1/p}(G) - \delta/2$ for some $\delta > 0$, contradicting to the fact that $w \in \mathcal{F}_{r,p}, w \ge 1 \ m-a.e.$ on G. Therefore u_G is unique.

To see the existence, take $u_n \in \mathcal{F}_{r,p}$ such that $u_n \ge 1$ on G and $\|u_n\|_{r,p}^p \to C_{r,p}(G)$. By Banach-Saks theorem for L^p-space, the Cesaro-mean of a subsequence of u_n converges strongly to some $w \in \mathcal{F}_{r,p}$, which obviously satisfies (7). If $u = V_r f, f \in L^p$, then $V_r f \le V_r f^+$ and $\|f^+\|_p \le \|f\|_p$ and hence $f = f^+$ by the uniqueness. q.e.d.

u_G of this lemma is called the $(r,p)-$capacitary potential of G.

Proof of Lemma 1.1(iv),(v). (iv): Suppose $p < p'$ and consider the (r,p')-capacitary potential $u_G = V_r f, f \in L_+^{p'}$ of open G. By the Hölder inequality $1 \le (V_r f)^{p'/p} \le V_r(f^{p'/p})$ holding $m-a.e.$ on G and noting $f^{p'/p} \in L^p$,

$$C_{r,p}(G) \le \|V_r(f^{p'/p})\|_{r,p}^p = \|f\|_{p'}^{p'} = C_{r,p'}(G).$$

(v): We may assume that E_n are open and the right hand side is finite. Let $u_{E_n} = V_r f_n, f_n \in L_+^p$ and $f = \sup_n f_n$. Since $f^p \le \sum_n f_n^p$, we have $f \in L^p$ and $\|V_r f\|_{r,p}^p \le \sum_n C_{r,p}(E_n)$. On the other hand, $C_{r,p}(\cup_n E_n) \le \|V_r f\|_{r,p}^p$ because $V_r f \ge 1 \ m-a.e.$ on $\cup_n E_n$. q.e.d.

A set N is called $(r,p)-$*polar* if $C_{r,p}(N) = 0$. Any $(r,p)-$polar set is m-negligible and get finer as r or p increases. "(r,p)- quasi-everywhere" or "(r,p)-q.e." will means "except for an $(r,p)-$polar set". An increasing sequence $\{F_n\}$ of closed sets is said to be an $(r,p)-$nest if $C_{r,p}(X - F_n) \to 0, n \to \infty$. If we reset $\tilde{F}_n = supp[I_{F_n} \cdot m] (\subset F_n)$, then $m(F_n - \tilde{F}_n) = 0$ and $C_{r,p}(X - \tilde{F}_n) = C_{r,p}(X - F_n)$ and consequently $\{\tilde{F}_n\}$ is again an $(r,p)-$ nest. A function u defined q.e. on X is called $(r,p)-$*quasi continuous* if there exists an $(r,p)-$nest $\{F_n\}$ such that $u|_{F_n}$ is continuous for each n. If u is $(r,p)-$ quasi continuous and $u \ge 0 \ m-a.e.$, then $u \ge 0 \ (r,p)-q.e.$ In fact take an $(r,p)-$ nest $\{F_n\}$ associated with u. Then $u(x) \ge 0$ for all $x \in \cup_n \tilde{F}_n$ for the redefined nest as above.

Remark 1.1 Consider the case that $p = 2$ and T_t are symmetric, namely, $\{T_t, t > 0\}$ is assumed to be a strongly continuous contraction semigroup of Markovian *symmetric* linear operators on $L^2(X;m)$. Then the infinitesimal generator L of $\{T_t, t > 0\}$ is a non-negative definite self-adjoint operator on $L^2(X;m)$ and admits a spectral representation: $-L = \int_0^\infty \lambda dE_\lambda$. Substitute $T_t f = \int_0^\infty e^{-\lambda t} dE_\lambda f$ into formula (1), we get $V_r f = \int_0^\infty (1+\lambda)^{-r/2} dE_\lambda f = (I-L)^{-r/2} f$ and consequently

$$\mathcal{F}_{r,2} = \mathcal{D}((I-L)^{r/2}),\ \|u\|_{r,2} = \|(I-L)^{r/2}u\|_2. \tag{8}$$

The Dirichlet form $(\mathcal{E}, \mathcal{F})$ of $\{T_t, t > 0\}$ on $L^2(X;m)$ is by definition given by

$$\mathcal{F} = \mathcal{D}(\sqrt{-L}), \mathcal{E}(u,v) = (\sqrt{-L}u, \sqrt{-L}v) \tag{9}$$

and accordingly we have the identification

$$\mathcal{F}_{1,2} = \mathcal{F}, \|u\|_{1,2}^2 = \mathcal{E}_1(u,u), \tag{10}$$

where $\mathcal{E}_\alpha(u,v) = \mathcal{E}(u,v) + \alpha(u,v)_2, \alpha > 0$ and $(\ ,\ )$ denotes the L^2-inner product. $C_{1,2}$ is exactly the capacity associated with the Dirichlet form of $\{T_t, t > 0\}$.

Remark 1.2 We call a function $p_t(x, E)$ of three variables $t > 0, x \in X, E \in \mathcal{B}(X)$ a transition function on X if it is measurable in (t, x), a measure in E with $p_t(x, X) \le 1$, and satisfies

$$p_t p_s = p_{t+s},\ \lim_{t \downarrow 0} p_t f(x) = f(x), x \in X, f \in C_b(X),$$

where $C_b(X)$ denotes the space of bounded continuous functions on X. Suppose that there are two transition functions p_t and $\hat{p}_t$ on X standing in duality with respect to m:

$$\int_X f(x) p_t g(x) m(dx) = \int_X \hat{p}_t f(x) g(x) m(dx)$$

for non-negative Borel f and g. Then $\{p_t\}$ decides uniquely a strongly continuous contraction semigroup of Markovian operators on L^p provided that either
$m(X) < \infty$ or
X is a locally compact separable metric space and m is a positive Radon measure on X.

In fact m is then p_t-excessive : $mp_t \le m$. By Hölder inequality we have $|p_t f|^p(x) \le p_t |f|^p(x)$ and the integration of both hand sides by m yields the contractivity. In the finite measure case, $p_t f \to f, t \to \infty$, in L^p for $f \in C_b(X)$ by the bounded convergence theorem and the strong continuity follows. To prove the strong continuity in the other case, we take $f \in C_b(X)$ with $supp(f) = K$ being compact and choose a relatively compact open set $G \supset K$. Further we take a function $\varphi \in C_b(X)$ with $\varphi \ge I_{X-G}$ and $\varphi = 0$ on K. Then

$$\|p_t f - f\|_p^p = \int_G |p_t f(x) - f(x)|^p m(dx) + \int_{X-G} |p_t f(x)|^p m(dx).$$

The first term of the right hand side tends to zero as $t \downarrow 0$. The second term is dominated by

$$\int_{X-G} p_t |f|^p(x) m(dx) = \int_K \hat{p}_t I_{X-G}(x) |f(x)|^p m(dx) \le \int_K \hat{p}_t \varphi(x) |f(x)|^p m(dx)$$

which tends to zero as $t \downarrow 0$.

Example 1.1 Let $X = R^d, m =$ the Lebesgue measure and

$$g_t(x) = (2\pi t)^{-d/2} exp\left(-\frac{|x|^2}{2t}\right).$$

The transition function of the d-dimensional standard Brownian motion is given by

$$p_t(x, dy) = g_t(y - x)dy$$

which determines a strongly continuous contraction semigroup $\{T_t, t > 0\}$ on $L^p(R^d) = L^p(R^d; m)$.

The Gamma transform V_r is represented as

$$V_r f = v_r * f,\ \ v_r(x) = \frac{1}{\Gamma(r/2)} \int_0^\infty t^{\frac{r}{2}-1} e^{-t} g_t(x) dt.$$

Denote by $\hat{u}$ the Fourier transform of u:

$$\hat{u}(\xi) = (2\pi)^{-d/2} \int_{R^d} e^{i(\xi,y)} u(y) dy.$$

Since $\hat{g}_t(\xi) = (2\pi)^{-\frac{d}{2}} e^{-\frac{t}{2}|\xi|^2}$, we have

$$\hat{v}_r(\xi) = (2\pi)^{-\frac{d}{2}} \left(1 + \frac{|\xi|^2}{2}\right)^{-\frac{r}{2}}$$

and we see that $v_r(x) = 2^{\frac{d}{2}} B_\alpha(\sqrt{2}x)$ where $B_\alpha(x)$ is the Bessel convolution kernel.

Accordingly $\mathcal{F}_{r,p}$ is just the space of Bessel potentials of L^p-functions. As is well known (cf.E.M.Stein (ref.[15]in §2)), the space $(\mathcal{F}_{r,p}, \|\ \|_{r,p})$ thus coincides with the Sobolev space $L^p_r(R^d)$ with equivalent norm if r is an integer.

1.2 Capacitary maximal inequalities - a reduction to L^p maximal inequalities

We make from now on an assumption of a regurality:

$$\text{there is a dense linear sublattice } \mathcal{D} \text{ of } L^p \text{ with } V_r(\mathcal{D}) \subset C(X), \tag{11}$$

where $C(X)$ denotes the set of all continuous (not necessarily bounded) functions on X. Then $\mathcal{F}_{r,p} \cap C(X)$ is dense in $\mathcal{F}_{r,p}$ and, just as in the case of Dirichlet spaces, we have the following:

Lemma 1.3 (i) *Each function $u \in \mathcal{F}_{r,p}$ admits an (r,p)-quasi continuous version $\tilde{u}$: $\tilde{u}$ is $(r,p)-$quasi continuous and $\tilde{u} = u$ $m - a.e.$*

(ii) $C_{r,p}(|\tilde{u}| > \lambda) \leq \frac{1}{\lambda^p} \|u\|^p_{r,p}, u \in \mathcal{F}_{r,p}$

(iii) *If $u_n, u \in \mathcal{F}_{r,p}$ and $\|u_n - u\|_{r,p} \to 0, n \to \infty$, then there exists a subsequaence n_k such that* $\lim_{n\to\infty} u_{n_k}(x) = \tilde{u}(x)$ $(r,p) - q.e.$ *for any choice of (r,p)-quasi continuous versions.*

We would like to know under what circumstances the $(r,p)-q.e.$ convergence takes place without taking a subsequence. The next lemma is taken from Albeverio-Ma[3]:

Lemma 1.4 *There exists a kernel $v_r(x,E)$ on $(X,\mathcal{B})$such that $v_r f(x) = \int_X v_r(x,dy)f(y)$ is an (r,p)-quasi continuous version of $V_r f$ for any non-negative Borel $f \in L^p$.*

Let $\{\theta_s, s \in I\}$ be a family of linear operators on $L^p(X;m)$ and let

$$M(f)(x) = \sup_{x \in I} |\theta_s f|(x). \tag{12}$$

We assume that either I is countable or $\theta_s f \in C(X), \forall s \in I,$ for all Borel $f \in L^p(X;m)$. $M(f)$ then makes sense as a Borel function.The following reduction theorem is an extension of Fukushima[7]:

Theorem 1.1 *Suppose that*

$$M(v_r f)(x) \leq v_r(Mf)(x), \forall x \in X \tag{13}$$

for any Borel $f \in L^p(X;m)$. We further assume that the operator M is of strong type p :There exists a positive constant A such that

$$\|M(f)\|_p \leq A\|f\|_p \text{ for Borel } f \in L^p(X;m). \tag{14}$$

Then M satisfies the next capacitary weak inequality with the same constant A:

$$C_{r,p}(M(u) > \lambda) \leq \frac{A^p}{\lambda^p}\|u\|_{r,p}^p \ \lambda > 0. \tag{15}$$

for any $u = v_r f \in \mathcal{F}_{r,p}$ with Borel $f \in L^p(X;m)$.

Proof Definition (4), Lemma 1.3 (ii) and (14) lead us to

$$\begin{aligned} C_{r,p}(M(u) > \lambda) &\leq C_{r,p}(v_r M(f) > \lambda) \leq \frac{1}{\lambda^p}\|V_r M(f)\|_{r,p}^p \\ &= \frac{1}{\lambda^p}\|M(f)\|_p^p \leq \frac{A^p}{\lambda^p}\|f\|_p^p = \frac{A^p}{\lambda^p}\|u\|_{r,p}^p. \end{aligned}$$

Remark 1.3 (i) Since v_r is positivity preserving, condition (13) is satisfied if

$$\theta_s v_r f(x) = v_r \theta_s f(x), \forall x \in X \forall \text{Borel } f \in L^p. \tag{16}$$

(ii) Under the condition (11), it was proven in Fukushima-Kaneko[9] that, for any $E \subset X$,

$$C_{r,p}(E) = \inf\{\|u\|_{r,p}^p : \tilde{u} \geq 1 \ (r,p) - q.e \text{ on } E\}$$

and $C_{r,p}$ satisfies the continuity:

$$E_n \uparrow E \Rightarrow C_{r,p}(E) = \sup_n C_{r,p}(E_n).$$

This implies the capacitability of a Borel set E (cf.[2]):

$$C_{r,p}(E) = \sup\{C_{r,p}(K) : E \supset K \text{compact}\}.$$

(iii) In the next section,we start with an a priori given specific transition function–the spatially homogeneous one. Kaneko[11] started with a general *analytic* Markovian semigroup and under the regularity condition (11) constructed a nice transition function. Kaneko[11] also constructed an associated Hunt process on X under a stronger regularity condition when X is a locally compact separable metric space and $r > 2$. The method is similar to the Dirichlet space case[6]. The analyticity of the Markovian semigroup is automatically satisfied when the semigroup on L^p is determined by the semigroup of symmetric operators on L^2 (cf. E.M.Stein[13]). In this case, we practically start with a Dirichlet form on L^2 but in many cases we then encounter difficulties in checking the regularity (11) for $(r,p) \neq (1,2)$. See Fukushima-Jacob-Kaneko[10] dealing with a pseudo-differential operator in this connection.
(iii) In the last decade, the theory of Dirichlet forms (which corresponds to the case $(r,p) = (1,2)$) has been well developed on a general (not necessarily locally compact) topological space X. This is what is presented in Röckner's lectures in this course. Here a tightness hypothesis of the capacity plays a crucial role.
(iv) As for further studies of the spaces $(\mathcal{F}_{r,p}, \| \|_{r,p})$ on general topologocal spaces X, we refer to Albeverio et al [4],Fukushima[8] and Kazumi-Shigekawa[12]. Here the following tightness of $C_{r,p}$ is playing also important roles :

$$\inf\{C_{r,p}(X - K) : K \text{ compact}\} = 0.$$

References in §1

[1] D.R.Adams, Maximal operators and capacity, Proc.Amer.Math.Soc. 34(1972),152-156.

[2] S.Albeverio and M.Röckner, Dirichlet forms on topological vector spaces- the construction of the associated diffusion process, Probab.Th.Rel.Fields 83(1989),405-434.

[3] S.Albeverio and Z.M.Ma, A note on quasi continuous kernels representing quasi linear maps, Forum Math., 3(1991),389-400.

[4] S.Albeverio, M.Fukushima, W.Hansen,Z.M.Ma and M.Röckner, Capacities on Wiener space : tihgtness and invariance, C.R.Acad.Sci.Paris, 312(1991),931-935.

[5] J.A.Clarkson, uniform convex spaces, TAMS, 40(1936)¡396-414.

[6] M.Fukushima, Dirichlet forms and Markov processes, Kodansha and North-Holland, 1980

[7] M .Fukushima, Capacitary maximal inequalities and an ergodic theorem, in "Probability Thoery and Mathematical Statistics" eds.K.Ito and I.V.Prokhorov, LNM 1021 Springer, 1983

[8] M.Fukushima, (r,p)-capacities and Hunt processes in infinite dimensions, in Proceedings of 6-th Japan Soviet Symp.at Kiev,1991, World Scientific, Singapore, to appear

[9] M.Fukushima and H.Kaneko, On (r,p)-capacities for general Markovian semigroups, in "Infinite dimensional analysis and stochastic processes", ed. S.Albeverio, Pitman, 1985

[10] M.Fukushima, N.Jacob and H.Kaneko, On $(r,2)-$capacities for a class of elliptic pseudo differential operators, Math. Ann.293(1992),343-348

[11] H .Kaneko, On (r,p)-capacities for Markov processes, Osaka J. Math.,23(1986), 325-336.

[12] T.Kazumi and I.Shigekawa, Measures of finite (r,p)-evergy and potentials on a separable metric space, Preprint

[13] E.M.Stein, Topics in harmonic analysis, Annals Math. Studies 63, Princeton Univ. Press, 1970.

2 Convolution semigroups and (r,p)-quasi everywhere convergences of functions

In this section, we consider (r,p)-capacities related to the convolution semigroups on R^d and the unit circle. We show that $(r,p)-$q.e. convergences take places in differentiability of functions of $\mathcal{F}_{r,p}$, non-tangential limits of Poisson integrals of boundary functions in $\mathcal{F}_{r,p}$ and in Fourier series of functions in $\mathcal{F}_{r,p}$.

2.1 Quasi everywhere convergences on R^d

We let $X = R^d$ and $m =$ the Lebesgue measure. The L^p-space is denoted by $L^p(R^d)$ with norm $\| \ \|_p$. A system of probability measures $\{\nu_t, t > 0\}$ on R^d is called a continuous convolution semigroup if

$$\nu_t * \nu_s = \nu_{t+s} \tag{17}$$

$$\lim_{t \downarrow 0} \nu_t = \delta \text{ weakly} \tag{18}$$

where $\nu_t * \nu_s(E)$ denotes the convolution $\int_{R^d} \nu_t(E-y)\nu_s(dy)$ and δ is the unit mass at the origin. Such a $\{\nu_t, t > 0\}$ is characterized by a triple (m, S, J) as follows(the Lévy-Khinchin formula):

$$\begin{aligned} \hat{\nu}_t(\xi) &= e^{-t\psi(\xi)} \\ \psi(\xi) &= i(m,\xi) + \tfrac{1}{2}(S\xi,\xi) + \int_{R^d}\left(1 - e^{i(\xi,y)} + \tfrac{i(\xi,y)}{1+|y|^2}\right) J(dy) \end{aligned} \tag{19}$$

where $\hat{\nu}_t(\xi) = \int_{R^d} e^{i(\xi,y)}\nu_t(dy)$ and

$$\begin{aligned} m \in R^d, \quad S &\text{ is a non-negative definite } n \times n \text{ symmetric matrix and} \\ J &\text{ is a measure on } R^d \text{ with } \textstyle\int \frac{|y|^2}{1+|y|^2} J(dy) < \infty. \end{aligned} \tag{20}$$

We shall fix a continuous convolution semigroup $\{\nu_t, t > 0\}$ on R^d. It defines a transition function $p_t(x, E)$ on R^d by

$$p_t(x,E) = \nu_t(E - x)$$

and accordingly, for non-negative Borel f

$$p_t f(x) = \int_{R^d} f(x+y)\nu_t(dy). \tag{21}$$

Then $\{p_t, t>0\}$ satisfies all properties stated in Remark 1.2 with the dual semigroup $\hat{p}_t$ with respect to the Lebesgue measure being given by the continuous convolution semigroup $\hat{\nu}_t(E) = \nu_t(-E)$. Therefore $\{p_t, t>0\}$ of (21) determines uniquely a strongly continuous contraction semigroup of Markovian operators on $L^p(R^d)$ and we can consider the associated objects $V_r, \mathcal{F}_{r,p}, \| \|_{r,p}, C_{r,p}$ and the associated notions"$(r,p)-$q.e.","$(r,p)-$quasi continuous" etc.

Let us introduce a probability measure γ_r on R^d by

$$\gamma_r(E) = \frac{1}{\Gamma(r/2)} \int_0^\infty t^{\frac{r}{2}-1} e^{-t} \nu_t(E) dt \tag{22}$$

the Gamma transform of $\{\nu_t, t>0\}$. We then define a kernel v_r by $v_r(x,E) = \gamma_r(E-x)$, namely, we set

$$v_r f(x) = \int_{R^d} f(x+y)\gamma_r(dy), \ f \geq 0, \text{ Borel.} \tag{23}$$

In view of (1) , $v_r f$ is a Borel version of $V_r f$ for non-negative Borel $f \in L^p(R^d)$. Denote by $C_0(R^d)$ the family of continuous functions on R^d with compact support. We then see from (23) that $v_r\left(C_0(R^d)\right) \subset C(R^d)$. Since $C_0(R^d)$ is a dense linear sublattice of L^p, the regularity condition (11)is valid. Moreover we can see that $v_r f$ defined by (23) is an $(r,p)-$quasi continuous version of $V_r f$ for any Borel $f \in L^p(R^d)$, because the convergence of f_n in L^p is equivalent to that of $v_r f_n$ in $\mathcal{F}_{r,p}$ and we can appeal to Lemma 1.3 (iii) and the monotone lemma.

We first prove the $(r,p)-$quasi everywhere differentiability of functions in $\mathcal{F}_{r,p}$. Denote by $B(x,\epsilon)$ the ball of radius $\epsilon > 0$, centered at $x \in R^d$. $m(B(x,\epsilon)) = \epsilon^d \Omega_d$ is the Lebesgue measure of $B(x,\epsilon)$ where Ω_d is that of the unit ball.

Theorem 2.1 *If $u \in \mathcal{F}_{r,p}$, then the limit*

$$\lim_{\epsilon \downarrow 0} \frac{1}{m(B(x,\epsilon))} \int_{B(x,\epsilon)} u(y)dy \tag{24}$$

exists for $(r,p)-$q.e. $x \in R^d$ and the limitting function gives us an $(r,p)-$quasi continuous version of u.

Proof We set

$$\theta_\epsilon f(x) = \frac{1}{m(B(x,\epsilon))} \int_{B(x,\epsilon)} f(y)dy \tag{25}$$

θ_ϵ transforms L^p-functions into continuous L^p-functions. Let $\phi(x)$ be the indicator function of the solid unit sphere $\{x \in R^d : |x| \leq 1\}$ divided by Ω_d and let $\phi_\epsilon(x) = \epsilon^{-d}\phi(x/\epsilon)$. Then we have

$$\theta_\epsilon f(x) = (f * \phi_\epsilon)(x)$$

and we get from (23) and the Fubini theorem that, for Borel $f \in L^p(R^d)$,

$$\theta_\epsilon v_r f(x) = v_r \theta_\epsilon f(x) \ \forall x \in R^d. \tag{26}$$

Therefore condition (16) of Remark 1.3 is fulfilled.

We next introduce the Hardy-Littlewood maximal function $\hat{M}(f)$ by

$$\hat{M}(f)(x) = \sup_{\epsilon>0} \frac{1}{m(B(x,\epsilon))} \int_{B(x,\epsilon)} |f(y)|dy.$$

It is well known that $\hat{M}(f)$ is of strong type p (cf.E.M.Stein [15]): if $f \in L^p(R^d)$, with $1 < p \leq \infty$, then $\hat{M}f \in L^p(R^d)$ and

$$\|\hat{M}(f)\|_p \leq A_p \|f\|_p$$

where A_p depends only on p and dimension d. Therefore our maximal function

$$M(f)(x) = \sup_{\epsilon>0} |\theta_\epsilon f|(x)$$

satisfies the same bound

$$\|M(f)\|_p \leq A_p \|f\|_p. \tag{27}$$

Now Theorem 1.1 implies that

$$C_{r,p}(M(u) > \lambda) \leq \frac{A_p^p}{\lambda^p} \|u\|_{r,p}^p, \; u \in \mathcal{F}_{r,p}, \lambda > 0. \tag{28}$$

To prove Theorem 2.1, we let

$$R(u)(x) = \lim_{n\to\infty} \sup_{0<\epsilon,\epsilon'<1/n} |\theta_\epsilon u(x) - \theta_{\epsilon'} u(x)|.$$

Then, for any $v \in \mathcal{F}_{r,p} \cap C(R^d), R(u)(x) = R(u-v)(x) \leq 2M(u-v)(x)$ and by (28)

$$C_{r,p}(R(u) > \lambda) \leq \frac{2^p A_p^p}{\lambda^p} \|u - v\|_{r,p}^p, \tag{29}$$

which can be made arbitrarily small by (11). Since $\lambda > 0$ is arbitrary, we get $R(u) = 0$ $(r,p) - q.e$ arriving at the first conclusion of the theorem.

On the other hand, we write $u = v_r f, f \in L^p$ to get from (4) and (26) that

$$\|\theta_\epsilon u - u\|_{r,p} = \|\theta_\epsilon f - f\|_p,$$

which converges to zero as $\epsilon \downarrow 0$ by virtue of the Lebesgue theorem. Hence the $(r,p)-$q.e. limit of $\theta_\epsilon u$ must be an $(r,p)-$quasi continuous version of f. q.e.d.

We turn to proving an $(r,p)-$q.e. non-tangential boundary limit theorem for boundary functions in $\mathcal{F}_{r,p}$. We shall think of R^d as the boundary hyperplane of the $(d+1)$-dimensional upper-half space R_+^{d+1}:

$$R_+^{d+1} = \{(x,y) : x \in R^d, y > 0\}.$$

Its boundary $\{(x,0)\}$ is identified with R^d. We consider the Poisson integral Hf of a function f given on R^d :

$$Hf(x,y) = \int_{R^d} P_y(t) f(x-t) dt \tag{30}$$

where

$$P_y(x) = \frac{c_d y}{(|x|^2 + y^2)^{\frac{d+1}{2}}}, \quad c_d = \frac{\Gamma\left(\frac{d+1}{2}\right)}{\pi^{\frac{d+1}{2}}}. \tag{31}$$

For any $x^o \in R^d$ and $\alpha > 0$, we let

$$\Gamma_\alpha(x^o) = \{(x, y) \in R_+^{d+1} : |x - x^o| < \alpha y\} \tag{32}$$

the infinite cone with vertex x^o. If $u(x, y)$ is defined at those points in R_+^{d+1} near a boundary point $(x^o, 0)$, then u has *non-tangential limit* (which equals l) at $(x^o, 0)$ if for every $\alpha > 0$ the conditions $(x, y) \in \Gamma_\alpha(x^o)$ and $(x, y) \to (x^o, 0)$ imply that $u(x, y) \to l$.

Theorem 2.2 *If $u \in \mathcal{F}_{r,p}$, then the Poisson integral $(Hu)(x, y)$ admits a non-tangential limit at $(x^o, 0)$ for $(r, p)-$q.e.$x^o \in R^d$. The limitting function is an $(r, p)-$quasi continuous version of u.*

Proof We fix $\alpha > 0$. By using the bound

$$P_y(x - t) \leq A_\alpha P_y(x) \text{ if } |t| < \alpha y,\ y > 0,$$

A_α being a positive constant independent of f, we get the following estimate (cf. E.M.Stein[15;pp197]) :

$$\sup_{(x,y) \in \Gamma_\alpha(x^o)} |Hu(x, y)| \leq A_\alpha \hat{M}(u)(x^o).$$

Without loss of generality, we take non-negative $u \in \mathcal{F}_{r,p}$. Then $\hat{M}(u) = M(u)$. We let

$$\Gamma_\alpha^n(x^o) = \{(x, y) \in \Gamma_\alpha(x^o) : |x - x^o|^2 + y^2 < \frac{1}{n}\}$$

and

$$R(u)(x^o) = \lim_{n \to \infty} \sup_{(x,y),(x',y') \in \Gamma_\alpha^n(x^o)} |Hu(x, y) - Hu(x', y')|.$$

Then, for $v \in \mathcal{F}_{r,p} \cap C(R^d)$,

$$R(u)(x^0) = R(u - v)(x^o) \leq 2A_\alpha M(u - v)(x^o)$$

and by (28)

$$C_{r,p}(x^o \in R^d : R(u)(x^o) > \lambda) < \frac{2^p A_\alpha^p A_p^p}{\lambda^p} \|u - v\|_{r,p}^p. \tag{33}$$

Hence we get the $(r, p)-$q.e.existence of the non-tangential limit of Hf just as in the proof of the preceding thoerem.

To get the last conclusion of the theorem, we write $Hg(x, y)$ as $(Hg)_y(x)$ by regarding it as a function of $x \in R^d$ with $y > 0$ fixed. Then, by the expressions (23) and (30), we have for $u = v_r f$, $f \in L^p$,

$$(Hu)_y = v_r[(Hf)_y]$$

and by (4)

$$\|(Hu)_y - u\|_{r,p} = \|(Hf)_y - f\|_p,$$

which converges to zero as $y \downarrow 0$ (cf. E.M.Stein[15;pp62]). This means that Hu is convergent in $\mathcal{F}_{r,p}$ to u along radials and consequently the $(r, p)-$q.e.non-tangential limit of Hu must be an $(r, p)-$quasi continuous version of u. q.e.d.

Example 2.1 (Classical case) Take $\nu_t(dx) = g_t(x)dx$ of Example 1.1 corresponding to the d-dimensional standard Brownian motion. $\psi(\xi) = \frac{1}{2}|\xi|^2$ in this case. The space $\mathcal{F}_{r,p}$ appearing in Theorem 2.1 and Theorem 2.2 is just the classical Sobolev space and $(r,p)-$capacity is also the classical one. When $rp > d$, then each function of $\mathcal{F}_{r,p}$ admits a continuous version by the Sobolev theorem and Theorem 2.1 and Theorem 2.2 become trivial.

Example 2.2 (trivial case) Take $\nu_t(dx) = \delta(dx)$ the unit mass at the orgin. This correspond to the trivial case that $m = S = J = 0$. Since $p_t f(x) = v_r f(x) = f(x)$, we have $\mathcal{F}_{r,p} = L^p(R^d)$ $\|u\|_{r,p} = \|u\|_p$ and $C_{r,p}(E) = m(E)$. Hence Theorem 2.1 and Theorem 2.2 reduce to the known almost everywhere convergence results.

Example 2.3 (symmetric case) Consider a continuous convolution semigroup $\{\nu_t, t > 0\}$ of symmetric probability measures:

$$\nu_t(E) = \nu_t(-E). \tag{34}$$

Then the exponent ψ in the Lévy-Khinchin formula (18) reduces to

$$\psi(\xi) = \frac{1}{2}(S\xi,\xi) + \int_{R^d}(1 - \cos(\xi,y))J(dy) \tag{35}$$

with a symmetric measure J.

The corresponding space $\mathcal{F}_{r,2}, \|\ \|_{r,2}$ can be expressed as follows:

$$\begin{aligned} \mathcal{F}_{r,2} &= \left\{u \in L^2(R^d) : \textstyle\int_{R^d} |\hat{u}(\xi)|^2(1+\psi(\xi))^r d\xi < \infty\right\} \\ \|u\|^2_{r,2} &= \textstyle\int_{R^d}(1+\psi(\xi))^r|\hat{u}(\xi)|^2 d\xi. \end{aligned} \tag{36}$$

In fact the Fourier transform $\hat{\gamma}_r$ of the measure γ_r is the Gamma transform $e^{-t\psi(\xi)}$ and equal to $(1+\psi(\xi))^{-\frac{r}{2}}$. If $u = v_r f, f \in L^2$, then $\hat{u} = \hat{\gamma}_r \cdot \hat{f}$ and by the Plancherel theorem

$$\begin{aligned} \|u\|^2_{r,2} &= \|f\|^2_2 = \|\hat{f}\|^2_2 = \|(1+\psi(\xi))^{\frac{r}{2}}\hat{u}(\xi)\|^2_2 \\ &= \textstyle\int_{R^d}(1+\psi(\xi))^r|\hat{u}(\xi)|^2 d\xi \end{aligned}$$

When $r = 1$, (36) is a translation invariant Dirichlet form $(\mathcal{F}, \mathcal{E}_1)$ on $L^2(R^d)$: if $u \in \mathcal{F}$, then for any $x \in R^d, (\tau_x u)(\cdot) = u(x+\cdot) \in \mathcal{F}$ and $\mathcal{E}_1(\tau_x u, \tau_x u) = \mathcal{E}_1(u,u)$. Any traslation invariant Dirichlet form on $L^2(R^d)$ arises in this way. In his celebrted article of a CIME course[7], J.Deny formulated this for a general abelien group X.

Example 2.4 (uniform motion to the right) The space $\mathcal{F}_{r,p}$ for non-symmetric ν_t is complicated. For instance consider the case that $d = 1, \nu_t(dx) = \delta_t(dx)$ corresponding to $m = 1, S = J = 0$. Then

$$v_r f(x) = \frac{1}{\Gamma(r/2)}\int_0^\infty t^{\frac{r}{2}-1}e^{-t}f(x+t)dt$$

is continuous in $x \in R^1$ and hence $\mathcal{F}_{r,p} \subset C(R^1)$. Theorem 2.1 and Theorem 2.2 are trivial in this case. But when $d = 2, \nu_t(dx_1 dx_2) = \delta_t(dx_1)\delta_0(dx_2)$ (which corresponds to $m = (1,0), S = J = 0$), $(v_r f)(x_1,x_2)$ is continuous only in x_1.

2.2 Quasi everywhere convergences on the unit circle

We next study the case that $X = T = [-\pi, \pi)$ the unit circle and m = the Lebesgue measure on T. We first formulate obvious analogues to Theorem 2.1 and Theorem 2.2. We fix a continuous convolution semigroup $\{\nu_t, t > 0\}$ of probability measures on T. $\{p_t, t > 0\}$ defined by (21) determines uniquely a strongly continuous contraction semigroup of Markovian operators on $L^p(T)$ and hence the associated objects $V_r, \mathcal{F}_{r,p}, \| \ \|_{r,p}, C_{r,p}$ and the associated notions "$(r,p)-$q.e." ,"$(r,p)-$quasi continuous" can be considered. In what follows, functions on T are extended periodically with period 2π to R^1. We introduce the Gamma transform of $\{\nu_t, t > 0\}$ by (22) and set

$$v_r f(x) = \int_T f(x+y)\gamma_r(dy), \ x \in T. \tag{37}$$

We can see as in the case of R^d that the space $\mathcal{F}_{r,p}$ satisfies the regularity (11) and that $v_r f$ is an $(r,p)-$quasi continuous version of $V_r f$ for Borel $f \in L^p(T)$.

Theorem 2.3 *If $u \in \mathcal{F}_{r,p}$, then the limit*

$$\lim_{\epsilon \downarrow 0} \frac{1}{2\epsilon} \int_{x-\epsilon}^{x+\epsilon} u(y)dy \tag{38}$$

exists for $(r,p)-q.e. x \in T$ and the limitting function is an $(r,p)-$quasi continuous version of u.

In fact we can proceed as in the proof of Theorem 2.1 using the Hardy- Littlewood maximal inequality(cf.Zygmund[16;pp33]):

$$\|\hat{M}(f)\|_p^p \le 4\left(\frac{p}{p-1}\right)^p \|f\|_p^p, \ p > 1,$$

where

$$\hat{M}(f) = \sup_{0<t\le\pi} \frac{1}{2t} \int_{x-t}^{x+t} |f(u)|du.$$

We next introduce the Poisson kernel

$$P_r(x) = \frac{1}{2} \frac{1-r^2}{1-2r\cos x + r^2}, \ 0 \le r < 1$$

and the Poisson integral

$$Hf(x,r) = \frac{1}{\pi} \int_{-\pi}^{\pi} P_r(t) f(x-t)dt, \ x \in T, 0 \le r < 1.$$

Exactly in the same way as in the proof of Theorem 2.2, we can prove

Theorem 2.4 *If $u \in \mathcal{F}_{r,p}$, then the Poisson integral $(Hf)(x,r)$ admits a non-tangential limit at $(x^\circ, 1) \in T$ for $(r,p) - q.e. x^\circ \in T$. The limitting function is an $(r,p)-$quasi continuous version of u.*

We can also formulate the $(r,p)-$quasi everywhere convergence of Fourier series. Let us consider the n-th partial sum of the Fourier series of function f on T:

$$S_n(f)(x) = \sum_{k=-n}^{n} \hat{f}(k)e^{ikx}, \ \hat{f}(k) = \frac{1}{2\pi}\int_{-\pi}^{\pi} e^{-ikx}f(x)dx. \tag{39}$$

Theorem 2.5 *If $u \in \mathcal{F}_{r,p}$, then its Fourier partial sum $S_n(u)(x)$ converges as $n \to \infty$ $(r,p) - q.e.$ on T to an $(r,p)-$quasi continuous version of u.*

Proof Since $S_n(f)$ is expressed as

$$S_n(f)(x) = \frac{1}{\pi}\int_{-\pi}^{\pi} f(t)D_n(t-x)dt, \ D_n(x) = \frac{\sin(n+\frac{1}{2})x}{2\sin\frac{1}{2}x},$$

by means of the Dirichlet kernel D_n, we get from the expression (37) the commutativity

$$S_n v_r f = v_r S_n f. \tag{40}$$

Hence the condition (13) of Theorem 1.1 is fulfilled by $S_n, n = 1,2,\cdots$. Further the maximal function $Mf(x) = \sup_n |S_n f(x)|, x \in T$, admits the Hunt estimates ([12],[11])

$$\|Mf\|_p \le A_p\|f\|_p, \ f \in L^p(T),$$

for some positive constant A_p. Therefore Theorem 1.1 applies and we get

$$C_{r,p}(|Mf| > \lambda) \le \frac{A_p^p}{\lambda^p}\|f\|_p^p, \ f \in L^p(T). \tag{41}$$

Denote by $C^1(T)$ the space of continuously differentiable functions on T. $S_n(f)$ converges uniformly on T to f for $f \in C^1(T)$. On the other hand, we notice that the transformation v_r of (37) makes the space $C^1(T)$ invariant and consequently $\mathcal{F}_{r,p} \cap C^1(T)$ is dense in $\mathcal{F}_{r,p}$. Now we can make use of (41) just as in the proof of Theorem 2.1 to derive the $(r,p)-$q.e convergence of $S_n(u)$ for $u \in \mathcal{F}_{r,p}$. Writing u as $u = v_r f, \ f \in L^p$, we get from (40) $S_n(u) = v_r S_n f$ and

$$\|S_n(u) - u\|_{r,p} = \|S_n f - f\|_p,$$

which converges to zero as $n \to \infty$. Hence the limitting function must be an $(r,p)-$quasi continuous version of u. q.e.d.

Example 2.5 (trivial case) If $\nu_t(dx) = \delta(dx)$,then $\mathcal{F}_{r,p} = L^p(T), \|u\|_{r,p} = \|f\|_p$ and $C_{r,p}(E) = m(E)$. Theorem 2.3,2.4 and 2.5 reduce to the classical almost everywhere convergence results.

Example 2.6 (symmetric case) If we require the symmetry of ν_t

$$\nu_t(E) = \nu_t(-E),$$

then the associated space $\mathcal{F}_{r,2}$ can be described explicitly. Denote by Λ the set of all real sequences $\lambda = \{\lambda_n\}_{n=-\infty}^{n}$ satisfying the following conditions:

$$\lambda_0 = 0, \lambda_n = \lambda_{-n}, \sum(\lambda_n + \lambda_m - \lambda_{n-m})\rho_n\rho_m \geq o$$

for any real sequence $\{\rho_n\}$ with finite support. $\lambda_n = |n|^\alpha, 0 < \alpha \leq 2, \lambda_n = \log(|n|+1)$ are elements of Λ. It is known (cf. [9: pp31]) that Λ stands in one to one correspondence with the family of continuous *symmetric* convolution semigroups $\{\nu_t\}$ by

$$\hat{\nu}_t(n) = e^{-t\lambda_n},\ \hat{\nu}_n(n) = \int_{-\pi}^{\pi} e^{in\pi}\nu_t(dx).$$

We have then the following expression:

$$\begin{aligned} \mathcal{F}_{r,2} &= \left\{u \in L^2(T) : \textstyle\sum_{k=-\infty}^{\infty} |\hat{u}(k)|^2 \lambda_k^r < \infty\right\} \\ \|u\|_{r,2}^2 &= 2\pi \textstyle\sum_{k=-\infty}^{\infty} |\hat{u}(k)|^2 (1+\lambda_k)^r. \end{aligned}$$

Since

$$\hat{\gamma}_r(n) = (1+\lambda_n)^{-\frac{r}{2}}, \widehat{v_r f}(n) = (1+\lambda_n)^{-\frac{r}{2}} \cdot \hat{f}(n),$$

we get for $u = v_r f, f \in L^2$

$$\|u\|_{r,2} = \|f\|_2 = \sum_{k=-\infty}^{\infty} |\hat{u}(k)|^2 (1+\lambda_k)^r,$$

by Perseval's formula.

In case that $\lambda_k = \frac{|k|}{2}$, the corresponding Dirichlet space($r = 1$)

$$\begin{aligned} \mathcal{F}_{1,2} &= \left\{u \in L^2(T) : \textstyle\sum_{k=-\infty}^{\infty} |\hat{u}(k)|^2 |k| < \infty\right\} \\ \mathcal{E}(u,v) &= \pi \textstyle\sum_{k=-\infty}^{\infty} \hat{u}(k)\bar{\hat{v}}(k)|k| \end{aligned}$$

is unitary eqivalent to the space of harmonic functions on the unit disk $D = \{|z| < 1\}$ with finite Dirichlet integrals(cf.[9; pp12]):

$$\frac{1}{2}\mathrm{D}(Hu, Hv) = \mathcal{E}(u,v),\ u, v \in \mathcal{F}_{1,2}.$$

It was Beurling [3] who first proved that any harmonic function on D with finite Dirichlet integral admits the limit along the radial to $x \in T$ except for a set of points of T whose planer logarithmic capacity is zero. Thus Theorem 2.4 may be thought of as an extension of Beurling's theorem. As for the identification of the set of zero logarithmic capacity and (1,2)- polar set with respect to the sequence $\lambda_k = \frac{|k|}{2}$, we refer to R.Bañuelos and B.Oksendal[1].

Actually Bañuelos and Oksendal[1] has proven the present Theorem 2.4 for $(r,p) = (1,2)$ (the Dirichlet space case) and under the symmetry requirement. Their method is similar to ours but different in that they employ, instead of the non-tangential limit, the limit along almost all conditional Brownian paths. Those two notions are known to be equivalent only when $d = 2$(cf.Durrett[8]). As is demonstrated in Theorem 2.2, our method works in higher dimensions as well.

As for related studies on q.e. convergences of harmonic functions and Fourier series, we refer to Carleson[4],Preston[13],Crezeiro[5],[6] and Bary[2], Zygmund[16]. Almost everywhere tangential limits were studied in Nagel, Rudin and Shapiro[14] in which a strong type capacitary inequality (but not a capacitary maximal inequality) was utilized.

References in §2

[1] R.Bañuelos and B.Oksendal, A stochastic approach to quasi everywhere boundary convergence of harmonic functions, J.Funct.Anal. 72(1987),13-27

[2] N.K.Bary, A treatise on trigonometric series, Pergammon, Oxford, 1964

[3] A. Beurling, Ensembles exceptinelles, Acta Math. 72(1940),1-13

[4] L. Carleson, Selected problem on exceptional sets, Van Nostrand, Princeton,1967

[5] Ana Bela Cruzeiro, Convergence quasi partout dans des domains paraboliques des fonctions d'intégrale de Dirichlet finie, C.R.A.S.Paris,t.294(1982),13-16

[6] Ana Bela Cruzeiro, Convergence au bord pour les fonctions harmoniques dans R^d de la class de Sobolev W_1^d, C.R.A.S.Paris, t294(1982),71-74

[7] J. Deny, Méthods Hilbertiennes et theorie du potentiel, Potential Theory, CIME, Edizioni Cremonese, Roma, 1970

[8] R.Durrett, Brownian motion and martingales in analysis,Wadsworth, Belmond,Calif. 1984

[9] M.Fukushima, Dirichlet forms and Markov processes, Kodansha and North Holland, 1980

[10] M. Fukushima, Capacitary maximal inequalities and an ergodic theorem, in "Probability theory and Mathematical statistics",eds.K.Ito and J.V.Prohorov, LNM 1021, Springer, 1983

[11] O.G.Jorsboe and L.Mejlbro,The Carleson-Hunt theorem on Fourier series, LNM 911, Springer, 1982

[12] R.A. Hunt, On the convergence of Fourier series,in Proc.S.I.U. Conf. on Orthogonal expansions, Southern Illinois Univ.Press,Carbondale,1968

[13] C.J.Preston, A theory of capacities and its application to some convergence results,Adv.Math. 6(1971),78-106

[14] A.Nagel, W.Rudin and J.H.Shapiro, Tangential boundary behavior of function in Dirichlet-type spaces, Ann.Math.116(1982),331-360

[15] E.M.Stein, Singular integrals and differentiability properties of functions, Princeton Univ. Press, 1970

[16] A.Zygmund, Trigonometric series, Cambridge Univ. Press, 1968

3 (r,p)–quasi everywhere convergences of martingales on the Wiener space

Thanks to specific structures of the Wiener space and the Ornstein-Uhlenbeck semigroup on it, we can give an application of Theorem 1.1 to the quasi everywhere convergences of martingales based on the Brownian filteration.

Consider the Wiener space

$$X = C_0([0,\infty), R) = \{w : [0,\infty) \to R \text{ continuous } w(0) = 0\}$$

equipped with the σ–field $\mathcal{B} = \sigma\{b_t; t \geq 0\}$, where $b_t = b_t(w)$ denotes the t–th coodinate map. Let m be the Wiener measure on X : the process $\{b_t, t \geq 0\}$ is of independent increments and $b_t - b_s, t > s$, has the Gaussian distribution $N(0, t-s)$.

The Ornstein-Uhelnbeck transition function on $(X, \mathcal{B})$ is given by the Mehler formula

$$p_t f(x) = \int_X f\left(e^{-t}x + \sqrt{1-e^{-2t}}y\right) m(dy) \tag{42}$$

for non-negative Borel f. p_t is m-symmetric and satisfies conditions in Remark 1.2. Therefore it decides uniquely a strongly continuous Markovian semigroup $\{T_t, t > 0\}$ on $L^p = L^p(X; m), p > 1$. The associated notions $V_r, \mathcal{F}_{r,p}, \| \|_{r,p}$ and $C_{r,p}$ are then introduced according to §1. $V_r f, f \in L^p$, has a representative $v_r f$ defined by

$$v_r f(x) = \frac{1}{\Gamma(r/2)} \int_0^\infty t^{\frac{r}{2}-1} e^{-t} p_t f(x) dt. \tag{43}$$

A function f on X of the form

$$f(x) = p(\ell_1(x), \cdots, \ell_n(x)),\ x \in X, n = 0, 1, \cdots,$$

is called a *polynomial* where p is a polynomial of n real variables and $\ell_1, \cdots, \ell_n \in X^*$ the topological dual space. The totality of polynomials is denoted by $\mathcal{P}$. Let

$$L^2 = \sum_{n=0}^{\infty} \oplus Z_n$$

be the Wiener chaos decomposition and J_n be the orthogonal projection on the n-th Wiener chaos Z_n. Since $p_t f = \sum_{n=0}^{\infty} e^{-nt} J_n f$ (finite sum) for $f \in \mathcal{P}$ (cf.S.Watanabe[5]), we have

$$v_r f = \sum_{n=0}^{\infty} (1+n)^{-\frac{r}{2}} J_n f \text{ (finite sum)},\ f \in \mathcal{P}. \tag{44}$$

Hence we see that v_r makes the space $\mathcal{P}$ invariant, $\mathcal{P}$ is a dense subset of $\mathcal{F}_{r,p}$ and accordingly $(\mathcal{F}_{r,p}, \| \|_{r,p})$ satisfies the regularity condition (11). $\mathcal{F}_{r,p}$ coincides with the space $\mathbf{D}_{p,r}$ appearing in S.Watanabe[5].

We let

$$\mathcal{B}_t = \sigma\{b_s; s \leq t\}, \tag{45}$$

$$f_t = E(f|\mathcal{B}_t),\ t \geq 0, f \in L^p, \tag{46}$$

where $E(f|\mathcal{B}_t)$ denotes the conditional expectation with respect to the Wiener measure m.

Theorem 3.1 *For any $u \in \mathcal{F}_{r,p}$, there exists a function $u(t,x), t \geq 0, x \in X$, such that $u(\cdot,x)$ is continuous for $(r,p)-$q.e. fixed $x \in X$,*
$u(t,\cdot)$ is an (r,p)-quasi continuous version of u_t for each $t \geq 0$ and
$\tilde{u}(x) = \lim_{t\to\infty} u(t,x)$ exists for $(r,p)-q.e.x \in X$, $\tilde{u}$ being an $(r,p)-$quasi continuous version of u.

Proof We first note that $v_r f(x), x \in X$, is an $(r,p)-$quasi continuous version of $V_r f$ for any Borel $f \in L^p$. This can be seen just as in the paragraph preceding Theorem 2.1. We further know that, for $f \in L^p$,

$$f_t \in L^p, \ \|f_t\|_p \leq \|f\|_p \tag{47}$$

$$M(f)(x) = \sup_{t\in Q_+} |f_t(x)| \in L^p, \ \|M(f)\|_p \leq \frac{p}{p-1}\|f\|_p, \tag{48}$$

where Q_+ denotes the set of positive rationals.

In view of the expression of elements of Z_n as the n-th repeated Ito integrals (multiple Wiener integrals), we easily derive from (44) the commutativity

$$(v_r f)_t \ = \ v_r(f_t) \tag{49}$$

for $f \in \mathcal{P}$ (cf. Bouleau-Hirsch[1 ;pp129]) and this is readily extended to $f \in L^p$.

Now take any $u \in \mathcal{F}_{r,p}$ and let $u = v_r f, f \in L^p$. We set

$$u(t,x) = v_r(f_t)(x), t \in Q_+, x \in X, \tag{50}$$

by choosing a Borel version of f_t for each $t \in Q_+$. Theorem 1.1 is then applicable and we get

$$C_{r,p}\left(\sup_{t\in Q_+} |u(t,x)| > \lambda\right) \leq \frac{1}{\lambda^p}\left(\frac{p}{p-1}\right)^p \|u\|^p_{r,p}. \tag{51}$$

On the other hand, if $v \in \mathcal{P}$ is expressed as

$$v(x) = p_1\left(b_{t_1}(x)\right) p_2\left(b_{t_2}(x)\right)\cdots p_n\left(b_{t_n}(x)\right), 0 < t_1 < t_2 < \cdots < t_n, x \in X,$$

for some polynomials $p_1, p_2, \cdots, p_n$ of one variable, then the conditional expectation v_t has a version $v(t,x), t \geq 0, x \in X$, which is uniformly continuous in $t \geq 0$ and constant for large t for each fixed $x \in X$. In fact, by virtue of the Markov property of the one dimensional Brownian motion, we may take

$$v(t,x) = \begin{cases} v_{t,t_1,\cdots,t_n}\left(b_t(x)\right) & 0 \leq t < t_1 \\ p_1\left(b_{t_1}(x)\right)\cdots p_{k-1}\left(b_{t_{k-1}}(x)\right) v_{t,t_k,\cdots,t_n}\left(b_t(x)\right) & t_{k-1} \leq t < t_k \\ v(x) & t \geq t_n \end{cases}$$

where

$$v_{t,t_k,\cdots,t_n}(\xi) = q_{t_k-t}\left[p_k\left\{q_{t_{k+1}-t_k}\cdots p_{n-1}\left(q_{t_n-t_{n-1}}p_n\right)\right\}\right](\xi)$$

with

$$q_s h(\xi) = \frac{1}{\sqrt{2\pi s}}\int_{R^1} e^{-\frac{(\xi-\eta)^2}{2s}} f(\eta)d\eta, \xi \in R^1.$$

Denote by $\mathcal{P}_0$ the linear span of polynomials of the above expression. We let

$$R_1(u)(x) = \lim_{n\to\infty} \sup_{|t-t'|<\frac{1}{n}, t,t'\in Q_+} |u(t,x) - u(t',x)|, x \in X$$

and

$$R_2(u)(x) = \lim_{n\to\infty} \sup_{t,t'>n, t,t'\in Q_+} |u(t,x) - u(t',x)|, x \in X.$$

Then, for $v \in \mathcal{P}_0$,

$$R_i(u)(x) = R_i(u-v)(x) \leq 2 \sup_{t\in Q_+} |u(t,x) - v(t,x)|, i = 1,2,$$

and by (51)

$$C_{r,p}(R_i(u) > \lambda) \leq \frac{2^p}{\lambda^p}\left(\frac{p}{p-1}\right)^p \|u-v\|_{r,p}^p, i = 1,2. \tag{52}$$

Since $\mathcal{P}_0$ is dense in $\mathcal{F}_{r,p}$, we have that $R_1(u) = 0$ and $R_2(u) = 0$ $(r,p) - q.e.$

Now it suffices to set

$$u(t,x) = \overline{\lim}_{t'\to t, t'\in Q_+} u(t',x), t \geq 0, x \in X.$$

$u(t,x)$ is then continuous in $t \geq 0$ for (r,p)–q.e. fixed $x \in X$ and admits an (r,p)–q.e. limit $\tilde{u}(x) = \lim_{t\to\infty} u(t,x)$. $u(t,x)$ and $\tilde{u}(x)$ are (r,p)–quasi continuous versions of u_t and u respectively because

$$\lim_{t'\to t, t'\in Q_+} f_{t'} = f_t, \lim_{t\to\infty} f_t = f \text{ in } L^p$$

and hence

$$\lim_{t'\to t, t'\in Q_+} v_r f_{t'} = v_r f_t = u_t, \lim_{t\to\infty} v_r f_t = v_r f = u \text{ in } \mathcal{F}_{r,p}.$$

Remark 3.1 Theorem 3.1 for $(r,p) = (1,2)$ has been proven by Bouleau-Hirsch[1;III.3] by a different method, where a stronger assertion that $u(t,x)$ can be taken to be jointly continuous in $(t,x) \in [0,\infty) \times F_k$ for some $(1,2)$–nest $\{F_k\}$.

In subsections 1.2 and 1.3, we have seen that Theorem 1.1 is well applied to the cases where θ_s commutes with V_r. Since T_s itself commutes with V_r in general, we can formulate (r,p)–q.e. convergence results for the semigroups as $t \downarrow 0$ and $t \to \infty$. See Fukushima[3] and Kaneko[4] in this connection.

References in §3

[1] N.Bouleau and F.Hirsch, Dirichlet forms and analysis on Wiener space, Walter de Gruyter, 1991

[2] C.Dellacherie and P.Meyer, Probability and potential B, North-Holland, 1982

[3] M. Fukushima,Capacitary maximal inequalities and an ergodic theorem,in "Probability Theory and Mathematical Statistics", eds.K.Ito and J.V.Prohorov,LNM 911, Springer,1982

[4] H.Kaneko, On (r,p)–capacities for Markov processes, Osaka J.Math.,23(1986),325-336

[5] S.Watanabe, Lectures on stochastic differential equations and Malliavin calculus,Tata Institute of Fundamental Research,vol.73, Springer-Verlag, 1984

Chapter 2 Additive functionals related to Dirichlet forms

4 Analytic characterizations of positive additive functionals

4.1 Revuz corespondence and Baxter-Dal Maso-Mosco correspondence

One of the most important concepts in the study of Markov processes is the positive additive functional. In this subsection, we shall give a brief historical survey on analytic characterizations of positive additive functionals of Markov processes. There have been two different but related approaches.

Let X be a locally compact separable metric space and $\mathbf{M} = (\Omega, X_t, \zeta, P_x)$ be a Hunt process on X : the sample function $X.$ is right continuous and has left limits on $[0, \infty)$ with values in the one point compactification $X_\Delta = X \cup \Delta$; $\mathbf{M}$ is strong Markov and quasi-left continuous. The filtration $\mathcal{F}_t, t \geq 0$, involved is taken to be the minimal completed σ-field containing $\sigma\{X_s : s \leq t\}$ (cf.[4]). The transition function of $\mathbf{M}$ is denoted by p_t:

$$p_t f(x) = E_x\left(f(X_t)\right), \; x \in X.$$

for any bounded Borel function f on X.

An $\mathcal{F}_t$-adapted extended real valued process $A_t(\omega), t \geq 0, \omega \in \Omega$, is called an *additive functional* (AF in abbreviation) if the following conditions are satisfied:

$$\text{there exists } \mathcal{L} \subset \Omega \text{ such that } P_x(\mathcal{L}) = 1 \text{ for every } x \in X, \tag{53}$$

and for each $\omega \in \mathcal{L}$

(A.1) $A_{t+s}(\omega) = A_s(\omega) + A_t(\theta_s\omega), A_t(\omega) = A_{\zeta(\omega)}(\omega) \; \forall t \geq \zeta(\omega)$,

(A.2) $A_t(\omega)$ is right continuous, has left limit for $t \geq 0$ and finite for any $t \in [0, \zeta(\omega))$.

Here θ_s is the transformation on Ω induced by the shift: $X_t(\theta_s(\omega)) = X_{t+s}(\omega), t \geq 0$. The set $\mathcal{L}$ in the above is called the *defining set* of the AF $A_t(\omega)$.

The notions of a *positive continuous additive functional* (PCAF in abbreviation) and a *positive additive functional* (PAF in abbreviation) are defined by replacing the condition (A.2) in the above with

(A.2)' $A_t(\omega)$ is continuous in $t \geq 0$ and non-negative finite for $t \in [0, \zeta(\omega))$,

(A.2)" $A_t(\omega)$ is right continuous in $t \geq 0$ and non-negative for $t \geq 0$. Further $A_{t-\epsilon}(\theta_\epsilon\omega)$ is right continuous in $\epsilon \in [0, t)$,

respectively. Notice that, in the condition (A.2)" for PAF, no finiteness of $A_t(\omega)$ is required. Hence the second condition in (A.2)" does not necessarily follow from the right continuity of $A_\epsilon(\omega)$.

Two AF's $A_t^{(1)}, A_t^{(2)}$ are regarded to be equivalent if

$$P_x\left(A_t^{(1)} = A_t^{(2)}, t \geq 0\right) = 1, \forall x \in X.$$

The simplest example is provided by a Borel function g on X if we let

$$A_t^g(\omega) = \int_0^{t\wedge\zeta(\omega)} g(X_s(\omega))ds, t > 0,$$

which is AF,PCAF and PAF respectively if g is bounded, bounded non-negative and non-negative. We fix a σ−finite Borel measure m on X which is p_t−excessive : $mp_t \leq m$. We look at the following correspondence between functionals and measures

$$A_t^g \longleftrightarrow g \cdot m$$

and try to extend this relation to a broader families of functionals and measures by continuity. Such a relation has been established first for the (equivalence classes of) family

$$\mathbf{A}_{c,1}^+ = \{A_t : PCAF of\ \mathbf{M}\}$$

by McKean-Tanaka([7],1961) and Wentzell-Dynkin([3],[12],1961) in the case that $\mathbf{M}$ is the multidimensional Brownian motion and by Revuz([8],1970) for a more general $\mathbf{M}$.

Indeed Revuz made a precise statement of the correspondence as follows. Suppose that $\mathbf{M}$ is in duality with another Hunt process $\hat{\mathbf{M}}$ with respect to the measure m in the sense of remark 1.2. Assume further the absolute continuity of the resolvent:

$$\exists g_\alpha(x,y), \alpha \geq 0, x, y \in X, e^{-\alpha t}p_t - \text{excessive in } x, e^{-\alpha t}\hat{p}_t - \text{excessive in } y,$$

such that for any non-negative Borel f

$$\begin{aligned} \int_0^\infty e^{-\alpha t}p_t f(x)dt &= \int_X g_\alpha(x,y)f(y)m(dy) \\ \int_0^\infty e^{-\alpha t}\hat{p}_t f(x)dt &= \int_X g_\alpha(y,x)f(y)m(dy). \end{aligned}$$

Then there is a one to one correspondence

$$\mathbf{A}_{c,1}^+ \ni A_t \longleftrightarrow \mu \in S_1$$

characterized by the relation

$$\lim_{t\downarrow 0} \frac{1}{t} E_m\left(\int_0^t f(X_s)dA_s\right) = \langle f, \mu\rangle, \forall f \geq 0, \tag{54}$$

where S_1 denotes the collection of all Borel measures μ on X satisfying the following conditions: μ charges no semipolar set and there exists Borel sets E_n increasing to X such that

$$\mu(E_n) < \infty, G_1\left(I_{E_n} \cdot \mu\right)(x) \text{ is bounded, } n = 1, 2, \cdots, \tag{55}$$

$$P_x\left(\lim_{n\to\infty} \sigma_{X-E_n} \geq \zeta\right) = 1, \forall x \in X. \tag{56}$$

In particular, if μ is finite, charges no semipolar set and $G_1\mu$ is bounded, then

$$E_x\left(\int_0^\infty e^{-t}dA_t\right) = G_1\mu(x), x \in X, \tag{57}$$

characterizes the associated $A \in \mathbf{A}^+_{c,1}$. (54) is called the *Revuz correspondence.*

The above description of the class S_1 of measures is rather intricate and this is caused by the strigent finiteness requirement imposed on PCAF. For instance $A_t = \int_0^t |X_s|^{-2}ds$ is not in $\mathbf{A}^+_{c,1}$ for the d-dimensional Brownian motion because $P_0(A_t = \infty) = 1$ on account of the law of the iterated logarithm, while $|x|^{-2}dx$ is a nice Radon measure on $R_d(d \geq 3)$ charging no polar set.

So far there have been two significant ways to relax the finiteness condition on PCAF:

1. PCAF admitting exceptional polar set on the Dirichlet space setting (Silverstein[9], 1974, Fukushima[4],1980)

2. HRM(homogeneous random measure) in the weak duality setting(Getoor- Sharpe[6], 1984).

In both settings, the Revuz correspondence (54) still makes sense. We denote by S (resp. $\tilde{S}$)the totality of the corresponding family of Revuz measures in the former (resp. latter) setting. S contains all positive Radon measures charging no polar set, while $\tilde{S}$ includes all $\sigma-$ finite Borel measures charging no semipolar set. Speaking of the above mentioned example, the former setting regards the one point set $\{0\}$ as exceptional for A_t and the latter regards A_t as a $\sigma-$finite positive measure on $(0, \infty)$ rather than on $[0, \infty)$. In both settings, $\mu = |x|^{-2}dx$ becomes the associated Revuz measure.

When $\mathbf{M}$ is the Brownian motion, a significant progress has been made recently by Baxter-Dal Maso-Mosco([1],1987) and Sturm([10],[11],1992) in establishing the correspondence between all Borel measures charging no polar set and all PAF's. Denote by $\mathcal{M}_0$ all positive Borel measures on R^d charging no polar set. For any $\mu \in \mathcal{M}_0$, let

$$\begin{aligned} \mathcal{F}^{(\mu)} &= H^1(R^d) \cap L^2(R^d : \mu) = \left\{u \in H^1(R^d) : \int \tilde{u}(x)^2\mu(dx) < \infty\right\} \\ \mathcal{E}^{(\mu)}(u,v) &= \frac{1}{2}\mathbf{D}(u,v) + \int_{R^d} \tilde{u}(x)\tilde{v}(x)\mu(dx),\ u, v \in \mathcal{F}^{(\mu)}. \end{aligned} \tag{58}$$

Here $\mathbf{D}$ denotes the Dirichlet integral and $\tilde{u}, \tilde{v}$ are the quasi-continuous versions. Let $L^2_\mu(R^d)$ be the closure of $\mathcal{F}^{(\mu)}$ in $L^2(R^d)$. $L^2_\mu(R^d)$ may be a proper subspace. Nevertheless $\left(\mathcal{F}^{(\mu)}, \mathcal{E}^{(\mu)}\right)$ is a closed form on $L^2_\mu(R^d)$. Denote by A^μ the corresponding self-adjoint operator on $L^2_\mu(R^d)$. A^μ may be thought of a realization of the Scrödinger operator

$$-\frac{1}{2}\Delta + \mu.$$

The resolvent $G^\mu_\alpha f = (\alpha - A^\mu)^{-1}f, f \in L^2_\mu(R^d)$, is characterized by

$$G^\mu_\alpha f \in \mathcal{F}^{(\mu)}, \mathcal{E}^{(\mu)}_\alpha(G^\mu_\alpha f, v) = \int_{R^d} fv dx,\ \forall v \in \mathcal{F}^{(\mu)}.$$

We introduce an equivalent relation $\sim$ in $\mathcal{M}_0$ by

$$\mu \sim \nu \Leftrightarrow (\mathcal{F}^{(\mu)}, \mathcal{E}^{(\mu)}) = (\mathcal{F}^{(\nu)}, \mathcal{E}^{(\nu)}). \tag{59}$$

It turns out that

$$\mu \sim \nu \Leftrightarrow \mu(F) = \nu(F) \text{ for any finely open Borel set } F \subset R^d.$$

For a PAF A_t of the Brownian motion on R^d, we set

$$\varphi^A f(x) = E_x\left(\int_0^\infty e^{-t-A_t} f(X_t)dt\right), x \in R^d, \tag{60}$$

which is well defined without finiteness condition on A_t.

The Baxter-Dal Maso-Mosco correspondence between all PAF's of the Brownian motion and the class $\mathcal{M}_0/\sim$ of measures is given by

$$\begin{gathered}\text{PCA } A_t \longleftrightarrow \mu \in \mathcal{M}_0/\sim \\ \varphi^A f = G_1^\mu P f, \forall f \in L^2(R^d),\end{gathered} \tag{61}$$

where P is the projection in $L^2(R^d)$ onto $L^2_\mu(R^d)$. Baxter,Dal Maso and Mosco[1] proved the correspondence $\longleftarrow$, while Sturm[10] provided a proof of the other correspondence $\longrightarrow$. (61) may be thought of the most general Feynman-Kac type realization of the solution of the equation

$$\left(-\frac{1}{2}\Delta + m + \mu\right)u = f.$$

A crucial observation in the proof is that any $\mu \in \mathcal{M}_0$ admits a $\nu \in S_1$ with $\mu \sim q \cdot \nu$ for some Borel $q \geq 0$. ν can be taken even to be of bounded potential. For $\mu \in \mathcal{M}_0$, one introduces its *permanent set* E^μ by

$$E^\mu = \left\{x \in R^d : \varphi^A 1(x) > 0\right\}, \tag{62}$$

where A is the PAF associated with μ by (61). An analytical description of E^μ is

$$E^\mu = \left\{x \in R^d : G_1(I_G \cdot \mu)(x) > 0 \text{ for some fine neighbourhood } G \text{ of } x\right\}.$$

It then turns out ([11])that

$$\begin{aligned} L^2_\mu(R^2) &= \left\{f \in L^2(R^d) : f = 0 \ a.e. \text{ on } R^d - E^\mu\right\}, \mu \in \mathcal{M}_0, \\ \mu \in S_1 &\Leftrightarrow E^\mu = R^d \\ \mu \in S &\Leftrightarrow E^\mu = R^d \ q.e. \end{aligned}$$

An extreme element $\overline{\infty} \in \mathcal{M}_0$ is defined by

$$\overline{\infty}(B) = \infty \cdot cap(B) = \begin{cases} 0 & \text{if } cap(B) = 0 \\ \infty & \text{otherwise.} \end{cases}$$

If $\mu = I_{G^c} \cdot \overline{\infty}$ with a Borel $G \subset R^d$, then $E^\mu = reg(G)$. Any measure $\mu \in \mathcal{M}_0$ and the associated PAF A_t is very smoothly behaved on the permanent set E^μ. in particular, any PAF A_t is continuous on $[0, s^A)$ and identically infinity on $[s^A, \infty)$ for some stopping time s^A ([10]).

So much for an account of analytical characterizations of positive additive functionals admitting their infinities. From the viewpoint of the transformations of the Brownian motion, the whole class $\mathcal{M}_0/\sim$ is now at our disposal in performing the associated killings of the Brownian motion. In dealing with random time changes,decompositions of not necessarily positive AF's,Girsanov type of trasformations and so on however, the finiteness of AF's at least for q.e. starting points should be required. The families S_1 and S of smooth measures thus retain their significances. In the next subsection, we shall treat those two families simultaneously under the Dirichlet form setting.

4.2 Dirichlet form setting and absolute continuity condition

Let X be a locally compact separable metric space and m be a positive Radon measure with supp$[m] = X$. Let $(\mathcal{E}, \mathcal{F})$ be a regular Dirichlet form on $L^2(X;m)$: $\mathcal{E}$ with domain $\mathcal{F}$ is a closed symmetric form on $L^2(X;m)$ on which the unit contraction operates. $\mathcal{F} \cap C_0(X)$ is assumed to be dense both in $\mathcal{F}$ and $C_0(X)$,$C_0(X)$ being the space of continuous functions on X with compact support. Consider a Hunt process $\mathbf{M} = (\Omega, X_t, \zeta, P_x)$ which is *associated with* $(\mathcal{E}, \mathcal{F})$ in the sense that the transition function $p_t f$ of $\mathbf{M}$ is a Borel version of the L^2−semigroup $T_t f$ for any $f \in L^2$. We then have automatically ([4])that

$$p_t f \text{ is a quasi continuous version of } T_t f, \forall f \in L^2. \tag{63}$$

A set B is called *properly exceptional* if B is Borel,$m(B) = 0$ and $X - B$ is $\mathbf{M}$-invariant. We say that $\mathbf{M}$ satisfies the absolute continuity condition if

$$p_t(x, \cdot) \prec m(\cdot), \forall t > 0. \forall x \in X. \tag{64}$$

Proposition 4.1 (i) *If two Hunt processes* $\mathbf{M}_1 = (X_t, P_x^1), \mathbf{M}_2 = (X_t, P_x^2)$ *are associated with the regular Dirichlet form* $(\mathcal{E}, \mathcal{F})$*, then there exists a common porperly exceptional set* B *for* $\mathbf{M}_1$ *and* $\mathbf{M}_2$ *such that* $P_x^1 = P_x^2$ *in law* $\forall x \in X - B$.

(ii) *If, in addition, both* $\mathbf{M}_1$ *and* $\mathbf{M}_2$ *satisfy the absolute continuity condition, then* $P_x^1 = P_x^2$ *in law* $\forall x \in X$.

(i) is a consequence of (63)(cf.[4; Theorem 4.3.6]). (ii) follows immediately from (i). The moral of this lemma is that what is associated with a regular Dirichlet form is not a single Hunt process but an equivalent family of Hunt processes. Accordingly in formulating some notions for an associated Hunt process, we have to admit exceptional sets of zero capacity in order to make the notions to be independent of the choice of the process from the equivalent family. The pointwise statement does not make sense unless either each singleton has a positive capacity (as in the cases of $H^1(R^1)$ and the Dirichlet forms on the nested fractals) or $\mathbf{M}$ satisfies the absolute continuity condition.

Fix now a Hunt process $\mathbf{M}$ associated with $(\mathcal{E}, \mathcal{F})$. From now on, we refer the notions of AF, PCAF and PAF introduced in the beginning of this section by adding the phrase "*in the strict sense*". We then call $A_t(\omega)$ simply an *AF* of $\mathbf{M}$ if there exists a properly exceptional set N such that $A_t(\omega)$ is an AF in the strict sense but with respect to the Hunt process $\mathbf{M}|_{X-N}$. In other words AF of $\mathbf{M}$ is now redefined by replacing X in condition (53) with $X - N$. N depends on A_t in general and called an *exceptional set for* A_t. *PCAF* of $\mathbf{M}$ is defined in the same way admitting its exceptional set. The set of all PCAF's of $\mathbf{M}$ is denoted by $\mathbf{A}_c^+$. We introduce an equivalence relation in $\mathbf{A}_c^+$ by

$$A_t^{(1)} \sim A_t^{(2)} \Leftrightarrow P_x\left(A_t^{(1)} = A_t^{(2)}, t \geq 0\right) = 1 \; q.e.x \in X.$$

A positive Radon measure μ on X is said to be *of finite energy integral* ($\mu \in S_0$ in notation) if

$$\int \varphi d\mu \leq C\sqrt{\mathcal{E}_1(\varphi, \varphi)}, \varphi \in \mathcal{F} \cap C_0(X).$$

The *1-potential* of μ is then well defined as an element $U_1\mu \in \mathcal{F}$ such that

$$\mathcal{E}_1(U_1\mu, \varphi) = \int \varphi d\mu, \; \varphi \in \mathcal{F} \cap C_0(X).$$

A Borel measure μ on X is said to be *smooth* ($\mu \in S$ in notation) if μ charges no set of zero capacity and there exists an increasing sequence of compact sets $\{F_n\}$ such that $\mu(F_n) < \infty$ for each n and $\lim_{n\to\infty} Cap(K - F_n) = 0$ for any compact set K. It is known ([4;theorem 3.2.3]) that $\mu \in S$ if and only if there exists an increasing sequence of closed set $\{F_n\}$ such that $I_{F_n} \cdot \mu \in S_0$ for each n, $\mu(X - \cup F_n) = 0$ and the equality (56) holds for q.e. $x \in X$. Hence $S_0 \subset S$ and S contains any positive Radon measures charging no set of zero capacity.

The next one to one correspondence has been established in [4; thoerem 5.1.3]:

$$\mathbf{A}_c^+/\sim \ni A_t \overset{(54)}{\longleftrightarrow} \mu \in S \tag{65}$$

In particular, when $\mu \in S_0$, the associated PCAF A_t is characterized by

$$E_x\left(\int_0^\infty e^{-t}dA_t\right) \text{ is a quasi continuous version of } U_1\mu. \tag{66}$$

In the rest of this subsection, we assume that $\mathbf{M}$ satisfies the absolute continuity condition (64). We denote by $\mathbf{A}_{c1}^+$ the family of all PCAF's in the strict sense of $\mathbf{M}$. We shall derive an analytical characterization of$\mathbf{A}_{c1}^+$ directly from the above correspondences (65) and (66).

Lemma 4.1 (i) *The resolvent $R_\alpha(x,E) = \int_0^\infty e^{-\alpha t}p_t(x,E)dt$ has a density function $g_\alpha(x,y)$ with respect to m which is symmetric in x,y and $\alpha-$excessive in x (and in y). Further, for $g_\alpha^y(\cdot) = g_\alpha(\cdot,y)$ and $0 < \alpha \le \beta$,*

$$g_\alpha^y = g_\beta^y + (\beta-\alpha)R_\alpha g_\beta^y = g_\beta^y + (\beta-\alpha)R_\beta g_\alpha^y, y \in X.$$

(ii) *For a positive Borel measure μ on X, $\mu \in S_0$ if and only if the double integral $\int\int g_1(x,y)\mu(dx)\mu(dy)$ is finite. In this case, $G_1\mu(x) = \int_X g_1(x,y)\mu(dy)$ is a quasi continuous version of $U_1\mu$.*

Proof (i) was proven in Blumenthal-Getoor[2;VI(1.4)] in a more general weak duality setting. To see (ii), suppose $\mu \in S_0$, then

$$(U_1\mu,\varphi) = \langle\mu, R_1\varphi\rangle = (G_1\mu,\varphi), \forall\varphi \in C_0(X).$$

Since $G_1\mu$ is 1-excessive and hence finely continuous, it is a quasi continuous version of $U_1\mu$ and $\langle G_1\mu,\mu\rangle = \mathcal{E}_1(U_1\mu, U_1\mu) < \infty$. Conversely assume that $\langle\mu, G_1\mu\rangle$ is finite. Then we have from (i) that for any $\alpha > 1$

$$G_\alpha\mu \in L^2(X;m),\ \beta(G_\alpha\mu - \beta R_{\beta+\alpha}G_\alpha\mu, G_\alpha\mu) \le \langle\mu, G_1\mu\rangle,$$

from which we conclude that

$$G_\alpha\mu \in \mathcal{F}, \mathcal{E}_1(G_\alpha\mu, G_\alpha\mu) \le \langle\mu, G_1\mu\rangle. \tag{67}$$

Since $G_\alpha\mu$ is easily seen to increase to $G_1\mu$ $m-a.e.$ as $\alpha \downarrow 1$, (67) holds for $\alpha = 1$ as well. Then, for any Borel $\varphi \in L^2(X;m)$,

$$\langle\mu, R_1\varphi\rangle = (G_1\mu,\varphi) = \mathcal{E}_1(G_1\mu, R_1\varphi).$$

In the above we may replace $R_1\varphi$ by $\alpha R_\alpha\varphi$ for non-negative $\varphi \in \mathcal{F} \cap C_0(X)$ and get by letting $\alpha \to \infty$

$$\langle \mu, \varphi \rangle \leq \sqrt{\langle \mu, G_1\mu \rangle}\sqrt{\mathcal{E}_1(\varphi, \varphi)}$$

arriving at $\mu \in S_0$. q.e.d.

Let us set

$$S_{01} = \{\mu \in S_0 : G_1\mu(x) < \infty, \forall x \in X\}.$$

A Borel measure μ on X is said to be *smooth in the strict sense* ($\mu \in S_1$ in notation) if there exists a sequence of Borel quasi-closed sets $\{E_n\}$ increasing to X such that

$$I_{E_n} \cdot \mu \in S_{01} \; n = 1, 2, \cdots, \tag{68}$$

and the equality (56) holds. Condition (68) can be replaced by an apparently stronger condition (55).

Given a PCAF A_t with defining set $\mathcal{L}$, the equality $P_x(\mathcal{L}) = 1$ holds only for q.e.$x \in X$ but

$$P_x\left(\theta_\epsilon^{-1}\mathcal{L}\right) = \int_{X_\Delta} p_\epsilon(x, dy) P_y(\mathcal{L}) = 1, \epsilon > 0,$$

for every $x \in X$ because of (64). Therefore if we let $\tilde{A}_t = \lim_{\epsilon_n \downarrow 0} A_{t-\epsilon_n}(\theta_{\epsilon_n}\omega)$, we may get a PCAF in the strict sense. In this way we can get from (66) (cf.[5] for the details)

Lemma 4.2 *For $\mu \in S_0$, let*

$$N_\mu = \{x \in X : G_1\mu(x) = \infty\}.$$

Then there is $A \in \mathbf{A}_c^+$ with exceptional set N_μ such that

$$E_x\left(\int_0^\infty e^{-t} dA_t\right) = G_1\mu(x), \forall x \in X - N_\mu.$$

If $A^{(1)}$ and $A^{(2)}$ are as above, then

$$P_x\left(A_t^{(1)} = A_t^{(2)}, t \geq 0\right) = 1, \forall x \in X - N_\mu.$$

This means that, if $N_\mu = \emptyset$, namely, $\mu \in S_{01}$, then we get $A \in \mathbf{A}_{c1}^+$ uniquely associated with μ. We then readily arrive at the next theorem([5]).

Theorem 4.1 *$\mathbf{A}_{c1}^+$ and S_1 stand in one to one correspondence by the Revuz relation (54). In particular, for $\mu \in S_{01}$, the associated $A \in \mathbf{A}_{c1}^+$ is characterized by the relation (57).*

References in §4

[1] J.Baxter, G.Dal Maso and U.Mosco, Stopping times and Γ– convergence, TAMS 303(1987),1-38

[2] R.M.Blumenthal and R.K.Getoor, Markov processes and potential theory, Acadmic press, 1968

[3] E.B. Dynkin, Markov processes, Springer, 1965

[4] M. Fukushima, Dirichlet forms and Markov processes, Kodansha and North-Holland, 1980

[5] M. Fukushima, On two classes of smooth measures for symmetric Markov processes, in "Stochastic Analysis", eds.M.Metivier and S.Watanabe, Lecture Notes in Math. 1322, Springer, 1988

[6] R.K.Getoor and M.J.Sharpe, Naturality, standardness and weak duality for Markov processes, Z.Wahrscheinlichleitsthoerie verw.Gebiete, 67(1984),1-62

[7] H.P.McKean and H.Tanaka, Additive functionals of the Brownian path,Memoire Coll.Sci.Univ.Kyoto 33(1961),479-506

[8] D. Revuz, Mésures associées aux fonctionelles additives de Markov I, TAMS 148(1970), 501-531

[9] M.L. Silverstein, Symmetric Markov processes, Lecture Notes in Math. 426, Springer, 1974

[10] K-T. Sturm, Measures charging no polar sets and additive functionals of Brownian motion, Forum Math. 4(1992),257-297

[11] K-T. Sturm,Schrödinger operators and Feynman-Kac semigroups with arbitrary nonnegative potentials, in "Operator Calculus and Spectral Theory",eds.Demuth and Schulze, Birkhäuser, to appear

[12] A.D. Wenzell, Nonnegative additive functionals of Markov processes, DAH 137(1961), 17-20

5 Decomposition of additive functionals of finite energy

As was stated in the preceding section, AF's in the Dirichlet space setting involve exceptinal starting point sets of zero capacity. Accordingly the representation of AF of the type $u(X_t) - u(X_0)$ as a sum of a martingale AF and AF of zero energy takes place P_x-almost surely for "quasi- every $x \in X$". It is important to know under what circumstances one can strengthen the above phrase inside the quatation marks into "every $x \in X$". In subsection 5.1, we give an answer to this question by using the results of the preceding subsection. In subsection 5.2, we apply it to the Skorohod type representation of the sample paths of the reflecting Brownian motion.

Thus, while the topics in Chapter 1 can be expressed as "from almost everywhere to quasi everywhere", the feature of Chapter 2 may be phrased as "from quasi everywhere to everywhere".

5.1 The decomposition and the absolute continuity condition

Let $X, m, \mathcal{E}, \mathcal{F}$ be as in the preceding subsection. Let $\mathbf{M} = (\Omega, X_t, \zeta, P_x)$ be a Hunt process associated with $(\mathcal{E}, \mathcal{F})$. $\mathbf{M}$ is assumed to be *of no killing inside* X :

$$P_x(X_{\zeta-} \in X, \zeta < \infty) = 0, x \in X, \tag{69}$$

or equivalently

$$\lim_{t \downarrow 0} \frac{1}{t} \int_X u(x)^2 (1 - p_t 1(x)) m(dx) = 0, u \in \mathcal{F}. \tag{70}$$

This amounts to the absence of the killing part in the Beurling-Deny expression of the form $\mathcal{E}$. We have then for $u \in \mathcal{F}$

$$\lim_{t \downarrow 0} \frac{1}{2t} E_m \left((u(X_t) - u(X_0))^2 \right) = \mathcal{E}(u, u). \tag{71}$$

In fact, the left hand side equals

$$\frac{1}{t}(u - p_t u, u) - \frac{1}{2t} \int u^2 (1 - p_t 1) dm.$$

For an AF A_t of $\mathbf{M}$, we set

$$e(A) = \lim_{t \downarrow 0} \frac{1}{2t} E_m \left(A_t^2 \right), \tag{72}$$

if the limit exists, and call it the *energy* of A_t. (71) means that

$$A_t^{[u]} = \tilde{u}(X_t) - \tilde{u}(X_0), \ u \in \mathcal{F} \tag{73}$$

is an AF of finite energy and

$$e\left(A^{[u]}\right) = \mathcal{E}(u, u). \tag{74}$$

We next set

$$\mathcal{M} = \{M : AF, \forall t > 0, E_x\left(M_t^2\right) < \infty, E_x(M_t) = 0 \ q.e. x \in X\}. \tag{75}$$

It is known that for any $M \in \mathcal{M}$, there exists $\langle M\rangle \in \mathbf{A}_c^+$ such that

$$E_x(\langle M\rangle_t) = E_x\left(M_t^2\right) \; q.e.x \in X. \tag{76}$$

$\langle M\rangle$ is called the *quadratic variation* of the martingale AF M.

Denote by $\mu_{\langle M\rangle}$ the Revuz measure of $\langle M\rangle$. Then by (76) and (54)

$$e(M) = \frac{1}{2}\mu_{\langle M\rangle}(X). \tag{77}$$

Let us introduce the spaces of AF's

$$\overset{\circ}{\mathcal{M}} = \{M \in \mathcal{M} : e(M) < \infty\} \tag{78}$$

$$\mathcal{N}_c = \{N : AF, continuous, e(N) = 0, E_x(|N_t|) < \infty \; q.e.x \in X\}. \tag{79}$$

The next three propositions are taken from [5].

Proposition 5.1 *For $u \in \mathcal{F}$, $A^{[u]}$ admits a unique decomposition*

$$A^{[u]} = M^{[u]} + N^{[u]}, \; \exists M^{[u]} \in \overset{\circ}{\mathcal{M}}, \; \exists N^{[u]} \in \mathcal{N}_c. \tag{80}$$

Furthermore

$$e\left(A^{[u]}\right) = e\left(M^{[u]}\right) = \mathcal{E}(u,u). \tag{81}$$

We proceed to computating $M^{[u]}$ and $N^{[u]}$. We denote by $\mathcal{F}_b$ the collection of m-essentially bounded functions in $\mathcal{F}$. For $u \in \mathcal{F}_b$,there exists a unique positive Radon measure $\mu_{\langle u\rangle}$ such that

$$\int_X \tilde{f}(x)\mu_{\langle u\rangle}(dx) = 2\mathcal{E}(uf,u) - \mathcal{E}(u^2,f) \; \forall f \in \mathcal{F}_b. \tag{82}$$

Indeed the right hand side equals

$$\lim_{t\downarrow 0}\frac{1}{t}\int(u(x)-u(y))^2 f(x)p_t(x,dy)m(dx) + \frac{1}{t}\int u(x)^2 f(x)(1-p_t1(x))m(dx)$$

defining a positive linear functional on $\mathcal{F} \cap C_0(X)$. $\mu_{\langle u\rangle}$ is called the *energy measure* of $u \in \mathcal{F}_b$.

For instance, when $X = R^d$ and the Dirichlet form $\mathcal{E}$ is given by

$$\mathcal{E}(u,v) = \frac{1}{2}\sum_{i,j=1}^{d}\int_{R^d}\frac{\partial u}{\partial x_i}\frac{\partial v}{\partial x_j}d\nu_{ij}, \; u,v \in C_0^\infty,$$

the energy measure of $u \in C_0^\infty$ is given by

$$\mu_{\langle u\rangle}(dx) = \sum_{i,j=1}^{d}\frac{\partial u}{\partial x_i}\frac{\partial u}{\partial x_j}\nu_{ij}(dx).$$

Proposition 5.2 *For $u \in \mathcal{F}_b$, the Revuz measure of $\langle M^{[u]}\rangle$ is equal to the energy measure of u: $\mu_{\langle M^{[u]}\rangle} = \mu_{\langle u\rangle}$.*

Proposition 5.3 *If* $\nu = \nu^{(1)} - \nu^{(2)}$ *for some* $\nu^{(1)}, \nu^{(2)} \in S_0$ *and*

$$\mathcal{E}(u,v) = \langle \nu, \tilde{v} \rangle \ \forall v \in \mathcal{F}, \tag{83}$$

then

$$N^{[u]} = -A^{(1)} + A^{(2)}, \tag{84}$$

where $A^{(1)}$ *(resp.*$A^{(2)}$*) is the PCAF with Revuz measure* $\nu^{(1)})$ *(resp.*$\nu^{(2)}$*).*

We now assume the absolute continuity condition (64). $M_t(\omega)$ is called an *martingale additive functional in the strict sense* if it is an AF in the strict sense and

$$E_x(M_t) = 0 \ E_x\left(M_t^2\right) < \infty \ \forall x \in X. \tag{85}$$

Theorem 5.1 *Suppose that a function u satisfies the following conditions:*

(i) $u \in \mathcal{F}, u$ *is bounded and finely continuous.*

(ii) $\mu_{\langle u \rangle} \in S_{01}$.

(iii) $\exists \nu = \nu^{(1)} - \nu^{(2)}$ *with* $\nu^{(1)}, \nu^{(2)} \in S_{01}$ *and*

$$\mathcal{E}(u,v) = \langle \nu, \tilde{v} \rangle \ \forall v \in \mathcal{F}.$$

Then

$$u(X_t) - u(X_0) = M_t^{[u]} + N_t^{[u]} \ P_x - a.e. \forall x \in X. \tag{86}$$

Here $M_t^{[u]}$ *is an MAF in the strict sense whose quadratic variation (in the strict sense) has the Revuz measure* $\mu_{\langle u \rangle}$*. Further*

$$N^{[u]} = -A^{(1)} + A^{(2)} \ P_x - a.e. \ \forall x \in X, \tag{87}$$

where $A^{(1)}$ *(resp.* $A^{(2)}$*) is a PCAF in the strict sense with Revuz measure* $\nu^{(1)}$ *(resp.* $\nu^{(2)}$*).*

Proof We define $N^{[u]}$ by (87) and set

$$M_t^{[u]} = u(X_t) - u(X_0) - N_t^{[u]}, \ t > 0.$$

Then $M_t^{[u]}$ is an AF in the strict sense and, by Proposition 5.3, $M^{[u]} \in \overset{\circ}{\mathcal{M}}$. In particular $E_x\left(M_t^{[u]}\right) = 0$ q.e. By noting that $E_x\left(A_t^{(i)}\right) < \infty \forall x \in X, i = 1,2$, we get

$$\begin{aligned} E_x\left(M_{t+\epsilon}^{[u]}\right) - E_x\left(M_\epsilon^{[u]}\right) &= E_x\left(E_{X_\epsilon}\left(M_t^{[u]}\right)\right) \\ &= \int_X P_\epsilon(x,dy) E_y\left(M_t^{[u]}\right) = 0, \ \forall x \in X, \end{aligned}$$

and by letting $\epsilon \downarrow 0$, we are led to

$$E_x\left(M_t^{[u]}\right) = 0 \ \forall x \in X.$$

Denote by B the PCAF in the strict sense associated with $\mu_{\langle u \rangle}$. Then by Proposition 5.2

$$E_x\left(\left(M_t^{[u]}\right)^2\right) = E_x(B_t) \; q.e.$$

Therefore as in the above

$$E_x\left(\left(M_{t+\epsilon}^{[u]} - M_\epsilon^{[u]}\right)^2\right) = E_x(B_{t+\epsilon} - B_\epsilon), \; \forall x \in X$$

and by the Fatou's lemma

$$E_x\left(\left(M_t^{[u]}\right)^2\right) \leq E_x(B_t) < \infty \; \forall x \in X.$$

We have proved that $M^{[u]}$ is a MAF in the strict sense. It is clear that

$$\langle M^{[u]} \rangle_t = B_t \; P_x - a.e. \forall x \in X.$$

5.2 Application to a reflecting Brownian motion

Let X be an arbitrary bounded domain D in R^d and m be the Lebesgue measure. We let

$$H^1(D) = \left\{ u \in L^2(D) : \frac{\partial u}{\partial x_i} \in L^2(D), i = 1, \cdots, d \right\}, \mathbf{D}(u,v) = \int_D \nabla u \cdot \nabla v dx.$$

Then $\left(\frac{1}{2}\mathbf{D}, H^1(D)\right)$ is a Dirichlet form on $L^2(D)$. But this Dirichlet form is not regular unless the complement of D is of zero capacity. A compact space D^* is called a *compactification* of D if D is densely and homeomorphically embedded into D^*. The Lebesgue measure on D is extended to D^* by setting $m(D^* - D) = 0$ and $L^2(D^*)$ is identified with $L^2(D)$. A *reflecting Brownian motion* is by definition a diffusion process on a compactification D^* of D which is m-symmetric and whose Dirichlet form on $L^2(D^*)$ is regular and coincides with $\left(H^1(D), \frac{1}{2}\mathbf{D}\right)$.

We consider the orthogonal decomposition $H^1(D) = H_0^1(D) \oplus \mathcal{H}_\alpha$ with respect to $\mathcal{E}_\alpha(u,v) = \frac{1}{2}\mathbf{D}(u,v) + \alpha(u,v)_{L^2(D)}$. $\mathcal{H}_\alpha$ is then the space of $\alpha-$harmonic functions of finite Dirichlet integrals and possesses the reproducing kernel $R_\alpha(x,y)$. We let

$$G_\alpha(x,y) = G_\alpha^0(x,y) + R_\alpha(x,y), x, y \in D,$$

where $G_\alpha^0(x,y)$ is the resolvent density associated with $\left(H_0^1(D), \frac{1}{2}\mathbf{D}\right)$. It is known [4] that there exists a function $p_t(x,y), t > 0, x, y \in D$ such that $p_t p_s = p_{t+s}, p_t 1 = 1$, and

$$\int_0^\infty e^{-\alpha t} p_t(x,y) dt = G_\alpha(x,y), \alpha > 0, x, y \in D.$$

We may well require for the reflecting Brownian motion to have these $p_t(x,y), G_\alpha(x,y)$ as its transition density and the resolvent density for $x, y \in D$. It was proven in [4] that arbitrary bounded domain D admits a compactification D^* called the Martin Kuramochi type compactification on which a reflecting Brownian motion satisfying all the requirements above can be constructed.

Suppose that D is a bounded Lipschitz domain. Then the Dirichlet form $\left(H^1(D), \frac{1}{2}\mathbf{D}\right)$ on $L^2(\bar{D})$ is known to be regular. Bass-Pei Hsu[2] extended $p_t(x,y)$ continuously to $\bar{D}\times\bar{D}$ and constructed a diffusion process $\mathbf{M}_{\bar{D}} = (X_t, P_x)$ on $\bar{D}$ associated with this transition density. $G_\alpha(x,y)$ is also extended and

$$G_1\sigma(x) = \int_{\partial D} G_1(x,y)\sigma(dy), x \in \bar{D}, \tag{88}$$

is bounded on $\bar{D}$, where σ is the surface measure on the boundary ∂D. In particular $\sigma \in S_{01}$ by Lemma 4.1.

$\mathbf{M}_{\bar{D}}$ may well be called a reflecting Brownian motion. By virtue of Theorem 4.1, there exists L_t a PCAF of $\mathbf{M}_{\bar{D}}$ in the strict sense such that

$$E_x\left(\int_0^\infty e^{-t}dL_t\right) = G_1\sigma(x), \forall x \in \bar{D}. \tag{89}$$

Now let $\varphi_i(x) = x_i, i = 1,\cdots,d$, and $\varphi(x) = (\varphi_1(x),\cdots,\varphi_d(x))$. Each φ_i is continuous and in $H^1(D)$. They have the properties

$$\mu_{\langle\varphi_i,\varphi_j\rangle}(dx) = \sum_{k=1}^d \frac{\partial\varphi_i}{\partial x_k}\frac{\partial\varphi_j}{\partial x_k}dx = \delta_{ij}dx, \tag{90}$$

$$\mathcal{E}(\varphi_i, v) = \frac{1}{2}\int_D \frac{\partial v}{\partial x_i}dx = \frac{1}{2}\int_{\partial D} \tilde{v}\nu_i(x)dx, v \in H^1, \tag{91}$$

where $\nu(x) = (\nu_1(x),\cdots,\nu_d(x))$ is the unit normal vector at $x \in \partial D$.

Applying Theorem 5.1 and using the martingale characterization of the multidimensional Brownian motion, we can now get a representation of the sample path of $\mathbf{M}_{\bar{D}}$:

$$X_t = x + B_t + \int_0^t \nu(X_s)dL_s\ P_x - a.s.\forall x \in \bar{D}. \tag{92}$$

Here B_t is a d-dimensional Brownian motion (but relative to the filtration generated by X_t) with $B_0 = 0$ under P_x for any $x \in R^d$. This Skorohod type representation of X_t is essentially due to Bass-Pei Hsu[1]. This formula being looked upon as an equation based on a Brownian motion B_t has been solved for an arbitrary convex domain by Tanaka[6].

Finally we mention a related recent work. Take an arbitrary bounded domain $D \subset R^d$ and consider the already mentioned reflecting Brownian motion $\mathbf{M}_{D^*} = (X_t^*, P_x)$ on the Martin-Kuramochi type compactification D^*. Since $\varphi_i \in H^1(D), \varphi$ has a quasi continuous version $\tilde{\varphi}$ on D^* and the R^d valued process $X_t = \tilde{\varphi}(X_t^*)$ is continuous $P_m - a.s.$ But $X_t \in D$ $P_m - a.s.$ for each fixed t and hence X_t is a $\bar{D}$ valued process under P_m. By making use of a variant of Proposition 5.3 concerning a characterization of AF $N^{[u]}$, Z.Q.Chen,Fitzsimmons and R.Williams [3] have given a necessary and sufficient condition for the domain D so that the process X_t becomes a quasi-martingale under P_m.

References in §5

[1] R.F.Bass and Pei Hsu, The semimartingale structure of reflecting Brownian motion, Proc.AMS 108(1990),1007-1010

[2] R.F.Bass and Pei Hsu, Some potential theory for reflecting Brownian motion in Hölder and Lipschitz domains, Ann.Prob.19(1991),486 -508

[3] Z.Q.Chen, P.J.Fitzsimmons and R.J. Williams, Reflecting Brownian motions : quasimartingales and strong Caccioppoli sets, Preprint

[4] M.Fukushima, A construction of reflecting barrier Brownian motions for bounded domains, Osaka J. Math. 4(1967),183-215

[5] M.Fukushima, Dirichlet forms and Markov processes, Kodansha/ North Holland 1980

[6] H.Tanaka, Stochastic differential equations with reflecting boundary condition in convex regions, Hiroshima Math.J. 9(1979),163-177

Logarithmic Sobolev Inequalities and Contractivity Properties of Semigroups

Leonard Gross

Department of Mathematics,
White Hall, Cornell University,
Ithaca, NY 14853, USA.

1 Introduction

Suppose that (X, μ) is a probability measure space and that H is a self-adjoint operator on $L^2(X, \mu)$ (which we shall take to be real functions only, in this introduction). Assume that H is bounded below. We will write, for $1 \leq q, p \leq \infty$, $\|e^{-tH}\|_{q \to p} = \sup\{\|e^{-tH} f\|_p : f \in L^2 \cap L^q, \|f\|_q \leq 1\}$ wherein $\|g\|_p$ denotes the $L^p(\mu)$ norm of g. Consider the following three questions concerning the relation between properties of H and properties of the semigroup e^{-tH} which it generates.

Question 1. Under what conditions on H is e^{-tH} a contraction semigroup on $L^2(X, \mu)$?

Answer (by the spectral theorem). $\|e^{-tH}\|_{2\to 2} \leq 1$ for all $t > 0$ if and only if

$$\langle Hf\,,\, f\rangle_{L^2(\mu)} \geq 0 \qquad \text{for all } f \text{ in } \mathcal{D}(H) \tag{1.1}$$

The semigroup e^{-tH} is called a *contraction semigroup* in L^p if $\|e^{-tH}\|_{p\to p} \leq 1$ for all $t > 0$ and is called *positivity preserving* if for all $t \geq 0$ $e^{-tH} f \geq 0$ a.e. whenever $f \geq 0$ a.e.. Both of these properties are important in probability theory because they are required for transition semigroups of Markov processes. By the Beurling–Deny theorem [D8 or Fu] the following question has an immediate answer. Write $g^+(x) = \max\{g(x)\,,\,0\}$.

Question 2. Assume H is a self-adjoint operator satisfying (1.1). Under what conditions on H is e^{-tH} a positivity preserving contraction semigroup on $L^p(X, \mu)$ for all p in $[1, \infty]$?

Answer (Beurling–Deny). e^{-tH} is positivity preserving and $\|e^{-tH}\|_{p\to p} \leq 1$ for all $t \geq 0$ and all p in $[1, \infty]$ if and only if

$$\langle Hf\,,(f-1)^+\rangle \;\geq\; 0 \qquad \text{for all } f \text{ in } \mathcal{D}(H) \tag{1.2}$$

In these notes we will be concerned with "smoothing" properties of the semigroup e^{-tH}, as represented by a statement that e^{-tH} is a contraction (more generally a bounded operator) from $L^q(\mu)$ to $L^p(\mu)$ for some $p > q$ wherein p may depend on q and t. In its simplest form the issue of concern to us is captured by the following question. The answer given below will be proven (in greater generality) in Sections 3, 4 and 5, and with quantitative relations between p, q and t.

Question 3. Assume that H is a self-adjoint operator in $L^2(X,\mu)$ which satisfies both (1.1) and (1.2). Under what conditions on H is e^{-tH} a contraction from $L^q(\mu)$ to $L^p(\mu)$ for some $t > 0$ and some q and p with $1 < q < p < \infty$?

Answer. $\|e^{-tH}\|_{q\to p} \le 1$ for some $t > 0$ and some q and p with $1 < q < p < \infty$ if and only if there is a constant $c > 0$ such that

$$c\langle Hf\,,\,f\rangle \;\ge\; \int_X f^2 \log|f|\,d\mu \;-\; \|f\|_2^2 \log\|f\|_2 \qquad \text{for all } f \text{ in } \mathcal{D}(H) \tag{1.3}$$

As we will see in the examples below the inequality (1.3) captures the spirit of Sobolev inequalities with a logarithm replacing a power. Hence (1.3) is generally referred to as a logarithmic Sobolev inequality.

In order to understand the relation between (1.3) and the classical Sobolev inequalities consider the following prototype example. Let $d\mu(x) = \omega(x)\,dx$ be a probability measure on R^n with a smooth strictly positive density ω. Define H by the equation

$$\langle Hf\,,\,f\rangle_{L^2(R^n,\mu)} \;=\; \int_{R^n} \nabla f(x)\cdot\nabla g(x)\;d\mu(x) \tag{1.4}$$

Here the dot on the right refers to the inner product on R^n. H is called the Dirichlet form operator for the measure μ. If we interpret the gradient operator ∇ as a closed densely defined operator from $L^2(R^n,\mu)$ to $L^2(R^n,\mu)\otimes R^n$ then H is simply given by $H = \nabla^*\nabla$. Conditions (1.1) and (1.2) are automatically satisfied for this operator. In order to establish bounds on $e^{-tH} : L^p \to L^q$ it suffices therefore to prove a logarithmic Sobolev inequality (similar to) (1.3). One can vary this typical example in many ways: one can replace R^n by a finite dimensional Riemannian manifold, by an infinite dimensional "Riemannian" manifold, by an open set in R^n or by a discrete set. Although, in the setting of R^n, the second order terms in $\nabla^*\nabla$ have constant coefficients one can replace them by bounded measurable (elliptic) coefficients. The measure μ on R^n may arise, for example, as the ground state measure for a Schrödinger operator $-\Delta + V$: one assumes that $-\Delta + V$ has a lowest eigenstate ψ with reasonable properties and puts $d\mu(x) = \psi(x)^2dx$. In this case $-\Delta + V$ is unitarily equivalent to $\nabla^*\nabla +$ constant. The underlying measure μ need not be finite for some of the techniques associated with logarithmic Sobolev inequalities to work, but in these notes we will largely consider only the finite measure case.

We will prove in Example 2.7 the following Gaussian logarithmic Sobolev inequality. Let

$$d\nu(x) \;=\; (2\pi)^{-n/2}\,e^{-|x|^2/2}\,dx, \qquad x \in R^n \tag{1.5}$$

Then

$$\int_{R^n} |f(x)|^2 \log|F(x)|\,d\nu(x) \;\le\; \int_{R^n} |\nabla f(x)|^2\,d\nu(x) \,+\, \|f\|_2^2 \log\|f\|_2 \tag{1.6}$$

(1.6) holds whenever f is in $L^2(\nu)$ and $|\nabla f(x)|$ (weak derivative) is in $L^2(\nu)$. (1.3) clearly reduces to (1.6) when $\mu = \nu$, $c = 1$ and H is given by (1.4). The inequality (1.6) has a dimension independent character. Not only are all the coefficients and powers independent of dimension but the inequality makes sense and remains correct for $n = \infty$ (ν is then simply an infinite product of probability measures on R^∞ and $|\nabla f(x)|^2 = \sum_{j=1}^{\infty} |\partial f(x_1, x_2, \dots)/\partial x_j|^2$). The theory of a Boson quantum field can be viewed as the theory of a certain elliptic differential operator in infinitely many variables. In view of the important role that the classical Sobolev inequalities play in the theory of finite dimensional elliptic operators it might come as no surprise that the dimension independent inequality (1.6) had its origin in constructive quantum field theory. Although logarithmic Sobolev inequalities continue to play an important role in the analysis of elliptic differential operators in infinitely many dimensions (see for example Stroock's Lectures on statistical mechanics in this volume) they have also found application in finite dimensions.

In contrast to logarithmic Sobolev inequalities the classical Sobolev inequalities are strongly dimension dependent. On R^n they may be written

$$\|f\|_{L^q(R^n,dx)} \le C_{p,n} \| |\nabla f| \|_{L^p(R^n,dx)} \quad f \in C_c^\infty(R^n) \tag{1.7}$$

wherein $C_{p,n}$ is a constant and

$$1/q = 1/p - 1/n, \quad 1 \le p < \infty. \tag{1.8}$$

(1.7) holds if $q < \infty$. If we complete $C_c^\infty(R^n)$ in the norm on the right of (1.7) then for any function f in the completion (1.7) tells us, among other things, that if the first derivatives of f are in L^p then f itself has milder local singularities than ∇f in the sense that f is in fact in L^q for some q strictly greater than p. Thus, on R^n (1.7) contains more information about the local singularities of f that (1.6) does. As $n \to \infty$, however, (1.8) shows that $q \downarrow p$. Information about the singularities of f compared to those of ∇f gets "lost" for large n. Of course for $n = \infty$ there is, on the one hand, no information gain at all about the nature of local singularities of f compared to those of ∇f while on the other hand the inequality (1.7) is meaningless because of the absence of a useful notion of infinite dimensional Lebesgue measure.

In Section 2 we describe three quite general properties of logarithmic Sobolev inequalities over an arbitrary finite measure space. There is first the Federbush semiboundedness theorem and the Faris converse, which together show the equivalence of logarithmic Sobolev inequalities with semiboundedness of $H + V$ for appropriate potentials V. Second, there is the Faris–Segal additivity theorem, a very neat and simple functorial property of logarithmic Sobolev inequalities which amounts to stability under products of measure spaces. Third, there is the Rothaus–Simon mass gap theorem. These three theorems concern properties of H in L^2.

In Section 3 we go into L^p and establish the basic relationship between logarithmic Sobolev inequalities generated by an operator H and L^p to L^q contraction properties of the associated semigroup e^{-tH}. In Section 4 we specialize to generalized Dirichlet forms. The connection between logarithmic Sobolev inequalities and L^p to L^q contraction properties of e^{-tH} simplifies in this case. In Section 5 we describe three kinds of contractivity properties for a semigroup e^{-tH}; hyper, super, and ultra contractivity, and their relation to logarithmic Sobolev inequalities. In Section 6 we give a survey of some of the contexts in which logarithmic Sobolev inequalities have been used or developed so far.

2 General properties of logarithmic Sobolev inequalities in L^2

Let (X, μ) be a finite measure space and $Q(f)$ a densely defined nonnegative quadratic form on $L^2(\mu)$ (real or complex functions). We say Q determines a logarithmic Sobolev inequality if

$$\text{(LS)} \qquad \int_X |f(x)|^2 \log|f(x)|\, d\mu(x) \le Q(f) + \|f\|_2^2 \log\|f\|_2 \quad \forall f \in \mathcal{D}(Q). \tag{2.1}$$

The following is the Federbush semiboundedness theorem [Fe, Grl] and the Faris converse [F]. Henceforth we write $\|f\|$ instead of $\|f\|_2$.

THEOREM 2.1 (Semiboundedness). *Assume Q satisfies* (LS) *and $V : X \to R$ is measurable. Suppose $\|e^{-V}\|_2 < \infty$. Then*

$$\text{(SB)} \qquad Q(f) + (Vf, f) \ge (-\log\|e^{-V}\|_2)\|f\|^2 \quad \textit{for } f \in \mathcal{D}(Q). \tag{2.2}$$

Conversely, if (SB) *holds whenever $\|e^{-V}\|_2 < \infty$ then Q satisfies* (LS).

PROOF. Assume $\|e^{-V}\| := \|e^{-V}\|_2 < \infty$ and that V is bounded above. Then $\int_X -V(x) |f(x)|^2\, d\mu(x)$ is well defined. Apply Young's inequality $st \le s\log s - s + e^t$, which is valid for $s \ge 0$ and all real t, to $s = |f(x)|^2$ and $t = -2V(x)$ to get

$$\begin{aligned} -(Vf, f) &= \frac{1}{2}\int |f(x)|^2(-2V(x))\, d\mu(x) \\ &\le \frac{1}{2}\int \{|f(x)|^2 \log|f(x)|^2 - |f(x)|^2\}\, d\mu(x) + \frac{1}{2}\int e^{-2V(x)}\, d\mu(x) \\ &\le Q(f) + \|f\|^2 \log\|f\| - \frac{1}{2}\|f\|^2 + \frac{1}{2}\|e^{-V}\|^2 \end{aligned}$$

which is finite if f is in $\mathcal{D}(Q)$. Thus $|(Vf, f)| < \infty$ if f is in $\mathcal{D}(Q)$ and moreover

$$Q(f) + (Vf, f) \ge -\|f\|^2 \log\|f\| + \frac{1}{2}(\|f\|^2 - \|e^{-V}\|^2). \tag{2.3}$$

Since (SB) is homogeneous in $\|f\|$ it suffices to verify it in case $\|f\| = \|e^{-V}\|$. But in this case (2.3) reduces to (SB). Now if V is not necessarily bounded above, but $\|e^{-V}\| < \infty$, put $V_n(x) = V(x)$ if $V(x) \le n$ and $V_n(x) = 0$ otherwise. Then $V_n \uparrow V$, $\|e^{-V_n}\| \to \|e^{-V}\|$ and $(V_n f, f) \uparrow (Vf, f)$ by the monotone convergence theorem. Apply (SB) to V_n and take the limit $n \to \infty$ to get (SB) for V.

For the converse suppose (SB) holds whenever $\|e^{-V}\| < \infty$. Let f be in $\mathcal{D}(Q)$ and put $V(x) = -\log|f(x)|$. Then $\|e^{-V}\| = \|f\| < \infty$. Inserting this choice of V into (SB) gives (LS). □

DEFINITION 2.2. Suppose that Q_i is a nonnegative quadratic form which is densely defined on $L^2(X_i, \mu_i)$ for $i = 1, 2$. We define

$$Q(f) = \int_{X_2} Q_1(f(\cdot, x_2))\, d\mu_2(x_2) + \int_{X_1} Q_2(f(x_1, \cdot))\, d\mu_1(x_1) \tag{2.4}$$

for f in $L^2(X_1 \times X_2, \mu_1 \times \mu_2)$ wherein we interpret $Q_i(u) = +\infty$ if u is not in $\mathcal{D}(Q_i)$. Then we take $\mathcal{D}(Q) = \{f \in L^2(\mu_1 \times \mu_2) : Q(f) < \infty\}$. We write $Q = Q_1 + Q_2$.

Next we will prove Faris' additivity theorem [F]. This is an infinitessimal version of the product theorem known as Segal's Lemma [S1, Lemma 1.4]. The hypotheses of the following theorem are too weak to imply the hypercontractivity of the associated semigroups (see Section 5) necessary to apply Segal's Lemma.

THEOREM 2.3 (Additivity). *If Q_i is a quadratic form on $L^2(X_i, \mu_i)$ which satisfies* (LS) *for $i = 1, 2$ then so does $Q_1 + Q_2$.*

PROOF. By Theorem 2.1 and Lemma 2.4 below it suffices to prove (SB) for bounded measurable V. Let V be bounded and write $W(x_2) = -\log \|e^{-V(\cdot, x_1)}\|$. Then W is also bounded and $\|e^{-W}\|_{L^2(\mu_2)} = \|e^{-V}\|$. Suppose that $f : X_1 \times X_2 \to R$ is square integrable and $Q(f) < \infty$. Then for almost all x_2 in X_2, $Q_1(f(\cdot, x_2)) < \infty$. Hence

$$\begin{aligned} Q_1(f(\cdot, x_2)) &+ \int_{X_1} V(x_1, x_2)|f(x_1, x_2)|^2 \, d\mu_1(x_1) \\ &\geq \left(-\log \|e^{-V(\cdot, x_2)}\|\right) \int_{X_1} |f(x_1, x_2)|^2 \, d\mu_1(x_1) \quad \text{a.e. } [\mu_2] \\ &= \int_{X_1} W(x_2)|f(x_1, x_2)|^2 \, d\mu_1(x_1) \quad \text{a.e. } [\mu_2]. \end{aligned} \tag{2.5}$$

Moreover, since $Q_2(f(x_1, \cdot)) < \infty$ for almost all x_1, we have

$$\begin{aligned} Q_2(f(x_1, \cdot)) &+ \int_{X_2} W(x_2)|f(x_1, x_2)|^2 \, d\mu_2(x_2) \\ &\geq (-\log \|e^{-V}\|) \int_{X_2} |f(x_1, x_2)|^2 \, d\mu(x_2) \quad \text{a.e. } [\mu_1]. \end{aligned} \tag{2.6}$$

Integrate (2.5) with respect to x_2 and (2.6) with respect to x_1 and add them to get (SB) for bounded V. □

LEMMA 2.4. *Suppose* (SB) *holds for all bounded measurable real valued functions V. Then it holds also whenever $\|e^{-V}\| < \infty$.*

PROOF. The issue here arises from the fact that if V is unbounded above and below then there exists f in $L^2(\mu)$ such that both the positive and negative parts of $V(x)|f(x)|^2$ are not integrable, so that (Vf, f) is undefined. Of course (SB) will force the negative part of $V(x)|f(x)|^2$ to be integrable once (SB) is known. Now suppose that $\|e^{-V}\| < \infty$. Put $V_k^{(n)}(x) = V(x)$ if $-k \leq V(x) \leq n$ and zero otherwise. Put $V^{(n)}(x) = V(x)$ if $V(x) \leq n$ and zero otherwise. Then $-V_k^{(n)}(x) \uparrow -V^{(n)}(x)$ for all x as $k \uparrow \infty$. By the monotone convergence theorem $\|e^{-V_k^{(n)}}\| \to \|e^{-V^{(n)}}\|$ and $(V_k^{(n)} f, f) \downarrow (V^{(n)} f, f)$. By the dominated convergence theorem $\|e^{-V^{(n)}}\| \to \|e^{-V}\|$. Applying (SB) to $V_k^{(n)}$ we may take the limit as $k \to \infty$ to get $Q(f) + (V^{(n)} f, f) \geq (-\log \|e^{-V^{(n)}}\|)\|f\|^2$ for f in $\mathcal{D}(Q)$. Hence $(V^{(n)} f, f)$ is finite. Thus the negative part of $V^{(n)}(x)|f(x)|^2$ is integrable and clearly independent of n for $n \geq 0$. We may therefore apply the monotone convergence theorem to the last inequality to derive (SB) for V itself. □

The third general theorem is the Rothaus–Simon mass gap theorem [Ro4, Si2].

THEOREM 2.5 (Mass gap). *Assume $\mu(X) = 1$ and also that*

a) *Q satisfies (LS) for real f*

b) $Q(1) = 0$

c) *$L^\infty(X, \mu) \cap \mathcal{D}(Q)$ is a core for Q.*

Then

$$\text{(MG)} \qquad Q(g) \geq \|g\|^2 \text{ if } g \perp 1 \quad (g \text{ real}). \qquad (2.7)$$

PROOF. Suppose g is in $L^\infty \cap \mathcal{D}(Q)$ and $g \perp 1$. If $B(u, v)$ is the symmetric bilinear form on $\operatorname{Re} L^2(\mu)$ associated to Q then the Schwarz inequality gives $|B(1, f)|^2 \leq Q(1)Q(f)$ for f in $\mathcal{D}(Q)$. Hence $B(1, f) = 0$ for f in $\mathcal{D}(Q)$. Thus $Q(1 + sg) = Q(1) + 2B(1, sg) + s^2Q(g) = s^2Q(g)$. Moreover $\|1 + sg\|^2 = 1 + s^2\|g\|^2$. For small real s, $1 + sg(x)$ is close to one uniformly in x. Hence we may expand $\log|1 + sg(x)| = \log(1 + sg(x))$ in a power series for small s. Put $f(x) = 1 + sg(x)$ in (LS) and expand both sides to order s^2. We have

$$\int_X (1 + sg(x))^2 \log(1 + sg(x))\, d\mu(x) \;\leq\; s^2Q(g) \;+\; (1 + s^2\|g\|^2)\tfrac{1}{2}\log(1 + s^2\|g\|^2)$$

and to order s^2 this reads

$$\int (1 + 2sg(x) + s^2g(x)^2)(sg(x) - s^2g(x)^2/2)\, d\mu(x)$$
$$\leq\; s^2Q(g) \;+\; \tfrac{1}{2}(1 + s^2\|g\|^2)s^2\|g\|^2 + O(s^3).$$

Hence

$$\int (sg(x) - s^2g(x)^2/2 + 2s^2g(x)^2)\, d\mu(x) \;+\; O(s^3) \;\leq\; s^2Q(g) \;+\; \tfrac{1}{2}s^2\|g\|^2 + O(s^3).$$

Use $\int g(x)\, d\mu(x) = 0$ to get

$$s^2\|g\|^2 \;\leq\; s^2Q(g) \;+\; O(s^3).$$

Divide by s^2 and let $s \to 0$ to get (MG) for g in $L^\infty \cap \mathcal{D}(Q)$. Now if g is in $\mathcal{D}(Q)$ and $g \perp 1$ then by c) there exists a sequence g_n in $L^\infty \cap \mathcal{D}(Q)$ such that $\|g - g_n\| \to 0$ while $Q(g - g_n) \to 0$. Since $(1, g_n) \to (1, g) = 0$ we may replace g_n by $g_n - (1, g_n)$ and maintain these limits. Therefore we may assume $(1, g_n) = 0$. Thus we may apply (MG) to g_n and let $n \to \infty$ to retrieve (MG) for arbitrary g in $\mathcal{D}(Q)$. □

EXAMPLE 2.6 (Two point inequality). Let $X = \{-1, 1\}$ and write $\mu(\{-1\}) = \mu(\{1\}) = 1/2$. If $f : X \to R$ define $(af)(x) = (f(1) - f(-1))/2$ for $x = \pm 1$. Define $Q(f) = \int_X (af)(x)^2 d\mu(x)$. Then Q is a quadratic form on $L^2(X, \mu)$. Of course since af is constant Q is also given by

$$Q(f) = (f(1) - f(-1))^2 / 4. \qquad (2.8)$$

We will prove that Q satisfies (LS). This is the simplest non trivial example. By the triangle inequality $Q(|f|) \le Q(f)$. Therefore it suffices to prove (LS) in case $f \ge 0$. Denote by x the coordinate function on X : $x(\pm 1) = \pm 1$. Every function on X is of the form $a + bx$ for some real a and b. This cannot be nonnegative unless $a > 0$. By homogeneity of (LS) we may therefore assume $f = 1 + bx$, which is nonnegative only for $-1 \le b \le 1$, and by symmetry we need only consider $0 \le b \le 1$. Let $f_s(x) = 1 + sx$, $0 \le s \le 1$ and put

$$h(s) = \int f_s^2 \log f_s \, d\mu - \|f_s\|^2 \log \|f_s\|$$

$$= \tfrac{1}{2}\left[(1+s)^2 \log(1+s) + (1-s)^2 \log(1-s)\right] - \tfrac{1}{2}(1+s^2)\log(1+s^2).$$

Since $af_s = s$ we have $Q(f_s) = s^2$. Thus it suffices to show that $h(s) \le s^2$ for $0 \le s \le 1$. Since $h(0) = 0$ it suffices to show that $h'(s)$, which equals $\left[(1+s)\log(1+s) - (1-s)\log(1-s)\right] - s\log(1+s^2)$, satisfies $h'(s) \le 2s$ on $[0,1)$. But since $h'(0) = 0$ it suffices to show that $h''(s)$, which equals $2 + \log[(1-s^2)/(1+s^2)] - 2s^2/(1+s^2)$, satisfies $h''(s) \le 2$ on $[0,1)$. This clearly holds since the second and third terms of $h''(s)$ are negative. This proves that Q satisfies (LS).

EXAMPLE 2.7 (Gauss Dirichlet form). Here we will prove (1.6). By the additivity theorem (Theorem 2.3) it suffices to prove (1.6) for $n = 1$. Let (X_j, μ_j), $j = 1, \dots, k$ be copies of the two point measure space of Example 2.6 with corresponding coordinate functions $x_1, \dots, x_k$. Put $\Omega = \prod_{j=1}^k X_j$ and $\mu = \prod_{j=1}^k \mu_j$. By the additivity theorem we have

$$\int_\Omega f(x)^2 \log|f(x)| \, d\mu(x) \le Q(f) + \|f\|^2 \log \|f\| \tag{2.9}$$

wherein

$$Q(f) = \sum_{j=1}^k \int_\Omega (a_j f)(x)^2 d\mu(x)$$

and

$$(a_j f)(x) = \frac{1}{2}\Big[f(x_1, \dots, x_{j-1}, 1, x_{j+1}, \dots, x_k) - f(x_1, \dots, x_{j-1}, -1, x_{j+1}, \dots, x_k)\Big].$$

Put $y = (x_1 + \cdots + x_k)k^{-1/2}$. Let us apply (2.9) to a function f of the form $f(x_1, \dots, x_n) = \varphi(y)$ with φ in $C_c^\infty(R)$. By the central limit theorem the left side of (2.9) converges to $\int_{-\infty}^\infty |\varphi(t)|^2 \log|\varphi(t)| \, d\nu(t)$ as $k \to \infty$ since the integrand is a continuous function with compact support. Similarly the second term on the right of (2.9) converges to the second term on the right of (1.6). We will show that $Q(f)$ converges to $\int_{-\infty}^\infty |\varphi'(t)|^2 d\nu(t)$. Since φ is in $C_c^\infty(R)$ we may write $(1/2)\left[\varphi(t - hx + h) - \varphi(t - hx - h)\right] - \varphi'(t)h = h^2 g(t, x, h)$ for t in R, $x = \pm 1$ and $0 < h \le 2$ where g is a bounded function on $R \times \{-1, 1\} \times (0, 2]$. Now put $h = k^{-1/2}$ and observe that $y|_{x_j = 1} = y - hx_j + h$ while $y|_{x_j = -1} = y - hx_j - h$. Hence

$$(a_j f)(x) = \tfrac{1}{2}\left[\varphi(y - hx_j + h) - \varphi(y - hx_j - h)\right]$$
$$= \varphi'(y)h + h^2 g(y, x_j, h).$$

Hence

$$\begin{aligned}\sum_{j=1}^{k}|(a_j f)(x)|^2 &= \sum_{j=1}^{k}|\varphi'(y)h + h^2 g(y,x_j,h)|^2 \\ &= kh^2|\varphi'(y)|^2 + \psi_k(x_1,\ldots,x_k,h) \\ &= |\varphi'(y)|^2 + \psi_k(x,h)\end{aligned}$$

where ψ_k is a sum of $2k$ terms of order h^3 or h^4 and is therefore of order $k^{-1/2}$ uniformly in x as $k \to \infty$. Hence

$$Q(f) = \int_X |\varphi'(y)|^2 d\mu(x) + \int_X \psi_k(x,h)\, d\mu(x),$$

which converges to $\int_{-\infty}^{\infty} |\varphi'(t)|^2 d\nu(t)$ by the central limit theorem as $k \to \infty$. This proves (1.6) in case f is in $C_c^\infty(R)$. But for any function f in $L^2(R,\nu)$ such that f' (weak derivative) is also in $L^2(R,\nu)$ there is a sequence f_n in $C_c^\infty(R)$ which converges to f in $\|f\|_{L^2(\nu)} + \|f'\|_{L^2(\nu)}$ norm. Since $t^2 \log t$ is bounded below on $[0,\infty)$ we may apply Fatou's Lemma to the left side of (1.6) for some a.e. convergent subsequence, obtaining (1.6) in general.

3 Logarithmic Sobolev inequalities and Contractivity Properties of Semigroups

In this section we consider a measure space (X,μ) and a semigroup of operators e^{-tH} acting on various function spaces over X. We will address the question "Given t, p and q, under what direct conditions on the infinitesimal generator H can one conclude that the operator e^{-tH} is a bounded operator from $L^p(\mu)$ to $L^q(\mu)$?" We will see that there is a simple equivalence between such boundedness properties of the semigroup e^{-tH} and logarithmic Sobolev inequalities determined by its infinitesimal generator H. We follow [Gr1]. For any complex valued function f and $p \in (1,\infty)$ we write $f_p = (\operatorname{sgn} f)|f|^{p-1}$ where $\operatorname{sgn} z = z/|z|$ if $z \neq 0$ and $\operatorname{sgn} z = 0$ if $z = 0$.

DEFINITION 3.1. Let (Ω,ν) be a probability space and let p be in $(1,\infty)$. An operator H in $L^p(\mu)$ is called a Sobolev generator of index p if it is the generator of a strongly continuous semigroup e^{-tH} in L^p and, for some real constants $c > 0$ and γ, there holds

$$\int |f|^p \log|f|\, d\mu \le c \operatorname{Re}\langle (H+\gamma)f, f_p\rangle + \|f\|_p^p \log\|f\|_p, \quad f \in \mathcal{D}(H). \tag{3.1}$$

c is called a principal coefficient of H and γ a local norm of H. Here and in the following $\langle f,g\rangle$ denotes $\int f\bar{g}\, d\mu$.

Suppose $1 \le a < b \le \infty$, $r \in (a,b)$ and e^{-tH} is a strongly continuous semigroup in $L^r(\mu)$ whose restriction to L^p, if $p > r$, or closure in L^p, if $p < r$, is a strongly continuous semigroup in L^p for all $p \in (a,b)$. We shall say that H is a Sobolev generator on (a,b) if there are continuous real valued functions $c(\cdot) > 0$ and $\gamma(\cdot)$ on (a,b) such that for all p

in (a,b) the generator in L^p of the semigroup e^{-tH} is a Sobolev generator of index p with principal coefficient $c(p)$ and local norm $\gamma(p)$.

REMARK 3.2. $\int |f|^p \log |f| \, d\mu - \|f\|_p^p \log \|f\|_p$ is positive homogeneous of degree p in f. Hence replacement of f by a non-zero scalar times f in (3.1) does not alter the inequality.

REMARK 3.3. The distinguished role played by the number $r \in (a,b)$ in the preceding definition is minor. For if $s \in (a,b)$ then the (restriction or closure of the) semigroup e^{-tH} in L^s has a generator H_s which is also a Sobolev generator on (a,b) with r in the definition replaced by s. Properly speaking the definition is concerned with a family of semigroups e^{-tH_s} on L^s, $s \in (a,b)$ such that $e^{-tH_s}|L^r = e^{-tH_r}$ whenever $s < r$. However it will be more convenient to use the definition as formulated. We shall refer to H_s as the L^s generator of e^{-tH}.

REMARK 3.4. Since $\varphi(x) = x \log x$ is convex in $[0,\infty)$ when $\varphi(0)$ is defined to be zero, Jensen's inequality shows that

$$\int |f|^p \log |f| \, d\mu - \|f\|_p^p \log \|f\|_p \equiv p^{-1}\Big\{ \int \varphi(|f|^p) - \varphi\Big(\int |f|^p\Big)\Big\} \geq 0.$$

Hence Equation (3.1) implies

$$\operatorname{Re}\langle (H+\gamma)f, f_p\rangle \geq 0, \quad f \in \mathcal{D}(H). \tag{3.2}$$

REMARK 3.5. The inequality (3.2) implies that $H+\gamma$ is a contraction semigroup generator. That is,

$$\|e^{-tH}\|_{p,p} \leq e^{t\gamma} \tag{3.3}$$

where, here and in the following, $\|A\|_{q,p}$ denotes the norm of an operator A from L^q to L^p. Indeed, Hölder's inequality and (3.2) yields $\|(H+\gamma+\lambda)f\|_p \|f\|_p^{p-1} \geq \operatorname{Re}\langle (H+\gamma+\lambda)f, f_p\rangle \geq \lambda \|f\|_p^p$ for all $\lambda \geq 0$. That is, $\|(H+\gamma+\lambda)f\|_p \geq \lambda\|f\|_p$, which shows that $\|e^{-t(H+\gamma)}\|_{p,p} \leq 1$ by the Hille–Yosida theorem. In particular, if H has local norm zero then e^{-tH} is a contraction semigroup in L^p.

REMARK 3.6. Write $p' = p/(p-1)$ for the conjugate exponent of p. We note that the map $f \to f_p$ is a homeomorphism of L^p onto $L^{p'}$ for $1 < p < \infty$, whose inverse is the map $g \to g_{p'}$.

THEOREM 3.7. *Let H be a Sobolev generator on (a,b) with principal coefficient function $c(\cdot)$ and local norm function $\gamma(\cdot)$. For each q in (a,b) let $p(t,q)$ be the solution of the initial value problem*

$$c(p)\, dp/dt = p, \quad p(0,q) = q \qquad t \geq 0 \tag{3.4}$$

and let

$$M(t,q) = \int_0^t \gamma(p(s,q))\, ds \tag{3.5}$$

$p(t,q)$ and $M(t,q)$ are both defined for as long as $p(t,q) < b$. Then

$$\|e^{-tH}\|_{q,p(t,q)} \leq e^{M(t,q)}. \tag{3.6}$$

LEMMA 3.8. *Let (Ω,μ) be a probability measure space. Suppose $1<p<\infty$, $\varepsilon>0$ and $q>p$. Let $s(t)$ be a real continuously differentiable function on $[0,\varepsilon)$ into $(1,\infty)$ such that $s(0)=p$ and let $f(t)$ be a continuously differentiable function on $[0,\varepsilon)$ into $L^q(\mu)$ with $f(0)=v\neq 0$. Then $\|f(t)\|_{s(t)}$ is differentiable at $t=0$ and*

$$\frac{d}{dt}\|f(t)\|_{s(t)}\Big|_{t=0} = \|v\|_p^{1-p}\Big[p^{-1}s'(0)\Big\{\int |v|^p\log|v|\,d\mu - \|v\|_p^p\log\|v\|_p\Big\} + \operatorname{Re}\langle f'(0),v_p\rangle\Big]. \tag{3.7}$$

PROOF. If $g:[0,\varepsilon)\to\mathbf{C}$ is continuously differentiable then a straightforward calculation shows that

$$d|g(t)|^{s(t)}/dt = s'(t)|g(t)|^{s(t)}\log|g(t)| + s(t)\operatorname{Re} g'(t)\overline{g_{s(t)}(t)} \tag{3.8}$$

and this is valid even when $g(t)=0$ for some t because $s(t)>1$. Proceeding informally for a moment, let us put $g(t)=f(t)(x)$ in (3.8), integrate with respect to x and interchange the t derivative on the left with the x integral to get

$$(d/dt)\int_\Omega |f(t)(x)|^{s(t)}d\mu(x) = \int_\Omega s'(t)|f(t,x)|^{s(t)}\log|f(t,x)|d\mu(x) + s(t)\operatorname{Re}\langle f'(t),f_{s(t)}\rangle. \tag{3.9}$$

Before addressing the technicalities involved in deriving (3.9) let us see that (3.9) proves the lemma. Put $V(t)=\int_\Omega |f(t)(x)|^{s(t)}d\mu(x)$ and observe that

$$\begin{aligned}(d/dt)\|f(t)\|_{s(t)} &= dV(t)^{s(t)^{-1}}/dt\\ &= s(t)^{-1}\big[V(t)^{s(t)^{-1}}/V(t)\big]V'(t) - (s'(t)/s(t)^2)\,V(t)^{s(t)^{-1}}\log V(t).\end{aligned}$$

Putting $t=0$ in this equality, $s(0)=p$, $V(0)=\|v\|_p^p$ and using (3.9) we get (3.7).

It remains to justify (3.9), which requires a little effort because $f(t,x)$ is not necessarily differentiable in t for a.e. x. Of course if Ω has only finitely many points then (3.9) holds. In the general case, if $A_1,\ldots,A_n$ is a measurable partition of Ω and $f_n(t)$ is the conditional expectation of $f(t)$ with respect to the σ-field generated by $A_1,\ldots,A_n$ then (3.9) holds for $f_n(t)$ because this is equivalent to the case of a finite set Ω again. Now write (3.9) for f_n in integral form by putting $\int_0^T dt$ on both sides and do the integral on the left. If the measure space (Ω,μ) is separable we may choose a sequence of partitions $\{A_1,\ldots,A_n\}$ such that $f_n(t)$ and $f_n'(t)$ converge in L^q to $f(t)$ and $f'(t)$, respectively, with good L^q boundedness in t, from which one can deduce (3.9) by standard limiting arguments. Finally, if (Ω,μ) is not separable there is a separable subfield with respect to which $f(t)(\cdot)$ and $f'(t)(\cdot)$, is measurable for all t in $[0,\varepsilon)$ (consider rational t first and then use continuity into L^q) and we are therefore in the previous case. A slightly different proof of this Lemma is given in [Gr1, Lemma 1.1]. A third way to deal with the technicalities involved in (3.9) is given in [D8, Lemmas 2.2.1 and 2.2.2]. □

PROOF OF THEOREM 3.7. Let g be a nonnegative function in $C_c^\infty(R)$ with support in $(0,\infty)$. Suppose that u is in $L^\infty(\mu)$. Then $v := \int_0^\infty g(s)e^{-sH}u\,ds$ exists as a Riemann integral in L^p for each p in (a,b) and is in $C^\infty(H_p)$. Put $f(t) = e^{-tH}v$ for $t \geq 0$. Then $f(\cdot)$ is a differentiable function into $L^r(\mu)$ for all r in (a,b). Hence by Lemma 3.8 the function $\alpha(t) = \|f(t)\|_{p(t,q)}$ is differentiable on its interval of definition and by (3.7), (3.4) and (3.1) we have

$$\begin{aligned} d\alpha(t)/dt &= \|f(t)\|_p^{1-p}\Big[c(p)^{-1}\Big\{\int |f(t)|^p \log|f(t)|\,d\mu - \|f(t)\|_p^p \log\|f\|_p\Big\} \\ &\qquad - \operatorname{Re}\langle Hf(t), f(t)_p\rangle\Big] \\ &\leq \gamma(p)\,\|f(t)\|_p \qquad p = p(t,q) \end{aligned} \tag{3.10}$$

That is, $d\log\alpha(t)/dt \leq \gamma(p(t,q))$. Thus $\log\alpha(t) \leq \log\alpha(0) + M(t,q)$. Since $\alpha(0) = \|v\|_q$ we have

$$\|e^{-tH}v\|_{p(t,q)} \leq e^{M(t,q)}\,\|v\|_q \tag{3.11}$$

Now fix t and let g run through a sequence g_n which converges to the Dirac measure $\delta(s)$ and such that the corresponding sequence v_n converges to u in L^q norm while $e^{-tH}v_n$, which converges to $e^{-tH}u$ in L^q, also converges pointwise almost everywhere. Apply (3.11) to v_n and apply Fatou's Lemma on the left to get (3.11) for u. Since $L^\infty(\mu)$ is dense in $L^q(\mu)$ we may apply Fatou's Lemma again to conclude that (3.11) holds for all v in $L^q(\mu)$. This proves (3.6). □

COROLLARY 3.9. *In case the local norm $\gamma(\cdot)$ is zero in Theorem 3.7 then $\|e^{-tH}\|_{q,p(t,q)} \leq 1$ where $p(\cdot,\cdot)$ is given by* (3.4).

For positivity preserving semigroups the hypotheses of Theorem 3.7 can be weakened in a very useful way.

COROLLARY 3.10. *Assume $\mu(X) < \infty$ and that e^{-tH} is a positivity preserving semigroup on $L^p(\mu)$ for $1 \leq a < p < b \leq \infty$. Suppose that $\gamma(p)$ and $0 < c(p)$ are continuous on (a,b) and that*

$$\int f^p \log f\,d\mu \leq c(p)\langle (H + \gamma(p))f, f^{p-1}\rangle + \|f\|_p^p \log\|f\|_p \tag{3.12}$$

holds for nonnegative f in $\mathcal{D}(H_p)$ for $a < p < b$. Then (3.6) *holds.*

PROOF. Note first that the inner product on the right of (3.12) is real for $f \geq 0$ because e^{-tH} and hence H is reality preserving. In the proof of Theorem 3.7 take $u \geq 0$ and bounded. Then since $g \geq 0$ and e^{-sH} is positivity preserving it follows that v as well as f(t) are nonnegative. Thus starting with (3.12) the proof of Theorem 3.7 goes as before and yields (3.11) for all nonnegative v in $L^q(\mu)$. Since $|(e^{-tH}(w))(x)| \leq (e^{-tH})|w|(x)$ a.e. for all complex functions w in L^q, (3.6) follows. □

COROLLARY 3.11 (Submarkovian case). *Assume $\mu(X) < \infty$ and that e^{-tH} is a positivity preserving semigroup on $L^2(\mu)$. Assume further that e^{-tH} is a contraction semigroup on L^∞. That is, $\|e^{-tH}u\|_\infty \le \|u\|_\infty$ for all u in $L^\infty(\mu)$ and all $t \ge 0$. Suppose that $\gamma(\cdot)$ and $0 < c(\cdot)$ are continuous on $[2,\infty)$ and that (3.12) holds for all bounded nonnegative functions f in the L^2 domain of H. Then (3.6) holds.*

PROOF. The proof is the same as that of Corollary 3.10. We need only note that now v and $f(t)$ are in $L^\infty(\mu)$ as well as nonnegative and in $\mathcal{D}(H_p)$ for $2 \le p < \infty$. Since, moreover $\mathcal{D}(H_p)$ is contained in the L^2 domain of H we can apply (3.12) and proceed as before. □

The next theorem is a converse of Theorem 3.7.

THEOREM 3.12. *Let (Ω, μ) be a probability space and let $1 \le a \le r \le b \le \infty$. Suppose that e^{-tH} is a strongly continuous semigroup in $L^r(\mu)$. Assume that for each q in (a,b) there are continuous functions $\rho(t,q)$ and $m(t,q)$ defined on an interval $[0,\varepsilon(q))$, $\varepsilon(q) > 0$, with $\rho(0,q) = q$ and $m(0,q) = 1$, such that*

$$\|e^{-tH}\|_{q,\rho(t,q)} \le m(t,q), \quad 0 \le t < \varepsilon(q), \quad q \in (a,b). \tag{3.13}$$

Assume that the right t derivatives of ρ and m exist at $t = 0$ and that $c(q)^{-1} \equiv q^{-1} d\rho(t,q)/dt\big|_{t=0}$ and $\gamma(q) \equiv dm(t,q)/dt\big|_{t=0}$ are continuous on (a,b) and that $c(\cdot)$ is strictly positive. Then H is a Sobolev generator on (a,b) with principal coefficient function $c(\cdot)$ and local norm function $\gamma(\cdot)$.

PROOF. Since $c(q) > 0$, $\rho(t,q) \ge q$ for small t. Hence $\|e^{-tH}\|_{q,q} \le m(t,q)$ for small t for each q in (a,b). Hence the restriction (or closure) of e^{-tH} in L^q is a semigroup of bounded operators and is uniformly bounded near $t = 0$. If f is in L^q and h is in $L^{q'}$ then $\langle (e^{-tH} - I)f, h\rangle$ is continuous at $t = 0$, as may be seen by approximating f in L^q norm by an element of L^r in case $q < r$ or by approximating h by an element of $L^{r'}$ in case $q > r$. Hence e^{-tH} determines a weakly and hence strongly continuous semi-group in L^q for all $q \in (a,b)$.

Let $\mathcal{D}$ be the linear span of the set of v constructed in the proof of Theorem 3.7, allowing g and u to vary over the indicated sets. Let v be a nonzero element of $\mathcal{D}$ and put $f(t) = e^{-tH}v$. Then for each t in $(0, \varepsilon(q))$ we have

$$t^{-1}(\|f(t)\|_{\rho(t,q)} - \|f(0)\|_q) \;\le\; \|v\|_q(m(t,q)-1)/t$$

by (3.13). By Lemma 3.8 we may take the limit as $t \downarrow 0$ in the last inequality to obtain

$$\|v\|_q^{1-q}\Big[q^{-1}qc(q)^{-1}\Big\{\int |v|^q \log|v|\,d\mu - \|v\|_q^q \ln\|v\|_q\Big\} \;-\; \mathrm{Re}\,\langle Hv, v_q\rangle\Big] \;\le\; \gamma(q)\|v\|_q.$$

Multiplying by $c(q)\|v\|_q^{q-1}$ yields (3.1) with $p = q$ since $\langle v, v_q\rangle \;=\; \|v\|_q^q$. Now $\mathcal{D}$ is a core for the L^p generator, H_p, of the semigroup e^{-tH}. If f is in $\mathcal{D}(H_p)$ there exists a sequence v_n in $\mathcal{D}$ such that $v_n \to f$ in H_p graph norm and such that $v_n \to f$ a.e.. Since $x^p \log x$ is bounded below on $[0,\infty)$ we may use Fatou's lemma on the left of (3.1) while on the right we observe that the map $f \to f_p$ is continuous from L^p to $L^{p'}$ so that the right side is a continuous function of f in H_p graph norm. Thus the validity of (3.1) for each v_n implies (3.1) for f. This concludes the proof. □

Davies and Simon [DS2, D8] have found it useful to solve the differential equation (3.4) for t in terms of p for each q instead of p in terms of t. The solution $p \to t(q,p)$ of (3.4) is clearly

$$t(q,p) = \int_q^p c(r)\, r^{-1}\, dr \tag{3.14}$$

Making the change of variable $r = p(s,q)$ in (3.5) we have $dr = p'(s,q)\, ds = r\, c(r)^{-1}\, ds$ by (3.4). So

$$M(t,q) = \int_q^p \gamma(r) c(r) r^{-1}\, dr\, . \tag{3.15}$$

Thus Theorem 3.7 gives

$$\|e^{-tH}\|_{q\to p} \le e^{M(t,q)} \quad \text{if} \quad t = t(q,p)\, . \tag{3.16}$$

$t(q,p)$ could be thought of as the smallest time t required for e^{-tH} to be a bounded operator from L^q to L^p (with the bound (3.6), anyway). In fact if H is the Dirichlet form operator associated to Gauss measure (1.5) then the function $t(q,p)$ that we will obtain in this way in Section 4 has been shown by E. Nelson [N4] to be indeed minimal in the sense that $\|e^{-tH}\|_{q\to p} = \infty$ if $t < t(q,p)$. In these considerations only $p \ge q$ is of interest because e^{-tH} is usually bounded from L^q to L^p for all $t \ge 0$ if $p < q$.

We can rephrase Theorem 3.7 as follows.

COROLLARY 3.13. *Assume the hypotheses of Theorem 3.7 and define $t(q,p)$ by (3.14) and $M(t,q)$ by (3.15). Then*

$$\|e^{-tH}\|_{q\to p} \le e^{M(t,q)} \quad \textit{if} \quad t = t(q,p) \tag{3.17}$$

4 Index 2 implies index p for generalized Dirichlet forms

In order to deduce contractivity properties of e^{-tH} from L^r to L^q via Theorem 3.7 it is necessary to know that H is a logarithmic Sobolev generator of index p for p running over some interval. In practice it is easiest to establish a logarithmic Sobolev inequality of index 2. It is fortuitous, therefore, that for an important class of operators — Dirichlet form operators — an index p logarithmic Sobolev inequality follows automatically from an index 2 logarithmic Sobolev inequality. The idea behind this can be understood easily in the following prototype case. We consider only real functions in this section for simplicity and without any real loss of generality. Suppose H is given by

$$(Hf,f)_{L^2(R^n,\mu)} = \int_{R^n} |\nabla f(x)|^2 d\mu(x)$$

wherein μ is a probability measure on R^n with, say, smooth density. We say H is the Dirichlet form operator for the measure μ. If ∇ has a closed version as a densely defined operator $\nabla : L^2(R^n,\mu) \to L^2(R^n,\mu)\otimes R^n$ then we may simply write $H = \nabla^*\nabla$. Ignoring domain issues for the moment consider a smooth bounded function $f : R^n \to (0,\infty)$. On the one hand, for $p > 1$ we have $|\nabla(f(x)^{p/2})|^2 = (p/2)^2(f(x)^{\frac{p}{2}-1})^2|\nabla f(x)|^2$ while on the other hand $\nabla f(x)\cdot\nabla(f(x)^{p-1}) = (p-1)f(x)^{p-2}|\nabla f(x)|^2$. Hence $|\nabla(f(x)^{p/2})|^2 = [(p/2)^2/(p-1)]\nabla f(x)\cdot\nabla(f(x)^{p-1})$. Therefore $(Hf^{p/2}, f^{p/2}) = [(p/2)^2/(p-1)]$ (Hf, f^{p-1}). Thus if H satisfies

$$\int_{R^n} f(x)^2\log f(x)\,d\mu(x) \;\le\; c(Hf,f) + \|f\|^2\log\|f\| \tag{4.1}$$

for smooth strictly positive f then, replacing f by $f^{p/2}$ and using the last identity we get

$$\int_{R^n} f(x)^p\log f(x)\,d\mu(x) \;\le\; \frac{cp/2}{p-1}(Hf,f^{p-1}) + \|f\|_p^p\log\|f\|_p. \tag{4.2}$$

By the Beurling–Deny theorem [D8, Fu] e^{-tH} is positivity preserving. Thus, except for problems concerning the domain of validity of these computations, we may apply Corollary 3.10 on $(1,\infty)$ with principal coefficient $c(p) = (p/2)(p-1)^{-1}c$ and local norm $\gamma \equiv 0$. There are domain questions here, however, which we will circumvent below.

A similar argument applies to the "discrete Dirichlet form" operator a^*a of Example 2.6 as was shown in [Gr1].

But D. Stroock has shown [St2] (see also [DeS1]) that the transition from (4.1) to (4.2) follows just from the fact that e^{-tH} is a positivity preserving contraction semigroup in $L^\infty(\mu)$. We will state and prove Stroock's theorem here in a slightly modified form, which is aimed at allowing the choice $H = \nabla^*\nabla + V$ wherein V is a positive potential. We follow [Gr7].

THEOREM 4.1 (Stroock). *Let (Ω,μ) be a finite measure space. Let L be a nonnegative self-adjoint operator on $L^2(\mu)$ such that e^{-tL} is positivity preserving and is a contraction on $L^\infty(\mu)$. Let $\mathcal{D}_1 = domain(L^{1/2})$ in $L^2(\mu)$. For f and g in $\mathcal{D}_1$ put*

$$Q(f,g) = (L^{1/2}f, L^{1/2}g)_{L^2(\mu)}.$$

If $0 \le f \in L^\infty \cap \mathcal{D}_1$ and $2 \le p < \infty$ then $f^{p/2}$ and f^{p-1} are in $\mathcal{D}_1$ and

$$Q(f^{p/2}, f^{p/2}) \;\le\; ((p/2)^2/(p-1))\;Q(f, f^{p-1}),\quad 2 \le p < \infty. \tag{4.3}$$

PROOF. Write $P_t = e^{-tL}$ and let $\sigma_t(y) = (P_t 1)(y)$. Then $1 - \sigma_t(y) \ge 0$ a.e. because P_t is a contraction on L^∞. If u and v are bounded measurable functions on Ω we write $u - u(x)1$ for the function $y \to u(y) - u(x)$. A straightforward expansion of the right side using the symmetry of P_t establishes the following well known equality.

$$\begin{aligned}((I-P_t)u,v) \;&=\; \frac{1}{2}\int P_t\{(u-u(x)1)(v-v(x)1)\}(x)\,d\mu(x)\\ &\quad + \int(1-\sigma_t(y))u(y)v(y)\,d\mu(y).\end{aligned} \tag{4.4}$$

Since $t^{-1}(1-e^{-t\lambda})$ increases to λ for $\lambda \geq 0$ as t decreases to zero the spectral theorem shows that for any functions u and v in $L^2(\mu)$

$$Q(u,v) = \lim_{t\downarrow 0} t^{-1}((I-P_t)u,v) \tag{4.5}$$

if u and v are in $\mathcal{D}_1$. Moreover u is in $\mathcal{D}_1$ if and only if $\lim_{t\downarrow 0} t^{-1}((I-P_1)u,u)$ is finite. Suppose $0 \leq f(y) \leq 1$ for all y in Ω and that $a \geq 1$. By the mean value theorem $(f(y)^a - f(x)^a)^2 \leq a^2(f(y)-f(x))^2$ while $(f(y)^a)^2 \leq f(y)^2 \leq a^2 f(y)^2$. Thus if we put $u = v = f^a$ in (4.4) we get $((I-P_t)f^a, f^a) \leq a^2((I-P_t)f,f)$ which implies that f^a is in $\mathcal{D}_1$ when f is in $\mathcal{D}_1$ and $0 \leq f \leq 1$. Since $p/2 \geq 1$ and $p-1 \geq 1$ the first assertion of the lemma follows for a bounded nonnegative function f by considering $C^{-1}f$ where $C = \sup f(y)$. To prove (4.3) we assume $0 \leq f$ is bounded. Now if $0 \leq \xi \leq \eta$ then

$$\begin{aligned}(\eta^{p/2}-\xi^{p/2})^2 &= \left(p/2\int_\xi^\eta s^{p/2-1}ds\right)^2 \\ &\leq (p/2)^2(n-\xi)\int_\xi^\eta s^{p-2}ds = (p/2)^2(p-1)^{-1}(\eta-\xi)(\eta^{p-1}-\xi^{p-1}).\end{aligned}$$

Hence

$$(f(y)^{p/2}-f(x)^{p/2})^2 \leq (p/2)^2(p-1)^{-1}(f(y)-f(x))(f^{p-1}(y)-f^{p-1}(x)).$$

Moreover $(p/2)^2(p-1)^{-1}$ is increasing on $[2,\infty)$ (take the derivative) and is therefore greater or equal to one on $[2,\infty)$. By (4.4) we therefore have

$$\begin{aligned}&((I-P_t)f^{p/2}, f^{p/2}) \\ &\quad= \frac{1}{2}\int P_t\{(f^{p/2}-f(x)^{p/2}1)^2\}(x)\,d\mu(x) + \int(1-\sigma_t(y))f^p(y)\,d\mu(y) \\ &\quad\leq (p/2)^2\,(p-1)^{-1}\frac{1}{2}\int P_t\left\{(f-f(x)1)(f^{p-1}-f(x)^{p-1}1)\right\}(x)\,d\mu(x) \\ &\qquad\qquad + \int(1-\sigma_t(y))f(y)f^{p-1}(y)\,d\mu(y) \\ &\quad\leq (p/2)^2\,(p-1)^{-1}\,((I-P_t)f, f^{p-1}).\end{aligned}$$

Inequality (4.3) now follows from (4.5). □

EXAMPLE 4.2 (Nelson's best Gaussian hypercontractive bounds, [N3, 4]). Denote by N the Dirichlet form operator for Gauss measure ν on R^n. See (1.5). Thus $(Nf,g)_{L^2(\nu)} = \int_{R^n} \nabla f(x)\cdot\nabla g(x)\,d\nu(x)$ for f and g in the domain of N. One can compute explicitly by an integration by parts that N is given by

$$(Nf)(x) = \sum_{j=1}^n\left(-\frac{\partial^2 f(x)}{\partial x_j^2} + x_j\frac{\partial f}{\partial x_j}\right) \tag{4.6}$$

for smooth f. Nelson proved [N3, 4] the bound in the following corollary and showed that it is best possible in the sense that if t is smaller than that allowed by (4.7) then e^{-tN} is unbounded from L^q to L^p!

COROLLARY 4.3 (Nelson [N4]). *If* $1 < q, p < \infty$ *and*

$$e^{-2t} \leq (q-1)/(p-1) \tag{4.7}$$

then

$$\|e^{-tN}\|_{q \to p} = 1 \tag{4.8}$$

PROOF. We have already proved the Gaussian logarithmic Sobolev inequality (1.6) in Example 2.7. Let $p \geq 2$ and suppose that f is a nonnegative bounded function in the L^2 domain of N. Replace f in (1.6) by $f^{p/2}$. Writing $Q(f,f) = \int |\nabla f(x)|^2 \, d\nu(x)$ and using (4.3) we obtain, upon canceling a factor $p/2$ in all terms, the inequality

$$\int_{R^n} f(x)^p \log f(x) \, d\nu(x) \;\leq\; [(p/2)/(p-1)] \, (Nf, f^{p-1}) \;+\; \|f\|_p^p \log \|f\|_p \tag{4.9}$$

Hence the inequality (3.12) holds for the functions f specified in Corollary 3.11 with $c(p) = (p/2)/(p-1)$ and $\gamma(p) = 0$. Since N is a Dirichlet form, e^{-tN} is a positivity preserving operator and also a contraction from L^p to L^p for all p in $[1, \infty]$. We may therefore apply Corollary 3.11. The solution for equation (3.4) is given by $p(t,q) = 1 + (q-1)e^{2t}$ for $q \geq 2$ and $t \geq 0$ (actually for all $q > 1$) as one verifies easily. By (3.5) $M(t,q) = 0$. Hence

$$\|e^{-tN}\|_{q \to p(t,q)} \leq 1 \quad \text{for all} \quad q \geq 2 \tag{4.10}$$

and all $t \geq 0$. Since N is a self adjoint operator which annihilates constant functions we have $e^{-tN}1 = 1$. Thus (4.10) is an equality. If we put $p = p(t,q)$ and solve for t in terms of p and q we get $e^{-2t} = (q-1)/(p-1)$. Thus we have proved that equality in (4.7) yields (4.8) in case $2 \leq q \leq p < \infty$.

Let us continue to consider the case of equality in (4.7) for other q and p. Define

$$t(q,p) = \frac{1}{2} \left[\log(p-1) - \log(q-1)\right] \qquad 1 < q \leq p < \infty. \tag{4.11}$$

This is the solution to (4.7) (equality case). It is of course also the value of the integral (3.14) is case $c(r) = (r/2)/(r-1)$, which is the case of interest to us in this example. It is clear from both (3.14) and (4.11) that $t(q,p)$ has the additivity property

$$t(a,c) = t(a,b) + t(b,c), \qquad 1 < a \leq b < \infty \tag{4.12}$$

It is clear from (3.14) that this property does not depend on $c(r)$. But our particular function (4.11) also relates well to the process of taking conjugate exponents: for $1 < p < \infty$ write $p' = p/(p-1)$ for the conjugate exponent. Then $p' - 1 = (p-1)^{-1}$. Hence by (4.11)

$$t(b', a') = t(a,b) \qquad 1 < a \leq b < \infty \tag{4.13}$$

Now we can conclude the proof of the corollary. Since e^{-tN} is a Hermitian operator we have

$$\|e^{-tN}\|_{q \to p} = \|e^{-tN}\|_{p' \to q'} \qquad t \geq 0 \tag{4.14}$$

Suppose that $1 < q \le p \le 2$. Then $2 \le p' \le q' < \infty$. Put $t = t(q,p)$ in (4.14). By (4.13) and what we have already proved the right side of (4.14) is one. Thus we now have

$$\|e^{-t(q,p)N}\|_{q\to p} = 1 \tag{4.15}$$

for $1 < q \le p < \infty$ provided both q and p are in $(1,2]$ or both are in $[2,\infty)$. If $1 < q \le 2 \le p < \infty$ then we may use the additivity (4.12) to get

$$\begin{aligned}\|e^{-t(q,p)N}\|_{q\to p} &= \|e^{-t_1N}e^{-t_2N}\|_{q\to p}\\ &= \|e^{-t_1N}\|_{2\to p}\,\|e^{-t_2N}\|_{q\to 2}\\ &\le 1\end{aligned}$$

wherein $t_1 = t(2,p)$ and $t_2 = t(q,2)$.

This completes the proof of (4.8) in case (4.7) is an equality. If $q > p$ then since $\|f\|_q \ge \|f\|_p$ we have $\|e^{-tN}f\|_p \le \|e^{-tN}f\|_q \le \|f\|_q$ which proves (4.8) in this case. (Of course we use $e^{-tN}1 = 1$ to get $\|e^{-tN}\|_{q\to p} = 1$.) If $q \le p$ and (4.7) holds put $s = t - t(q,p)$. Then $s \ge 0$. Hence $\|e^{-tN}\|_{q\to p} = \|e^{-sN}e^{-t(q,p)N}\|_{q\to p} \le \|e^{-sN}\|_{p\to p}\|e^{-t(q,p)N}\|_{q\to p} \le 1$. This completes the proof of Corollary 4.3. □

REMARK 4.4. Although we have used the rather sophisticated Theorem 4.1 in the proof of Corollary 4.3 to get from (4.1) to (4.2) (and only for $p \ge 2$) one should not lose sight of the fact that the elementary calculus derivation of (4.2) from (4.1) given at the beginning of this section captures the essence of (4.3) in our case and does so for all p in $(1,\infty)$. However there are technical problems for $p < 2$ which must be dealt with. For example if $p = 3/2$ and $f \ge 0$ then $f^{p-1} = f^{1/2}$ and differentiating this fucntion at a point x where $f(x) = 0$ can lead to trouble in our elementary derivation of (4.2). This can be circumvented for the operator N by considering only those functions u in the proof of Theorem 3.7 satisfying $u \ge \varepsilon > 0$. Since $e^{-tN}1 = 1$ the functions v and $f(t)$ in that proof also satisfy this restriction if $\int_0^\infty g(s)\,ds = 1$. See [DS2, D8, DeS1, St2] for various other ways to deal with this issue.

5 Hypercontractive, supercontractive and ultracontractive semigroups

DEFINITION 5.1. Let (X,μ) be a measure space and let H be a nonnegative self-adjoint operator on $L^2(S,\mu)$. The semigroup e^{-tH} is called *hypercontractive* if

(a) $e^{-tH} : L^p \cap L^2 \to L^p$ is a contraction in L^p norms for each p in $[1,\infty]$

and

(b) $\|e^{-TH}\|_{2\to 4} = b < \infty$ for some $T > 0$.

This definition is taken from [SHK].

The generator H of a hypercontractive semigroup is a Sobolev generator on $(1,\infty)$ as we will now show. Let us assume $\mu(X) < \infty$ to avoid technical issues concerning the possible ill definedness of $\int_X |f(x)|^2 \log|f(x)|\,d\mu$ when $\mu(X) = \infty$. One can compute a principal coefficient $c(\cdot)$ and a local norm $\gamma(\cdot)$ by using the Stein interpolation theorem [Sn] as follows (cf. [Gr1]).

Applying the Stein interpolation theorem to the analytic family $T_z = e^{-zTH}$, $\operatorname{Re} z \geq 0$, with $\|T_{iy}\|_{2,2} = 1$ and $\|T_{1+iy}\|_{2,4} = b$ one obtains $\|e^{-sTH}\|_{2,4/(2-s)} \leq b^s$ for $0 \leq s \leq 1$. By (a) we have $\|e^{-tTH}\|_{\infty,\infty} \leq 1$ for all $t \geq 0$. Hence, applying the Riesz–Thorin theorem to T_s with indices $(2, 4/(2-s))$ and (∞,∞), one obtains, for $2 < r < \infty$, $\|e^{-sTH}\|_{r,2r/(2-s)} \leq b^{2s/r}$ for $0 \leq s \leq 1$. Using the Hermiticity of e^{-sTH} one obtains a similar inequality for $1 < r \leq 2$. A principal coefficient function $c(\cdot)$ and a local norm function $\gamma(\cdot)$ can now be computed by Theorem 3.12. The result is

$$c(p) = \begin{cases} 2T, & 2 \leq p < \infty, \\ 2T/(p-1), & 1 < p < 2, \end{cases} \tag{5.1}$$

$$\gamma(p) = \begin{cases} (2\ln b)/(pT), & 2 \leq p < \infty, \\ (2\ln b)(p-1)/(pT), & 1 < p < 2, \end{cases} \tag{5.2}$$

which are continuous.

REMARK 5.2. The number operator N (alias Ornstein–Uhlenbeck operator) discussed in Section 4 is the prototype and first studied generator of a hypercontractive semigroup. The exact (i.e., smallest) principal coefficients for N are those given in Section 4. The principal coefficients given by (5.1) are strictly larger than the known smallest ones because there is some information lost in using only knowledge of the time T for which $e^{-TN} : L^2 \to L^4$ is a contraction ($2T = \log 3$ which is greater than one) and then using the Stein interpolation theorem.

DEFINITION 5.3. Let (X,μ) be a measure space and let H be a nonnegative self-adjoint operator on $L^2(X,\mu)$. We say that the semigroup is *ultracontractive* if

(a) $e^{-tH} : L^p(\mu) \to L^p(\mu)$ is a contraction for each $t > 0$ and each p in $[1,\infty]$,

(b) $e^{-tH} : L^2(\mu) \to L^\infty(\mu)$ is bounded for each $t > 0$.

It is clear that ultracontractivity is a much stronger notion than hypercontractivity if $\mu(X) < \infty$. We could repeat the interpolation argument that we used above for hypercontractivity but we will instead use only a small part of that argument to derive information about a principal coefficient and local norm just at $p = 2$. Unlike the hypercontractive case we now have t at our disposal. Write

$$e^{M(t)} = \|e^{-tH}\|_{2\to\infty}, \quad t > 0.$$

Put $t = \varepsilon$ and $A(z) = e^{-z\varepsilon H}$ for $0 \leq \operatorname{Re} z \leq 1$. Then $A(iy) : L^2 \to L^2$ is a unitary operator and has norm one while $A(1+iy) : L^2 \to L^\infty$ has norm $e^{M(\varepsilon)}$. We may apply the Stein interpolation theorem [Sn] to conclude that

$$\|e^{-s\varepsilon H}\|_{2\to p(s)} \leq e^{sM(\varepsilon)}, \quad 0 \leq s \leq 1, \tag{5.3}$$

where $p(s) = 2/(1-s)$. Now we may apply the method of Theorem 3.12 to the semigroup $s \to e^{-s\varepsilon H}$, which has generator εH. We take $q = 2$, $\rho(s,2) = p(s)$, and $m(s,2) = e^{sM(\varepsilon)}$. Since $c(2)^{-1} = 2^{-1}dp(s)/ds|_{s=0} = 1$ and $dm(s,2)/ds|_{s=0} = M(\varepsilon)$ we get, for $\varepsilon > 0$ and $f \in \mathcal{D}(H)$,

$$\int_X |f(x)|^2 \log|f(x)|\, d\mu(x) \le \varepsilon(Hf,f) + M(\varepsilon)\|f\|^2 + \|f\|^2 \log\|f\|^2. \tag{5.4}$$

The distinction between hypercontractivity and ultracontractivity is thus already visible at the quadratic form level in that the principal coefficient, ε, in (5.4) can be made arbitrarily small. This contrasts sharply with the number operator N for which the coefficient of the gradient term in (1.6) cannot be reduced at all even if one adds a multiple of $\|f\|^2$ to the right side. We will now show that for Dirichlet forms (5.4) yields ultracontractivity if $M(\varepsilon)$ does not grow too rapidly as $\varepsilon \downarrow 0$.

THEOREM 5.4 (Davies, Simon [DS2, D8]). *Let $\beta : (0,\infty) \to [0,\infty)$ be continuous. Suppose (X,μ) is a finite measure space and H is a self-adjoint operator in $L^2(X,\mu)$. Assume that e^{-tH} is a positivity preserving semigroup for $t \ge 0$ and is a contraction on $L^p(\mu)$ for $1 \le p \le \infty$. (Equivalently, H satisfies the Beurling-Deny condition* [Fu, D8].) *Assume further that*

$$\int_X |f(x)|^2 \log|f(x)|\, d\mu(x) \le \varepsilon(Hf,f) + \beta(\varepsilon)\|f\|_2^2 + \|f\|_2^2 \log\|f\|_2, \quad f \in \mathcal{D}(H), \tag{5.5}$$

and that

$$M(t) = \frac{1}{t}\int_0^t \beta(\varepsilon)\, d\varepsilon < \infty \quad \text{for all } t > 0. \tag{5.6}$$

Then e^{-tH} is an ultracontractive semigroup and

$$\|e^{-tH}\|_{2\to\infty} \le e^{M(t)}. \tag{5.7}$$

PROOF. By Corollary 3.11 it suffices to consider nonnegative bounded f in $\mathcal{D}(H)$. For $p \ge 2$ apply (5.5) to $f^{p/2}$ and use Theorem 4.1 to get

$$\int f(x)^p \log f(x)\, d\mu \le \varepsilon\,[(p/2)/(p-1)]\,(Hf, f^{p-1}) + (2\beta(\varepsilon)/p)\|f\|_p^p + \|f\|_p^p \log\|f\|_p.$$

Since $p \ge 2$ we have $2p - 2 = p + p - 2 \ge p$. So $(p/2)/(p-1) \le 1$. Hence

$$\int f(x)^p \log f(x)\, d\mu \le \varepsilon(Hf, f^{p-1}) + 2\beta(\varepsilon)p^{-1}\|f\|_p^p + \|f\|_p^p \log\|f\|_p$$

for all $\varepsilon > 0$ and all $p \ge 2$. Fix $T > 0$ and put $\varepsilon = 2Tp^{-1}$. Then we have (3.12) holding for nonnegative f with $c(p) = 2Tp^{-1}$ and $c(p)\gamma(p) = 2\beta(2Tp^{-1})p^{-1}$. From (3.14) we have

$$t(2,p) = \int_2^p 2Tr^{-2}\, dr = 2T(2^{-1} - p^{-1}) = T(1 - 2p^{-1}) \le T$$

and from (3.15) we have, with $t = t(2,p)$,

$$M(t,2) \;=\; \int_2^p 2\beta(2Tr^{-1})r^{-2}\,dr \;=\; T^{-1}\int_{2Tp^{-1}}^T \beta(\varepsilon)\,d\varepsilon \;\le\; M(T).$$

Hence by Corollaries 3.11 and 3.9 we have, for f in L^2 and $t = t(2,p)$,

$$\|e^{-TH}f\|_p \le \|e^{-t(2,p)H}f\|_p \le e^{M(t,2)}\|f\|_2 \le e^{M(T)}\|f\|_2$$

for all $p < \infty$. Hence (5.7) holds with $t = T$. □

REMARK 5.5. Since e^{-tH} is Hermitian in Theorem 5.4, (5.7) implies $\|e^{-tH}\|_{1\to 2} \le e^{M(t)}$. Hence

$$\|e^{-tH}\|_{1\to\infty} \le \|e^{-tH/2}\|_{2\to\infty}\|e^{-tH/2}\|_{1\to 2} \le e^{2M(t/2)}.$$

Suppose that X is a locally compact Hausdorf space and μ is a regular Borel measure which assigns strictly positive measure to non empty open sets. Suppose further that e^{-tH} is given by an integral operator with continuous kernel; $(e^{-tH}f)(x) = \int_X K_t(x,y)f(y)\,d\mu(y)$. Then the inequality $\|e^{-tH}\|_{1\to\infty} \le e^{2M(t/2)}$ is easily seen to imply the uniform bound

$$|K_t(x,y)| \le e^{2M(t/2)}, \quad t > 0, \quad x,y \in X. \tag{5.8}$$

We will sketch later (Section 6.(xii)) how E.B. Davies obtained pointwise bounds on heat kernels from (5.8).

The definition of ***supercontractivity*** is the same as that for hypercontractivity (see Definition 5.1) except that condition b) is required to hold for all $T > 0$. It is clear from the derivation of (5.1) that the generator H of a supercontractive semigroup satisfies a logarithmic Sobolev inequality of the form (5.5) for all $\varepsilon > 0$. The distinction between supercontractive and ultracontractive appears to lie in the rate at which $\beta(\varepsilon) \to \infty$ as $\varepsilon \downarrow 0$.

6 Survey of related topics and applications

The Gaussian logarithmic Sobolev inequality (1.6) can be derived in a number of differeent ways. Its equivalent form, Corollary 4.3, which is E. Nelson's hypercontractivity bound can also be derived in several essentially distinct ways. Of course a direct proof of Corollary 4.3 is also an indirect proof of (1.6) and vice versa. Logarithmic Sobolev inequalities for the energy form of other measures either on R^n or on some other finite or infinite dimensional manifold have been derived with specific applications in mind. At the present time there are approximately 150 papers known to the author which derive new logarithmic Sobolev inequalities, give new derivations of old logarithmic Sobolev inequalities or corresponding hypercontractivity, give applications of the logarithmic Sobolev inequalities or the equivalent hypercontractive or ultracontractive inequalities, or do several of the above. In this section we will give a survey of many of these topics.

All of the references for these notes except [Fu], [Sn] and [S3] deal directly with these topics. The list of references represents, I believe, all of the papers on these topics that have come to my attention up to the present time, August 1992.

(i) *Direct proofs of Gaussian hypercontractivity*

The entire subject of this lecture was started in the paper [N1] of Edward Nelson. It was the goal of [N1] to prove that the $(\varphi^4)_2$ quantum field Hamiltonian is bounded below when one approximates space, which is one dimensional in this model, by a large circle. The constants in this paper are not best possible. J. Glimm [Gl] improved on this by showing that e^{-tN} is not only bounded from L^2 to L^4 for large t, but is also a contraction, for larger t, thereby allowing the important replacement of the large circle by a large interval. I. E. Segal [S1] gave an independent proof of this contractivity. In two following papers [N3, 4] Nelson gave two different proofs of contractivity of e^{-tN} from L^p to L^q. These papers give for the first time the best constants. The equivalence of hypercontractivity bounds with logarithmic Sobolev inequalities [Gr1] yields a different route to the proof of the best constants result. See Section 4. However, some very novel direct proofs (i.e., without going through logarithmic Sobolev inequalities) have been discovered which are very different from one another. See [Bnr2, 4], [BrL], [C1], [CLo], [Ep], [Li2], [MM], [Ne]. We have at the present time (at least) nine *direct* proofs of Nelson's Gaussian hypercontractivity bounds given in Corollary 4.3.

(ii) *Direct proofs of the Gaussian logarithmic Sobolev inequality*

The first paper known to us which contains an inequality similar to (1.6) and capturing some of the spirit of (1.6) is the paper [Fe] of Paul Federbush. His goal there was to prove semiboundedness of the $(\varphi^4)_2$ quantum field Hamiltonian without using Nelson's hypercontractivity explicitly. In the process he proves a combination of the Federbush semiboundedness theorem (see Theorem 2.1 above) and a slightly weaker version of (1.6) in infinite dimensions (cf. [Fe, Equation (14)].) Although he is not concerned with separating these ingredients in his proof of semiboundedness or in getting optimal constants, the two key ideas: a) use Young's inequality (as in the proof of Theorem 2.1) and b) differentiate $(d/dt)\int|e^{-tH}f|^{2+\lambda t}\,d\mu|_{t=0}$ and control this derivative, are in this paper. Our proof of (1.6) in Example 2.7 above is a slight simplification of our proof in [Gr1]. [Gr1] contains also an attempted derivation of (1.6) from the classical Sobolev inequality (cf. [Gr1, Theorem 5]). It is essentially a failure because it leads to an n dependent local norm. R. Seneor [Se] has pointed out, however, that if one starts with the best known constants in the classical Sobolev inequalities then the method works. Faris [F] proved the converse of the Federbush semiboundedness theorem and used it to give another proof of the 2-point inequality [Example 2.6]. Unknown to most of the workers in this field for many years including myself, the 2-point inequality had already been established quite early by Bonami [Bo]. A very different direct proof of (1.6) (for $n = 1$, which is all that's needed in view of the additivity theorem) was given by O. Rothaus [Ro1]. His very simple proof just uses Jensen's inequality and the fact that the lowest eigenfunction for a Sturm–Liouville boundary value problem with

Dirichlet boundary conditions is positive. R. A. Adams and F. H. Clarke [AC] have given another short proof of (1.6). Their proof formulates (1.6) as a problem in the calculus of variations. Finally we have the method of D. Bakry and M. Emery [BE1, 2]. This method has proven itself to be very powerful in the sense that it works well for other manifolds (e.g., S^n) and measures other than Gauss measure. This method will be described in detail in lectures of D. Bakry in the 1992 St. Flour summer school. See [AK1, 2], [B1, 2, 3], [De], [DeS2] and [Gel] for recent developments of this technique. We have at the present time (at least) five *direct* proofs of (1.6) which are very different from one another. We note also that E. Carlen [C1] has shown that one can add to the left side of (1.6) another term involving the Fourier transform of f.

(iii) *Higher order logarithmic Sobolev inequalities*

We know from (1.6) that if f and $|\nabla f|$ are in $L^2(R^n, \nu)$ then f is in the Orlicz space $L^2 \log L(\nu)$. Suppose that all the derivatives of f up to order k are in $L^2(\nu)$. What can one say about the "growth rate" of f? G. Feissner [Fe 1] and independently R. A. Adams [A] showed that f is in the Orlicz space $L^2(\log L)^k$. More generally if $1 < p < \infty$ and if f and $N^{k/2} f$ are in $L^p(R^n, \nu)$ then f is in $L^p(\log L)^k$. The associated inequalities are independent of the dimension, n.

(iv) *Logarithmic Sobolev inequalities on spheres*

If μ denotes normalized Lebesgue measure on S^n then the Dirichlet form operator for μ is the Laplacian, Δ. Although $e^{t\Delta} : L^p(\mu) \to L^q(\mu)$ is a bounded operator for all $t > 0$ and all p and q in $[1, \infty]$ there is a minimal time $t(p, q)$ for which it is a contraction. Correspondingly there is a smallest principle coefficient $c(2)$ in (3.12) for which the local norm is zero. C. Mueller and F. Weissler [MW] have determined the coefficient $c(2)$ on S^n for $n \geq 2$. By projecting S^n onto one coordinate axis they reduce the problem to a study of $L^2([-1, 1], w_\alpha(x)\,dx)$ where w_α is the induced density from μ. Instead of just allowing integer α they study densities w_α parametrized by $\alpha \in (0, \infty)$. They obtain thereby a continuum of logarithmic Sobolev inequalities in which the S^n cases are embedded. In the limit $\alpha \downarrow 0$ their inequalities reduce to the 2-point inequality (Example 2.6) and in the limit $\alpha \to \infty$ they obtain the Gaussian inequality (1.6) on R! Another derivation of best constant for S^2 is given in [J]. The circle S^1 has some special features which have been studied separately in [BnrJJ], [EmY], [Ro2] and [W3]. Finally, the very general method of Bakry and Emery [BE1, 2] works on S^n also, for $n \geq 2$. For an even stronger but related type of inequality see [Bnr4].

(v) *Logarithmic Sobolev inequalities and hypercontractivity over Clifford algebras*

There is a well known unitary map U between the space of symmetric tensors over $\mathbb{C}^n$ and $L^2(R^n, \nu)$ wherein ν is given by (1.5). The Dirichlet form operator $\nabla^*\nabla$ is diagonalized by U in the sense that $U\nabla^*\nabla U^{-1}$ multiplies a k-tensor by k. There is a quite precise analog for the space A of anti-symmetric tensors over $\mathbb{C}^n$. One replaces

$L^2(R^n, \nu)$ by the complexified Clifford algebra $\mathcal{C}$ over R^n. There is then a unitary map $U : \mathcal{A} \to L^2(\mathcal{C})$ which intertwines the natural actions of $O(n)$ on these two spaces and maps the Fermionic number operator to a "Clifford–Dirichlet" form operator N on $L^2(\mathcal{C})$. If one interprets $L^p(\mathcal{C})$ in the sense of I. E. Segal's noncommutative integration theory [S3] then it happens [Gr2] that $e^{-tN} : L^2(\mathcal{C}) \to L^4(\mathcal{C})$ is a contraction for exactly the same value of t, (4.7), as in the Gaussian case. Moreover a link between hypercontractivity and logarithmic Sobolev inequalities over $\mathcal{C}$ was developed in [Gr3] similar to that in Section 3 above. However two issues were not settled in [Gr3]. The best principal coefficient seemed likely to be one, but I could only derive a coefficient of log 3 (from interpolation of the previous $L^2 \to L^4$ exact time). J. M. Lindsay [Lin] extended the $L^2 \to L^4$ result by showing that $e^{-tN} : L^{2^k} \to L^{2^j}, 1 \le k < j$ is a contraction if t is given by (4.7). Lindsay and Meyer [LinM] extended this further by allowing even integers instead of powers of two. I. Wilde [Wi] emphasized the "Itô expansion" aspects of Fermionic hypercontractivity. The second issue left unresolved in [Gr3] was the transition from a logarithmic Sobolev inequality to contractivity of e^{-tN} on non-Hermitian elements of $L^2(\mathcal{C})$. These two issues have now been completely resolved in a recent paper by E. A. Carlen and E. Lieb [CL]. Fermionic hypercontractivity holds for exactly the same times (4.7) as in the Gaussian case.

(vi) *Existence of ground states*

Suppose that e^{-tH} is a positivity preserving hypercontractive semigroup on $L^2(X, \mu)$ with $\mu(X) < \infty$. If $\lambda = \inf \text{spectrum}\, H$, is λ an eigenvalue of H? If so any corresponding eigenstate ψ is called a ground state of H. Positivity preserving operators such as $A := e^{-H}$ have a long history, going back to the Perron–Frobenius theorem for matrices with nonnegative entries. Of course, λ is an eigenvalue for H if and only if $e^{-\lambda} \equiv \|A\|$ is an eigenvalue for the Hermitian operator A. If X were a finite point set then the Perron–Frobenius theorem would ensure that $\|A\|$ is an eigenvalue belonging to a nonnegative eigenfunction. But if $L^2(X, \mu)$ is infinite dimensional one needs some additional condition on the positivity preserving Hermitian operator A to ensure that $\|A\|$ is an eigenvalue. For example compactness of A is more than enough, as is spectral isolation of $\|A\|$. But hypercontractivity by itself is also enough and plays therefore a kind of compactness role. The idea behind the proof of this existence theorem is very simple. If $\{M_1, \ldots, M_n\}$ is a partition of X into measurable sets of positive measure and E_n is the corresponding conditional expectation projection in $L^2(X, \mu)$ then $E_n A E_n$ is identifiable in an obvious way with an $n \times n$ symmetric matrix A_n with nonnegative entries. By the Perron–Frobenius theorem this operator has a normalized nonnegative eigenfunction ψ_n which is constant on each M_j. Any weak limit ψ of the ψ_n would be an eigenfunction for A provided that ψ is not zero. This is the key point. Since the ψ_n's do not oscillate (being nonnegative) and since $\mu(X) < \infty$ the only way that ψ_n can possibly go weakly to zero is for the measure $\psi_n^2(x)\, d\mu(x)$ to "concentrate" on a set of μ measure zero. But this forces $\|\psi_n\|_p \to \infty$ for any $p > 2$, which is impossible because $M \equiv \|A\|_{2 \to p} < \infty$ for some $p > 2$ so that $\|A_n\|\|\psi_n\|_p = \|E_n A \psi_n\|_p \le M \|\psi\|_2$. For details see [Gr2] which contains also a similar theorem for noncommutative gauge spaces. The preceding discussion hinges on the use of hypercontractivity. But I. Shigekawa has shown [Sh] that a logarithmic Sobolev inequality can also be used directly as a "compactness replacement". He proved the existence of invariant measures for certain diffusion processes with infinite dimensional state spaces! See also [Gr7, 8] for application to existence of ground states for Schrödinger operators over homotopy classes in loop groups.

(vii) *Connection with isoperimetric inequalities*

See the work of M. Ledoux [L1] and O. Rothaus [Ro5].

(viii) *Semiboundedness of Hamiltonians*

We have already noted in topic (i) the motivating role that the semiboundedness problem for quantum field Hamiltonians played in E. Nelson's first step toward hypercontractivity while the same motive led P. Federbush to an inequality capturing the spirit of (1.6). The exact form of (2.2) is in [Gr1, Theorem 7] and the Faris converse in in [F]. The inequality (2.2) with these same best constants was also derived directly from Nelson's best hypercontractivity bounds by Guerra, Rosen and Simon [GuRS2] without going through logarithmic Sobolev inequalities.

(ix) *Intrinsic hyper- and ultra-contractivity*

Suppose μ is a measure on R^n with a smooth strictly positive density and $V : R^n \to R$ is a potential. For example $\mu =$ Lebesgue measure is of principal interest. If $H_0 := \nabla^*\nabla$ is the Dirichlet form operator for μ and $H = H_0 + V$ then e^{-tH} may not have any kind of $L^p(\mu) \to L^q(\mu)$ boundedness properties. But suppose further that $-\infty < \lambda :=$ inf spectrum H is an eigenvalue for H and that H has a corresponding normalized eigenfunction ψ which is strictly greater than zero everywhere. This is often the case. Let $d\nu(x) = \psi(x)^2\, d\mu(x)$. Then ν is a probability measure on R^n and the map $U : L^2(R^n, \mu) \to L^2(R^n, \nu)$ given by $(Uf)(x) = f(x)/\psi(x)$ is clearly unitary. Moreover one can compute that $U(H-\lambda)U^{-1}$ is the Dirichlet form operator for ν. Thus if D is given by $(Df, g)_{L^2(\nu)} = \int_{R^n} \nabla f(x) \cdot \overline{\nabla g(x)}\, d\nu(x)$ with appropriate domain (see the lectures of M. Fukushima and M. Röckner in this volume) then we have $U \exp(-t(H-\lambda))U^{-1} = e^{-tD}$ which automatically is a contraction from $L^p(\nu)$ to $L^p(\nu)$ for all $t > 0$ and all p in $[1, \infty]$. The unitary map $U : L^2(\mu) \to L^2(\nu)$ is called the ***ground state representation*** for H. Both U and D are entirely determined by H. One says that the semigroup e^{-tH} is ***intrinsically hypercontractive*** if e^{-tD} is a hypercontractive semigroup. Similarly e^{-tH} is ***intrinsically ultracontractive*** if e^{-tD} is ultracontractive. (See Section 5.) Of course one can replace R^n by a Riemannian manifold in these concepts. For R^n with $\mu =$ Lebesgue measure the operator H is simply the Schrödinger operator $H = -\Delta + V$. For what potentials V is e^{-tH} intrinsically hyper or ultracontractive? This and related questions are addressed by R. Carmona [Ca], B. Simon [Si1], E. B. Davies and B. Simon [DS1, 2, 3], E. B. Davies [D1, 2, 3, 8], R. Banuelos [Bn], B. Davis [Da] and F. Cipriani [Ci1, 2].

(x) *Statistical mechanics*

Let M be a compact Riemannian manifold, let Δ denote the Laplacian on M and let μ be normalized Riemann–Lebesgue measure on M. Then Δ determines a logarithmic Sobolev inequality on M with local norm zero:

$$\int_M f(x)^2 \log |f(x)|\, d\mu(x) \le c \int_M |\nabla f(x)|^2\, d\mu(x) + \|f\|^2_{L^2(\mu)} \log \|f\|^2_{L^2(\mu)} \tag{6.1}$$

for some constant c. By the additivity theorem, Theorem 2.3, (6.1) holds with the same constant c if M is replaced by the product M^n with the product metric, $n = 1, 2, \ldots$. Let L be a countable set, typically $L = Z^d$ for lattice statistical mechanics. Then (6.1) clearly continues to hold for the infinite product M^L with the product measure μ^L. However, the statistical mechanical formalism of Dobrushin, Lanford and Ruelle leads naturally to other interesting probability measures (Gibbs states) on M^L. If one replaces μ^L by a Gibbs state measure does one still obtain a logarithmic Sobolev inequality on M^L? The usefulness of an affirmative answer and conditions under which the answer is affirmative have been described by R. Holley, D. Stroock and B. Zegarlinski [HS1, 2], [StZ1, 2, 3] and [Z1, 2, 3]. Although I have described these questions in the context of a Riemannian manifold M the previous references address also the case in which M is a finite set. See the lectures of D. Stroock in this volume for more details.

(xi) *Large deviations*

D. Stroock has used logarithmic Sobolev inequalities as an infinite dimensional replacement for compactness in this context. See [St2] and the joint work with J. Deuschel [DeS1]. See also the work by G. Grillo [Gri] and S. Jacquot [Ja].

(xii) *Ultracontractivity, heat kernel bounds and boundary behavior*

Suppose that Ω is a bounded open set in R^n. If Δ denotes the Dirichlet Laplacian for Ω and ψ is an eigenfunction for Δ then ψ approaches zero at the boundary of Ω in a manner that depends on the local shape of Ω. At a point x in $\partial\Omega$ at which $\partial\Omega$ is smooth $\psi(y)$ goes to zero like $|y - x|$. But if Ω has a corner at x then the asymptotic behavior of $\psi(y)$ as $y \to x$ is more complicated. E. B. Davies and B. Simon [DS2] showed that the boundary behavior of ψ is controlled by the boundary behavior of the ground state, ψ_0, for Δ. Their technique, in part, consists in studying the semigroup $e^{t\Delta}$ in its ground state representation and showing ultracontractivity using the general methods described in Section 5 above. The notions of ultracontractivity and intrinsic ultracontractivity were first introduced in [DS2]. In this paper they also studied related questions for Schrödinger operators, $-\Delta + V$, using similar techniques. See also item (ix) above — intrinsic ultracontractivity — for some further discussion and references. In another direction E. B. Davies supplemented these techniques to obtain very detailed information on upper and lower bounds for heat kernels of elliptic operators. In Section 5 we saw how one could get uniform bounds $|K_t(x, y)| \le c_t$ on a heat kernel by using the equivalence of ultracontractive bounds with certain logarithmic Sobolev inequalities. If φ is a strictly positive function then the kernel $\varphi(x)^{-1} K_t(x,y)\varphi(y)$ is again the (informally at this level of exposition) kernel of a semigroup. By choosing φ cleverly Davies showed that the resultant semigroup generator satisfies the kind of logarithmic Sobolev inequalities which lead to the ultracontractive bounds $\varphi(x)^{-1} K_t(x,y)\varphi(y) \le c_t$. This gives a pointwise bound $K_t(x,y) \le c_t\varphi(y)^{-1}\varphi(x)$, which by suitable specialization of φ (depending even on x and y) yields bounds similar to the prototype Gaussian case $(2\pi t)^{-n/2} e^{-|x-r|^2/2t}$ [D4]. For an exposition of the work of Davies and his collaborators on heat kernel bounds see his recent book [D8].

(xiii) *Heritability of logarithmic Sobolev inequalities by submanifolds and loop groups*

Suppose that P is a probability measure on a Riemannian manifold Ω such that

$$\int_\Omega f^2 \log|f|\,dP \le \int_\Omega |\operatorname{grad} f(x)|^2\,dP + \|f\|_2^2 \log\|f\|_2 \tag{6.2}$$

Suppose further that $v : \Omega \to R^n$ is a smooth function which is nondegenerate on $M := v^{-1}(0)$ and let μ be the conditioned measure on M; $\mu = P(\ \mid v = 0)$. One can ask whether (M,μ) "inherits" the logarithmic Sobolev inequality (6.2) or some variant of it. It was shown in [Gr7] that under suitable conditions there is a potential V on M and a constant $\alpha > 0$ such that

$$\int_M f^2 \log|f|\,d\mu \le \alpha \int_M |\nabla_0 f|^2\,d\mu + (Vf,f)_{L^2(\mu)} + \|f\|_2^2 \log\|f\|_2$$

for functions $f : M \to R$ wherein ∇_0 is the tangential gradient on M. The question, which is also meaningful if Ω is infinite dimensional, is motivated by an application to pinned Brownian motion on loop groups. See [Gr7, 8] and [Gel, 2].

References

[A] Adams, R. A., *General logarithmic Sobolev inequalities and Orlicz imbeddings*, J. Funct. Anal. **34** (1979), 292–303.

[AC] ——— and Clarke, F. H., *Gross's logarithmic Sobolev inequality: a simple proof*, Amer. J. Math. **101** (1979), 1265–1270.

[AK1] Antonjuk, A. and Kondratiev, J., *Log-Sobolev inequality for Dirichlet operators on Riemannian manifold and its applications*, BiBoS preprint (1991).

[AK2] ———, *Log-concave smooth measures on Hilbert space and some properties of corresponding Dirichlet operators*, BiBoS preprint (1991).

[AKR] Albeverio, S., Kondratiev, Yu. G., and Röckner, M., *Dirichlet operators and Gibbs measures*, preprint (1992).

[Av] Avrim, Joel, *Perturbation of Log. Sob. Generators by electric and magnetic potentials*, Berkeley Thesis, 1982.

[B1] Bakry, D., *Une remarque sur les diffusions hypercontractives*, preprint, Nov. (1987), 3pp.

[B2] ———, *Weak Sobolev inequalities*, preprint (1990), Toulouse, 19pp.

[B3] ———, *Inégalités de Soboloev faibles: un critère* Γ_2, preprint (1990), Toulouse, 27pp.

[B4] ———, *Lectures in St. Flour summer school 1992.* (to appear).

[BE1] ——— and Emery, M., *Hypercontractivité de semi-groupes de diffusion*, Comptes-Rendus **299** (1984), 775–778.

[BE2] ——— and Emery, M., *Diffusions hypercontractives*, Sem. de Probabilités XIX, Lecture Notes in Mathematics 1123 (Azema, J. and Yor, M., eds.), Springer–Verlag, New York, 1985, pp. 177–207.

[BM] ——— and Meyer, P. A., *Sur les inequalitíes de Sobolev logarithmic I & II*, preprints, (1980/81), Univ. de Strasboug.

[BMi] ——— and Michel, D., *Inégalités de Sobolev et minorations du semigroupe de la chaleur*, preprint, Oct. (1989), Univ. de Toulouse, 37pp.

[Bn] Banuelos, R., *Intrinsic ultracontractivity and eigenfunction estimates for Schrödinger operators*, J. Funct. Anal. **100** (1991), 181–206.

[Bnr1] Beckner, W., *Inequalities in Fourier Analysis on* R^n, Proc. Nat. Acad. Sci. U.S.A. **72** (1975), 638–641.

[Bnr2] ———, *Inequalities in Fourier Analysis*, Ann. of Math. **102** (1975), 159–182.

[Bnr3] ———, *A generalized Poincaré inequality for Gaussian measures*, Proc. Amer. Math. Soc. **105** (1989), 397–400.

[Bnr4] ———, *Sobolev inequalities, the Poisson semigroup and analysis on the sphere* S^n, Proc. Nat. Acad. Sci. **89** (1992), 4816–4819.

[BnrJJ] ———, Janson, S., and Jerison, D., *Convolution inequalities on the circle*, Conference on harmonic analysis in honor of Antoni Zygmund, vol. 1, Wadsworth, Belmont, 1983, pp. 32–43.

[Bi] Bialynicki-Birula, Iwo, *Entropic uncertainty relations in quantum mechanics*, Quantum Probability and Applications II, Lecture Notes in Mathematics 1136 (Accardi, L. and von Waldenfels, W., eds.), Springer–Verlag, 1985, p. 90.

[BiM1] ——— and Mycielski, Jerzy, *Uncertainty relations for information entropy in wave mechanics*, Comm. Math. Phys. **44** (1975), 129–132.

[BiM2] ——— and Mycielski, Jerzy, *Wave equations with logarithmic nonlinearities*, Bull. Acad. Polon. Sci. Cl. III **23** (1975), 461.

[BiM3] ——— and Mycielski, Jerzy, Ann. Phys. **100** (1976), 62.

[Bo] Bonami, A., *Étude des coefficients de Fourier des fonctions de $L^p(G)$*, [Contains "two point" inequalities, see pp. 376–380], Ann. Inst. Fourier **20:2** (1970), 335–402.

[Br1] Borell, Christer, *On polynomial chaos and integrability*, preprint, (1980), Univ. of Göteborg, Prob. & Math. Statistics.

[Br2] ———, *A Gaussian correlation inequality for certain bodies in R^n*, Math. Ann. **256** (1981), 569–573.

[Br3] ———, *Convexity in Gauss space*, Les aspects statistiques et les aspects physiques des processus gaussiens, No. 307, CNRS, Paris, 1981, pp. 27–37.

[Br4] ———, *Positivity improving operators and hypercontractivity*, Math. Z. **180** (1982), 225–234.

(This prolific author has many other papers tangentially related to hypercontractivity.)

[BrL] Brascamp, H. J. and Lieb, E. H., *Best constants in Youngs inequality, its converse, and its generalization to more than three functions*, Advances in Math. **20** (1976), 151–173.

[BuJ] Buchwalter, Henri and Chen Dian Jie, *Une liason entre une inégalité logarithmique de Weissler sur le cercle et une question d'extremum dans le problème des moments de Stieltjes*, preprint, U. E. R. de Mathématiques, Univ. Claude-Bernard-Lyon 1, 43, bd. du 11 Nov. 1918; 69622 Villeurbanne Cedex.

[C1] Carlen, Eric, *Superadditivity of Fisher's information and logarithmic Sobolev inequalities*, J. Funct. Anal. **101** (1991), 194–211.

[C2] ———, *Some integral identities and inequalities for entire functions and their application to the coherent state transform*, J. Funct. Anal. (to appear).

[CK] ——— and Kreé, P., *On sharp L^p estimates for iterated stochastic integrals and some related topics*, White Noise Analysis (Hida, T., Kuo, H. H., Potthoff, J., Streit, L., eds.), World Scientific, NJ, 1990, pp. 43–48.

[CL] ——— and Lieb, E. H., *Optimal hypercontractivity for Fermi fields and related non-commutative integration inequalities*, Comm. in Math. Phys. (to appear).

[CLo] ——— and Loss, Michael, *Extremals of functionals with competing symmetries*, J. Funct. Anal. **88** (1990), 437–456.

[CSo] ——— and Soffer, A., *Entropy production by block variable summation and central limit theorems*, Comm. Math. Phys. **140** (1991), 339–371.

[CS1] ——— and Stroock, Daniel, *An application of the Bakry–Emery criterion to infinite dimensional diffusions*, Strasbourg Sem. de Probabilité (Azima and Yor, eds.), vol. XX, 1986.

[CS2] ——— and Stroock, Daniel, *Ultracontractivity, Sobolev inequalities, and all that*, preprint (June 1985), M. I. T.

[Ca] Carmona, R., *Regularity properties of Schrödinger and Dirichlet semigroups*, J. Funct. Anal. **33** (1979), 259–296.

[Ci1] Cipriani, F., *Intrinsic ultracontractivity of Schrödinger operators with deep wells potentials*, preprint (April 1992), SISSA.

[Ci2] ———, *Intrinsic ultracontractivity of Dirichlet Laplacians in non-smooth domains*, preprint (April 1992), SISSA.

[CM] Cowling, M. and Meda, S., *Harmonic analysis and ultracontractivity*, preprint (Oct. 1991), Trento.

[D1] Davies, E. B., *Spectral properties of metastable Markov semigroups*, J. Funct. Anal. **52** (1983), 315–329.

[D2] ———, *Hypercontractive and related bounds for double well Schrödinger operators*, Quart. J. Math. Oxford Ser. (2) **34** (1983), 407–421.

[D3] ———, *Perturbations of ultracontractive semigroups*, Quart. J. Math. Oxford Ser. (2) **37** (1986), 167–176.

[D4] ———, *Explicit constants for Gaussian upper bounds on heat kernels*, Amer. J. Math. **109** (1987), 319–334.

[D5] ———, *The equivalence of certain heat kernel and Green function bounds*, J. Funct. Anal. **71** (1987), 88–103.

[D6] ———, *Heat kernel bounds for second order elliptic operators on Riemannian manifolds*, Amer. J. of Math. **109** (1987), 545–570.

[D7] ———, *Gaussian upper bounds for heat kernels of some second order operators on Riemannian manifolds*, J. Funct. Anal. **80** (1988), 16–32.

[D8] ———, *Heat kernels and spectral theory*, Cambridge Univ. Press, 1989, pp. 197.

[DGS] ———, Gross, L. and Simon, B., *Hypercontractivity: a bibliographic review*, Proc. Hoegh-Krohn Memorial Conference (Alberverio, S., ed.) (to appear).

[DM1] ——— and Mandouvalos, N., *Heat kernel bounds on manifolds with cusps*, J. Funct. Anal. **75** (1987), 311–322.

[DM2] ——— and Mandouvalos, N., *Heat kernel bounds on hyperbolic space and Kleinian groups*, Proc. London Math. Soc. **57** (1988), 182–208.

[DP] ——— and Pang, M. M. H., *Sharp heat kernel bounds for some Laplace operators*, Quart. J. Math. Oxford Ser. (2) **40** (1989), 281–290.

[DR] ——— and Rothaus, O. S., *Markov semigroups on C^*-bundles*, J. Funct. Anal. **85** (1989), 264–286.

[DS1] ——— and Simon, B., *Ultracontractive semigroups and some problems in analysis*, Aspects of Mathematics and its Applications, North-Holland, Amsterdam, 1984, pp. 18.

[DS2] ______ and Simon, B., *Ultracontractivity and the heat kernel for Schrödinger operators and Dirichlet Laplacians*, J. Funct. Anal. **59** (1984), 335–395.

[DS3] ______ and Simon, B., *L^1 properties of intrinsic Schrödinger semigroups*, J. Funct. Anal. **65** (1986), 126–146.

[Da] Davis, Burgess, *Intrinsic ultracontractivity and the Dirichlet Laplacian*, J. Funct. Anal. **100** (1991), 162–180.

[De] Deuschel, J. D., *Logarithmic Sobolev inequalities of symmetric diffusions*, Seminar on Stoch. Processes, San Diego 1989, Prog. in Probability **18**, Birkhauser, Boston, pp. 17–22.

[DeS1] ______ and Stroock, D. W., *Large deviations*, Pure Appl. Math., vol. 137, Academic Press, Boston, 1989.

[DeS2] ______ and Stroock, D. W., *Hypercontractivity and spectral gap of symmetric diffusions with application to the stochastic Ising model*, J. Funct. Anal. **92** (1990), 30–48.

[Ec] Eckmann, J. P., *Hypercontractivity for anharmonic oscillators*, J. Funct. Anal. **16** (1974), 388–406.

[EmY] Emery, M. and Yukich, J. E., *A simple proof of the logarithmic Sobolev inequality on the circle*, Seminaire de Probabilités XXI, Lecture Notes in Math. 1247, Springer–Verlag, Berlin, 1987, pp. 173–176.

[Ep] Epperson, J., *The hypercontractive approach to exactly bounding an operator with complex Gaussian kernel*, J. Funct. Anal. **87** (1989), 1–30.

[F] Faris, W., *Product spaces and Nelson's inequality*, Helv. Phys. Acta **48** (1975), 721–730.

[Fe] Federbush, P., *A partially alternate derivation of a result of Nelson*, J. Math. Phys. **10** (1969), 50–52.

[Fei] Feissner, G., *Hypercontractive semigroups and Sobolev's inequality*, T. A. M. S. **210** (1975), 51–62.

[Fu] Fukushima, M., *Dirichlet Forms and Markov Processes*, North Holland, New York, 1980.

[Ge1] Getzler, E., *Dirichlet forms on loop space*, Bull. Sc. Math. 2ᵉ ser. **113** (1989), 151–174.

[Ge2] ______, *An extension of Gross's log-Sobolev inequality for the loop space of a compact Lie group*, Probability Models in Mathematical Physics (Morrow, G., Yang, W-S., eds.), World Scientific, NJ, 1991, pp. 73–97.

[Gl] Glimm, J., *Boson fields with nonlinear self-interaction in two dimensions*, Comm. Math. Phys. **8** (1968), 12–25.

[Gri] Grillo, G., *Logarithmic Sobolev inequalities and Langevin algorithms in R^n*, preprint (January 1991), Trieste.

[Gr1] Gross, L., *Logarithmic Sobolev inequalities*, Amer. J. of Math. **97** (1975), 1061–1083.

[Gr2] ———, *Existence and uniqueness of physical ground states*, J. Funct. Anal. **10** (1972), 52–109.

[Gr3] ———, *Hypercontractivity and logarithmic Sobolev inequalities for the Clifford-Dirichlet form*, Duke Math. J. **42** (1975), 383–396.

[Gr4] ———, *Logarithmic Sobolev inequalities — a Survey*, Vector Space Measures and Applications I (Aron, R. and Dineen, S., eds.), Lecture Notes in Math. 644, 1977, pp. 196–203.

[Gr5] ———, *Logarithmic Sobolev inequalities on Lie groups*, Ill. J. Math. **36** (1992), 447–490.

[Gr6] ———, *Logarithmic Sobolev inequalities for the heat kernel on a Lie group*, White Noise Analysis (Hida, T., Kuo, H. H., Potthoff, J., Streit, L., eds.), World Scientific, NJ, 1990, pp. 108–130.

[Gr7] ———, *Logarithmic Sobolev inequalities on loop groups*, J. Funct. Anal. **102** (1991), 268–313.

[Gr8] ———, *Uniqueness of ground states for Schrödinger operators over loop groups*, Cornell preprint October 1991, J. Funct. Anal. (to appear).

[Gu] Guennoun, O., *Inégalités logarithmiques de Gross-Sobolev et hypercontractivité*, Thèse 3ème cycle, Université Lyon 1, 1980, pp. 90.

[GuRS1] Guerra, F., Rosen, L. and Simon, B., *Nelson's symmetry and the infinite volume behavior of the vacuum in $P(\varphi)_2$*, Comm. Math. Phys. **27** (1972), 10–22.

[GuRS2] ———, *The vacuum energy for $P(\varphi)_2$: infinite volume limit and coupling constant dependence*, Comm. Math. Phys. **29** (1973), 233–247.

[GuRS3] ———, *The $P(\varphi)_2$ Euclidean quantum field theory as classical statistical mechanics*, Ann. of Math. **101** (1975), 111–259.

[H] Herbst, Ira, *Minimal decay rate for hypercontractive measures*, private communication (November 1975).

[HS1] Holley, R. and Stroock, D., *Diffusions on an infinite dimensional torus*, J. Funct. Anal. **42** (1981), 29–63.

[HS2] ———, *Logarithmic Sobolev inequalities and stochastic Ising models*, J. of Statistical Physics **46** (1987), 1159–1194.

[HS3] ———, *Uniform and L^2 convergence in one dimensional stochastic Ising models*, Comm. Math. Phys. **123** (1989), 85–93.

[Hol] Hooton, J. G., *Dirichlet forms associated with hypercontractive semigroups*, Trans. A. M. S. **253** (1979), 237–256.

[Ho2] ———, *Dirichlet semigroups on bounded domains*, Rocky Mountian J. Math. **12** (1982), 283–297.

[Ho3] ———, *Compact Sobolev imbeddings on finite measure spaces*, J. Math. Anal. and Applications **83** (1981), 570–581.

[J] Janson, S., *On hypercontractivity for multipliers on orthogonal polynomials*, [Best constant for Laplacian on S^2], Ark. Mat. **21** (1983), 97–110.

[Ja] Jacquot, S., *Résultats d'ergodicité sur l'espace de Wiener et application au recuit simulé*, Doctoral Thesis (Jan 1992), Université d'Orleans.

[Ke] Kesten, H., *An absorption problem for several Brownian motions*, preprint (1991), Cornell.

[Kl] Klein, Abel, *Selfadjointness of the locally correct generators of Lorentz transformations of $P(\varphi)_2$*, Math. of Contemporary Physics (Streater, R. F., ed.), Academic Press, London, 1972, pp. 227–236.

[KL1] ——— and Landau, L., *Construction of a unique self-adjoint generator for a symmetric local semigroup*, J. Funct. Anal. **44** (1981), 121–137.

[KL2] ——— and Landau, L., *Singular perturbations of positivity preseving semigroups via path space techniques*, J. Funct. Anal. **20** (1975), 44–82.

[KL3] ———, Landau, L. and Shucker, D., *Decoupling inequalities for stationary Gaussian processes*, Ann. of Probability **10** (1982), 702–708.

[KR] ——— and Russo, Bernard, *Sharp inequalities for Weyl operators and Heisenberg groups*, Math. Ann. **235** (1978), 175–194.

[Ko1] Korzeniowski, Andrzej, *On hypercontractive estimates for the Laguerre semigroup*, preprint (Feb 1984), U. of Texas, Arlington.

[Ko2] ———, *On logarithmic Sobolev constant for diffusion semi-groups*, J. Funct. Anal. **71** (1987), 363–370.

[KS] ——— and Stroock, D. W., *An example in the theory of hypercontractive semigroups*, Proc. A. M. S. **94** (1985), 87–90.

[KrS] Krakowiak, W. and Szulga, J., *Hypercontraction principle and random multilinear forms*, Wroclaw Univ. preprint #34 (1985), Prob. Theory and Related Fields (to appear).

[KuS] Kusuoka, S. and Stroock, D., *Some boundedness properties of certain stationary diffusion semigroups*, J. Funct. Anal. **60** (1985), 243–264.

[KwS] Kwapien, S. and Szulga, J., *Hypercontraction methods for comparison of moments of random series in normed spaces*, preprint (Sept. 1, 1988), Case Western Reserve, 13pp.

[L1] Ledoux, M., *Isopérimétrie et inégalités de Sobolev logarithmiques Gaussiennes*, C. R. Acad. Sci. Paris. ser. 1 **306** (1988), 79–82.

[L2] ———, *On an integral criterion for hypercontractivity of diffusion semigroups and extremal functions*, J. Funct. Anal. **105** (1992), 444–465.

[Lia] Lianantonakis, M., *Ultracontractive heat kernel bounds for singular second order elliptic operators*, preprint (Aug. 1991), King's College, London, 31pp.

[Li1] Lieb. E. H., *Proof of an entropy conjecture of Wehrl*, Comm. Math. Phys. **62** (1978), 35–41.

[Li2] ———, *Gaussian kernels have only Gaussian maximizers*, Invent. Math. **102** (1990), 179–208.

[Lin] Lindsay, M., *Gaussian hypercontractivity revisited*, J. Funct. Anal. **92** (1990), 313–324.

[LinM] ——— and Meyer, P. A., *Fermionic hypercontractivity*, preprint (1991).

[M] Meyer, Paul, *Some analytical results on the Ornstein–Uhlenbeck semigroup in infinitely many dimensions*, preprint, IFIP Working Conference on Theory and Applications of Random Fields, Bangalore, Jan. 4–9, 1982, 10pp.

[MW] Mueller, C. E. and Weissler, F., *Hypercontractivity for the heat semigroup for ultraspherical polynomials and on the n-sphere*, J. Funct. Anal. **48** (1982), 252-283.

[MM] Müller, P. H. and Müllner, A., *A representation formula in a space of functionals of the Wiener process and some applications*, preprint (May 1987), Tech. Univ. Dresden.

[N1] Nelson, E., *A quartic interaction in two dimensions*, Mathematical Theory of Elementary Particles (Goodman, R. and Segal, I., eds.), M. I. T. Press, Cambridge, Mass., 1966, pp. 69–73.

[N2] ———, *Quantum fields and Markov fields*, AMS Summer Institute on Partial Differential Equations, Berkeley, 1971.

[N3] ———, *Probability theory and Euclidean field theory*, Constructive Quantum Field Theory, Lecture Notes in Physics (Velo, G. and Wightman, A., eds.), vol. 25, Springer, 1973.

[N4] ———, *The free Markov field*, J. Funct. Anal. **12** (1973), 211–227.

[Ne] Neveu,. J., *Sur l'espérance conditionnelle par rapport à un mouvement brownien*, [Contains a probabilistic proof of hypercontractivity], Ann. Inst. H. Poincaré **12** (1976), 105–109.

[P1] Pang, M. M. H., L^1 *properties of two classes of singular second order elliptic operators*, J. London Math. Soc. (2) **38** (1988), 525–543.

[P2] ———, *Heat kernels of graphs*, preprint (Aug. 1991), King's College, London, 24pp.

[Pe] Peetre, J., *A class of kernels related to the inequalities of Beckner and Nelson*, A tribute to Ake Pleijel, Uppsala, 1980, pp. 171–210.

[PT] Pelliccia, E. and Talenti, G., *A proof of a logarithmic Sobolev inequality*, preprint (June 1992), Firenze.

[Pi] Pinsky, Ross, *Hypercontractivity estimates for nonselfadjoint diffusion semigroups*, Proc. of A. M. S. **104** (1988), 532–536.

[R] Rosen, J., *Sobolev inequalities for weight spaces and supercontractivity*, Trans. A. M. S. **222** (1976), 367–376.

[Ro1] Rothaus, O. S., *Lower bounds for eigenvalues of regular Sturm–Liouville operators and the logarithmic Sobolev inequality*, Duke Math. Journal **45** (1978), 351–362.

[Ro2] ———, *Logarithmic Sobolev inequalities and the spectrum of Sturm–Liouville operators*, J. Funct. Anal. **39** (1980), 42-56.

[Ro3] ———, *Logarithmic Sobolev inequalities and the spectrum of Schrödinger operators*, J. Funct. Anal. **42** (1981), 110–120.

[Ro4] ———, *Diffusion on compact Riemannian manifolds and logarithmic Sobolev inequalities*, J. Funct. Anal. **42** (1981), 102–109.

[Ro5] ———, *Analytic inequalities, isoperimetric inequalities and logarithmic Sobolev inequalities*, J. Funct. Anal. **64** (1985), 296–313.

[Ro6] ———, *Hypercontractivity and the Bakry–Emery criterion for compact Lie groups*, J. Funct. Anal. **65** (1986), 358–367.

[S1] Segal. I. E., *Construction of nonlinear local quantum processes, I*, Ann. of Math. **92** (1970), 462–481.

[S2] ———, *Construction of nonlinear local quantum processes, II*, Inventiones Math. **14** (1971), 211–241.

[S3] ———, *A non-commutative extension of abstract integration*, Ann. of Math. **57** (1953), 401–457.

[Se] Seneor, R., *Logarithmic Sobolev inequalities from Sobolev inequalities*, private communication (November 1975).

[Sh] Shigekawa, I., *Existence of invariant measures of diffusions on an abstract Wiener space*, Osaka J. Math. **24** (1987), 37–59.

[Si1] Simon, B., *Schrödinger semigroups*, Bull. A. M. S. **7** (1982), 447-526.

[Si2] ———, *A remark on Nelson's best hypercontractive estimates*, Proc. Amer. Math. Soc. **55** (1976), 376–378.

[Si3] ———, *The $P(\varphi)_2$ Euclidean (Quantum) field theory*, Princeton Univ. Press, Princeton, N. J., 1974.

[SHK] ——— and Hoegh-Krohn, R., *Hypercontractive semi-groups and two dimensional self-coupled Bose fields*, J. Funct. Anal. **9** (1972), 121–180.

[Sn] Stein, E. M., *Interpolation of linear operators*, Trans. Amer. Math. Soc. **83** (1956), 482–492.

[St1] Stroock, D. W., *Some remarks about Beckner's inequalitiy*, Proc. of Symposia in Pure Math. **35** (1979), Part I, 149–157.

[St2] ——, *An Introduction to the Theory of Large Deviations*, Universitext, Springer–Verlag, 1984, pp. 196.

[StZ1] —— and Zegarlinski, B., *The logarithmic Sobolev inequality for continuous spins on a lattice*, J. Funct. Anal. **104** (1992), 299–326.

[StZ2] —— and Zegarlinski, B., *The equivalence of the logarithmic Sobolev inequality and the Dobrushin–Shlosman mixing condition*, Comm. Math. Phys. (to appear).

[StZ3] —— and Zegarlinski, B., *The logarithmic Sobolev inequality for discrete spin systems on a lattice*, Comm. Math. Phys. (to appear).

[Sz] Szulga, Jerzy, *On hypercontractivity of α-stable random variables*, $0 < \alpha < 2$, preprint (July 1987), U. of North Carolina Statistics Dept..

[V1] Varopoulos, N. Th., *Iteration de J. Moser, Perturbation de semigroupes sous-markoviens*, C. R. Acad. Sci. Paris Ser. 1 **300** (1985), 617–620.

[V2] ——, *Hardy-Littlewood theory for semigroups*, J. Funct. Anal. **63** (1985), 240–260.

[W1] Weissler, F., *Logarithmic Sobolev inequalities for the heat-diffusion semigroup*, Trans. A. M. S. **237** (1978), 255–269.

[W2] ——, *Two-point inequalities, the Hermite semigroup, and the Gauss–Weierstrass semigroup*, J. Funct. Anal. **32** (1979), 102–121.

[W3] ——, *Logarithmic Sobolev inequalities and hypercontractive estimates on the circle*, J. Funct. Anal. **37** (1980), 218–234.

[Wi] Wilde, I., *Hypercontractivity for Fermions*, J. Math. Phys. **14** (1973), 791–792.

[Z1] Zegarlinski, B., *On log-Sobolev inequalities for infinite lattice systems*, Lett. Math. Phys. **20** (1990), 173–182.

[Z2] ——, *Log-Sobolev inequalities for infinite one-dimensional lattice systems*, Comm. Math. Phys. **133** (1990), 147–162.

[Z3] ——, *Dobrushin uniqueness theorem and logarithmic Sobolev inequalities*, J. Funct. Anal. **105** (1992), 77–111.

[Zh] Zhou, Z., *The contractivity of the free Hamiltonian semigroup in the L_p space of entire functions*, J. Funct. Anal. **96** (1991), 407–425.

Potential Theory of Non-Divergence form Elliptic Equations

by
Carlos E. Kenig[1]
Department of Mathematics
University of Chicago
Chicago, IL 60637 USA

§I Introduction

In these notes we develop (giving complete proofs) those aspects of the theory of non-divergence form elliptic equations, with bounded measurable coefficients, which we believe are most important towards understanding the potential theoretic properties of the subject. Immediately after saying this we must become precise, in order to clarify an important proviso, pervasive throughout these notes.

Let $A(X) = (a_{ij}(X))$, $1 \leq i, j \leq n$ be a symmetric real matrix, with $\lambda|\xi|^2 \leq \langle A(X)\xi, \xi\rangle \leq \lambda^{-1}|\xi|^2$ for all $X, \xi \in \mathbb{R}^n$. We will also make the a priori assumption that $A \in C^\infty(\mathbb{R}^n)$, and we will study solutions to $Lu = 0$, where $L = \sum_{i,j=1}^{n} a_{ij}(X)\frac{\partial^2}{\partial X_i \partial X_j}$. Our aim is to study the potential theoretic bounds associated with the operator L depending only on λ and n, and independent of the smoothness of the coefficients. The hope is that, eventually, we will be able to 'make sense' of the expression '$Lu = 0$' for A merely bounded and measurable, and that all the estimates explained here will then hold in this case. (See (4.4) and (4.5) for an explicit discussion in this direction).

In section §2 we prove the maximum principle of Alexandrov-Bakelman-Pucci ([1], [2] and [41]) and the celebrated Harnack inequality of Krylov and Safonov ([35], [44]). These results have revitalized the subject, and have led to numerous applications to non-linear partial differential equations (See for example [11], [12], [15], [25], and [33]).

In section §3 we study properties of non-negative solutions to the adjoint equation $L^* = \sum_{i,j=1}^{n} \frac{\partial^2 a_{ij}(X)}{\partial X_i \partial X_j}$. We present work of P. Bauman [7] and Fabes and Stroock [28] on doubling and reverse Hölder inequalities. This can be interpreted as 'localized' versions of the dual formulation of the maximum principle in §2. Applications to interior L^p estimates ($p < 1$ in general!) for gradients and second derivatives of solutions to $Lu = 0$, due to L.C. Evans ([26]) and F.H. Lin (37) are also presented. It is known ([42], [50]) that no local L^p estimate $p \geq 1$ for second derivatives of solutions is valid. What the 'best' L^p estimates for gradients of solutions are, remains an open problem (see (2.4), (3.12), (6.9) and (6.10)).

In section §4 we study the boundary behavior of solutions to $Lu = 0$ in smooth (C^2) domains, using versions of the classical Hopf maximum principle and the results of section §2. We then introduce a natural notion of 'good solution' to '$Lu = 0$' when the coefficients are merely bounded and measurable ([18]). The usefulness of this notion is still unclear. It depends on a positive solution to conjecture (4.5). Next, using the Hopf type maximum principle, we present the boundary version, on C^2 domains, of the estimates of Evans [26] and Lin [37]. These are due to Lin [37].

---

[1]Supported in part by the NSF

Examples of Bauman [6] show that adjoint solutions need not be locally bounded. To overcome this, Bauman [7] introduced the important concept of 'normalized adjoint solution.' A normalized adjoint solution is basically simply the ratio between any adjoint solution and a fixed, non-negative adjoint solution. We next present proofs of basic results of Bauman [7] on the Dirichlet problem, maximum principle, Harnack inequality and interior Hölder continuity of normalized adjoint solutions. Section §4 concludes with applications, again due to Bauman [8], of the concept of 'normalized adjoint solution' to size estimates for Green's functions, and a 'Wiener test,' valid for 'good solutions.'

In section §5 we study the boundary behavior of solutions and normalized adjoint solutions on Lipschitz domains. We present results of Fabes, Garofalo, Marin Malave, and Salsa [27]. The main result is that ratios of solutions or normalized adjoint solutions which vanish on a piece of the boundary are Hölder continuous up to the closure. This fact can be regarded as a version of the Hopf maximum principle, valid in non-smooth domains. It has many applications, such as the doubling property of L-harmonic measure and normalized adjoint L-harmonic measure, Fatou theorems, extensions up to the boundary of non-smooth domains of the gradient and second derivative estimates of Evans [26] and Lin [37], etc.

In section §6 we present (without proof) recent square function estimates for solutions and normalized adjoint solutions, due to L. Escauriaza and C. Kenig [24]. We isolate an important step in the proof, a weighted Poincaré inequality for solutions (6.9) whose proof we present. This inequality (and possible improvements of it (6.12), (6.13)) seems to be connected to the uniqueness of 'good solutions,' conjecture (4.5).

The material in these notes stems from a graduate course that I taught at the University of Chicago, in the winter of 1990, and from the C.I.M.E. course that I taught in Varenna in 1992. I would like to thank the scientific organizers of the C.I.M.E. course on Dirichlet Forms, Professors J.F. Dell' Antonio and U. Mosco for their kind invitation to participate in this course, and also Professor P. Zucca of C.I.M.E. for the excellent organization of the course. I would also like to thank Professor E. B. Fabes for many illuminating conversations on the subject matter of these notes.

§2. Main interior estimates for solutions

As mentioned in the introduction, we will make the a priori assumption that our coefficients are of class C^∞, but the constants appearing in our estimates will only depend on ellipticity and dimension. We start out with the well-known maximum principle of Alexandrov-Bakelman-Pucci ([1], [2] and [4]). Let $B = B_1(0) = \{X \in \mathbb{R}^n : |X| < 1\}$.

THEOREM 2.1. *If* $u \in C^\infty(\overline{B}), u|_{\partial B} \equiv 0$, *then*

$$|u(X)| \le C_\lambda \left(\int_B |Lu|^n \right)^{\frac{1}{n}} .$$

PROOF. We first introduce some notation. For $X_0 \in \mathbb{R}^n, t > 0$, we let $\Omega(X_0, t) = \{Y \in \mathbb{R}^n : Y \cdot (\xi - X_0) + t > 0, \text{ for all } \xi \in \overline{B}\}$. Thus, $\Omega(X_0, t)$ is the set of gradients of linear functions, which are positive in $\overline{B}$, and take the value t at X_0. The following properties of $\Omega(X_0, t)$ are easily verified:

(i) $$\Omega(X_0, t) = t\,\Omega(X_0, 1)$$

(ii) $$X_0 \in \overline{B} \Rightarrow B_{\frac{1}{4}} \subset \Omega(X_0, 1) \qquad \text{(here } Br = Br(0) = \{X \in \mathbb{R}^n : |X| < r\}\text{).}$$

(iii) $\Omega(X_0, t)$ is open and bounded.

If $u \in C^\infty(\overline{B})$, $\Omega \subset \overline{B}$, we will denote by $\nabla u(\Omega)$, the set of gradients of u at points in Ω. Also, $\Gamma = \{v : v \geq u \text{ in } B$, and $-v$ is convex in $B\}$, and $\hat{u}(X) =$ the upper convex envelope of $u = \inf\{v(X) : v \in \Gamma\}$, so that $\hat{u} \in \Gamma$ and is minimal with that property. Let $\Omega_u = \{X \in B : Hu = \left(\frac{\partial^2 u}{\partial X_i \partial X_j}\right) \leq 0\}$, so that $\Omega_u = \{X \in B : \hat{u} = u\}$, and $\nabla u(\Omega_u) = \nabla \hat{u}(\Omega_u)$. The proof will follow from the following claims:

CLAIM 1:. *If* $u \in C^\infty(\overline{B}), u \leq 0$ *on* $B, u(X) > 0$, *then* $\Omega(X, u(X)) \subset \nabla u(\Omega_u)$.

CLAIM 2:. $|\nabla u(\Omega_u)| \leq \int_{\Omega_u} |det\, H(u)|$.

Let us take the claims for granted, and show how to finish the proof. Let $f = Lu$. By writing $f = f_1 - f_2$, where each $f_i \geq 0$, $f_i \in C^\infty(\overline{B}), |f_i| \leq 2|f|$, we see that we can assume that $f \leq 0$, and hence, by the maximum principle, $u \geq 0$ on $\overline{B}$. Let X_0 be a point on $\overline{B}$ where u takes its maximum value, and use Claim 1 at X_0. Hence,

$$u(X_0)\,\Omega(X_0, 1) = \Omega(X_0, u(X_0)) \subset \nabla u(\Omega_u),$$

by (i). Using (ii) and Claim 2, we see that

$$C_n\, u(X_0)^n \leq |\Omega(X_0, u(X_0))| \leq |\nabla u(\Omega_u)| \leq \int_{\Omega_u} |\det Hu|.$$

All that remains to show, then, is that, on Ω_u, $|\det Hu| \leq C_\lambda (Lu)^n$. But, if $A(X) = (a_{ij}(X)), Lu(X) = trA(X)Hu(X)$, and if $Hu(X) \leq 0$ and $C_n \leq C_{n-1} \leq \cdots \leq C_1 \leq 0$ are its eigenvalues, then, $trA(X)Hu(X) = \sum_{i=1}^{n} \alpha_i C_i$, where $\alpha_i \geq \lambda^{-1}$. In fact, $Hu(X) = 0C0^t$, where 0 is an orthogonal matrix, and $(C)_{ij} = \delta_{ij} C_i$, where $\delta_{ij} =$ *Kronecker delta*. Hence $trAHu = trAOCO^t = tr(O^t AO)C$, and the diagonal elements of $O^t AO$ are the α_i. But then,

$$|Lu| = |\sum_{i=1}^{n} \alpha_i\, C_i| = \sum \alpha_i |C_i| \geq \lambda^{-1} \sum |C_i| =$$
$$= n\lambda^{-1} \frac{\sum |C_i|}{n} \geq n\lambda^{-1} (\Pi |C_i|)^{\frac{1}{n}}.$$

Since $|det\, Hu| = \Pi |Ci|$, our inequality is verified.

To finish the proof we want to establish the claims.

PROOF OF CLAIM 1. If $Y \in \Omega(X, u(X))$, then $Y \cdot (\xi - X) + u(X) > 0 \,\forall \xi \in \overline{B}$. Let $\lambda_0 = \inf\{\lambda : \lambda + Y \cdot (\xi - X) \geq u(\xi) \text{ for } \xi \in \overline{B}\}$. By continuity, $\lambda_0 + Y \cdot (\xi - X) \geq u(\xi) \,\forall \xi \in \overline{B}$, and there exists at least one $\overline{\xi} \in \overline{B}$ for which equality holds. If $\overline{\xi} \in B$, we have $\nabla u(\overline{\xi}) = Y$, and so $\overline{\xi} \in \Omega_u$, since the function $\lambda_0 + Y \cdot (\xi - X) - u(\xi)$ achieves a minimum at $\overline{\xi} \in B$. We will now show that $\overline{\xi}$ must belong to B. Since $\lambda_0 + Y \cdot (\xi - X) \geq u(\xi)$ for all $\xi \in \overline{B}$, we take $\xi = X$ to see that $u(X) \leq \lambda_0$, and since $Y \cdot (\xi - X) + u(X) > 0, \lambda_0 + Y \cdot (\xi - X) > 0$ for $\xi \in \overline{B}$. Since $u|_{\partial B} \leq 0$, $u(\overline{\xi}) = \lambda_0 + Y \cdot (\overline{\xi} - X), \overline{\xi} \in B$, and the claim is established.

PROOF OF CLAIM 2. To each point in $\nabla u(\Omega_u)$ there is at least one point in Ω_u mapped to it by the mapping $X \mapsto \nabla u(X)$, and $H(u)$ is the Jacobian of this mapping. The change of variables formula establishes the claim.

COROLLARY 2.2. *Suppose that* $u \in C^\infty(\overline{B}r), u|_{\partial Br} \equiv 0$, *then* $|u(X)| \leq C_{\lambda,n}\, r \left(\int_{Br} |Lu|^n\right)^{1/n}$.

The corollary follows from the theorem by a dilation, and the observation that our class of equations is dilation invariant.

COROLLARY 2.3. *Let* $G(X,Y)$ *be the Green's function for* L *in* B, *i.e.* $L_X G(-,Y) = -\delta_Y, G(-,Y)|_B \equiv 0, L_Y^* G(X,-) = -\delta_X$, $G(X,-)|_{\partial B} \equiv 0$. *(Here* $L^* = \sum_{i,j=1}^{n} \frac{\partial^2}{\partial Y_i\, \partial Y_j} (a_{ij}(Y) \cdot -)$*). Then,*

$$\sup_{X \in B} \int_B G(X,Y)^{n/n-1}\, dY \leq C_{\lambda,n}$$

PROOF. Since the solution to $Lu = f$, $u|_{\partial B} = 0$ is given by

$$u(X) = -\int_B G(X,Y)\, f(Y) dY$$

, the corollary follows from Theorem 2.1 by duality.

In order to establish local L^∞ estimates and the Harnack inequality for solutions we will use a version of the Cacciopoli estimate, valid in the non-divergence from case.

LEMMA 2.4 ([26], [28]). *Suppose that* $Lu = f$ *on* B, $f \in C^\infty(B)$, $\Psi \in C_0^\infty(B)$, $v \geq 0$, *and* $L^* v \leq 0$ *on a neighborhood of the support of* Ψ. *Then,*

$$\int A \nabla u \nabla u\, v \Psi^2 \leq C_{\lambda,n} \int v u^2 \{|L\Psi^2| + |\nabla \Psi|^2\} + C_{\lambda,n} \int v|u||f|\Psi^2$$

PROOF. We use the readily checked identity

$$2\, A\nabla u \nabla u \Psi^2 = L(u^2\Psi^2) - 2u\, f\Psi^2 - u^2 L\Psi^2 - 8\, A\nabla u\, \nabla\Psi\, \Psi u$$

multiply it by v, integrate and use the Cauchy-Schwarz inequality.

We are now ready for the L^∞ estimate

LEMMA 2.5 ([45], [28]). *Assume that $Lu = 0$ in B. Then,*

$$\|u\|_{L^\infty(B_{1/2})} \le C_{\lambda,n} \|u\|_{L^{2n}(B)}$$

PROOF. Choose $\varphi \in C_0^\infty(B)$, $\varphi \ge 0$, $\varphi \equiv 1$ on $B_{1/2}$. For

$$X \in B_{1/2}, u(X) = u(X)\varphi(X) = -\int_B G(X,Y)\, L(u\varphi)(Y)dY =$$

$$= -\int_B G(X,Y)\, u(Y) L\varphi(Y)\, dY +$$

$$+2\int_B G(X,Y)\, A(Y)\nabla u(Y)\nabla\varphi(Y)dY =$$

$$= \mathrm{I} + \mathrm{II}$$

$|\mathrm{I}| \le C_{\lambda,n}\|u\|_{L^n(B)}$ by (2.3). To bound II, choose $\Psi \in C_0^\infty(B\backslash B_{1/2}), \Psi \equiv 1$ on $\operatorname{supp} \nabla\varphi$, so that

$$|\mathrm{II}| = 2\left|\int_B G(X,Y)\, A(Y)\nabla u(Y)\nabla\varphi(Y)\Psi^2(Y)\, dY\right| \le$$

$$\le 2\left(\int_B G(X,Y)\, A(Y)\, \nabla u(Y)\nabla u(Y)\Psi^2(Y)dY\right)^{1/2} \cdot$$

$$\cdot\left(\int_B G(X,Y)\, A(Y)\nabla\varphi(Y)\nabla\varphi(Y)\Psi^2(Y)\, dY\right)^{1/2} \quad \text{by}$$

Cauchy-Schwarz. The second factor is bounded by $C_{\lambda,n}$ by (2.3). For the first factor, we apply (2.4) with $v = G(X,Y)$ which shows that it is less than or equal to

$$C_{\lambda,n}\left(\int_B G(X,Y)\, u^2(Y)dY\right)^{1/2}$$

Another application of (2.3) finishes the proof.

REMARK 2.6. *An examination of the above proof shows that there exists $N = N(\lambda, n)$ so that, for $1/2 \le \rho < r < 1$, we have*

$$\|u\|_{L^\infty(B_\rho)} \le \frac{C_{\lambda,n}}{(r-\rho)^N} \|u\|_{L^{2n}(B_r)}$$

COROLLARY 2.7. *Let u be as in (2.5). Then, for any $p > 0$, we have*

$$\|u\|_{L^\infty(B_{1/2})} \leq C_{p,\lambda,n} \|u\|_{L^p(B)}$$

PROOF. The corollary is a direct consequence of (2.6) as we will now show. We can assume, without loss of generality that $p < 2n$ and that $\|u\|_{L^p(B)} = 1$. Using (2.6) we see that

$$\|u\|_{L^\infty(B_\rho)} \leq \frac{C_{\lambda,n}}{(r-\rho)^N} \|u\|_{L^\infty(B_r)}^{1-p/2n} \|u\|_{L^p(B_r)}^{p/2n}$$

Let $\mathrm{I}(\rho) = \|u\|_{L^\infty(B_\rho)}$, so that, $\mathrm{I}(\rho) \leq \frac{C_{\lambda,n}}{(r-\rho)^N} \cdot \mathrm{I}(r)^\theta$, where $\theta = 1 - p/2n < 1$, $\theta > 0$. Hence,

$$\log \mathrm{I}(\rho) \leq \log C_{\lambda,n} + N \log \frac{1}{r-\rho} + \theta \log \mathrm{I}(r).$$

Choose now $\rho = r^a$, $a > 1$, a to be determined, and integrate against $\frac{dr}{r}$ for $\frac{2}{3} < r < 1$, to obtain,

$$\int_{2/3}^{1} \log \mathrm{I}(r^a) \frac{dr}{r} \leq Ca + \theta \int_{2/3}^{1} \log \mathrm{I}(r) \frac{dr}{r}$$

Changing variables in the first integral, we see that

$$\frac{1}{a} \int_{(\frac{2}{3}^a)}^{1} \log \mathrm{I}(r) \frac{dr}{r} \leq Ca + \theta \int_{\frac{2}{3}}^{1} \log \mathrm{I}(r) \frac{dr}{r}$$

Assume now that a is so close to 1 that $(\frac{2}{3})^a > \frac{1}{2}$. If $\mathrm{I}\,(1/2) \leq 1$, there is nothing to prove. If, on the other hand $\mathrm{I}\,(1/2) > 1$, $\mathrm{I}\,(r) > 1$ for $(\frac{2}{3})^a < r < (\frac{2}{3})$ and hence

$$\frac{1}{a} \int_{(\frac{2}{3})}^{1} \log \mathrm{I}(r) \frac{dr}{r} \leq Ca + \theta \int_{(\frac{2}{3})}^{1} \log \mathrm{I}(r) \frac{dr}{r}.$$

If we now choose a so close to 1 that $(\frac{1}{a} - \theta) > 0$, the Lemma follows.

COROLLARY 2.8. *Let $Lu = 0$ in B_{2r}. Then, for $p > 0$,*

$$\|u\|_{L^\infty(B_r)} \leq C_{p,\lambda,n} \left(\fint_{B_{2r}} |u|^p \right)^{1/p},$$

where $\fint_{B_{2r}} = \frac{1}{|B_{2r}|} \int_{B_{2r}}$.

PROOF. Use dilation, and the fact that our class of equations is dilation invariant.

In order to establish the Harnack inequality, we now establish lower bounds for non-negative solutions.

LEMMA 2.9. *Let $Lu = 0, u \geq 0$ in B_{2r}. Then, there exists ξ_0, $C > 0$, $0 < \xi_0 < 1$ (depending only on λ, n) such that if $|\{X \in B_r : u > z\}| \geq \xi_0 |B_r|$, then*

$$\inf_{B_{r/2}} u \geq Cz$$

PROOF. We can assume, without loss of generality, that $r = 1, z = 1$. For a set $\Gamma \subset B_1$, and G the Green's function of L in B_1, we set $W_\Gamma(X) = \int_\Gamma G(X,Y)dY$. We now choose $\Gamma = \{X \in B_1 : u \geq 1\}$. On ∂B_1, $u \geq 0$, $W_\Gamma \equiv 0$, on Γ $u \geq 1$, and, by (2.3) $W_\Gamma(X) \leq C|\Gamma|^{1/n} \leq C$. Thus, by the maximum principle $u(X) \geq CW_\Gamma(X)$ for $X \in B_1$. Moreover, (2.3) shows that $W_{B_1 \setminus \Gamma} \leq C(1-\xi_0)^{1/n}$, and $W_\Gamma(X) = W_{B_1} - W_{B_1\setminus\Gamma}$. The lemma will then follow if we can show that

$$\text{(2.10)} \qquad \inf_{B_{1/2}} W_{B_1}(X) \geq C_{\lambda,n}$$

(2.10) follows from a barrier argument. Let $\varphi(X) = [1 - |X|^2]$, so that $|L\varphi| \leq C_{\lambda,n}$. Now, let $h(X) = W_{B_1}(X) - C\varphi(X)$. We can then choose C so small that $Lh \leq 0$ on B_1. On ∂B_1, we have $h \geq 0$, and hence, by the maximum principle, $h \geq 0$ in B_1 and (2.10) follows.

LEMMA 2.11. *Let $Lu = 0$, $u \geq 0$ in B_{2r}. Assume that $\inf_{B_{\varepsilon r}} u(X) \geq 1$, for $0 < \varepsilon \leq 1$. Then, $u(X) \geq C\varepsilon^m$ on $B_{r/2}$, where C, m depend only on λ, n.*

PROOF. Without loss of generality $r = 1$. Let now $\varphi(X) = |X|^{-m}$, $X \in \overline{B}_1 \setminus B_\varepsilon$. A computation shows that $L[\varphi] \geq 0$ for m large. Thus, if $h(X) = u(X) - \varepsilon^m[\varphi(X) - 1]$, we have $Lh \leq 0$ on $\overline{B}_1 \setminus B_\varepsilon$. Also, on $\partial B_1 \cup \partial B_\varepsilon$ we have $h \geq 0$, and so $h \geq 0$ on $\overline{B}_1 \setminus B_\varepsilon$. Since $\varphi - 1 \geq 2^m - 1 \geq 1$ on $\overline{B}_{1/2}$, 2.11 follows.

In order to deduce Harnack's inequality from these results, we need to invoke a variant of the Calderón-Zygmund [46] decomposition.

LEMMA 2.12. *Let $R > 0$, $0 < \xi < 1$ be given, and $\Gamma \subset B_R$, $|\Gamma| > 0$ be given. Assume that $|\Gamma| < \xi|B_R|$. Then, we can find a countable collection $B_j = B_{r_j}(X_j)$ of balls, with center X_j, and radius r_j such that each B_j is contained in B_R, and*

(i) $$|\Gamma \cap B_j| \geq \xi |B_j|$$

(ii) $$|\Gamma \setminus \cup B_j| = 0$$

(iii) $$|\cup B_j| \geq [1 + (1-\xi)/5^n]\,|\Gamma|$$

Let us postpone the proof of (2.12), and use it to establish the Harnack inequality.

LEMMA 2.13. *Let $Lu = 0, u \geq 0$ in B_{2r}. Then, there exists $p_0 = p_0(\lambda, n) > 0$ and $C_{\lambda,n} > 0$ such that*

$$\left(\fint_{B_r} u^{p_0} \right)^{1/p_0} \leq C_{\lambda,n} \inf_{B_{r/2}} u$$

PROOF. Without loss of generality, $r = 1$, $\inf_{B_{1/2}} u = 1$. Let $z > 0$ be given, and let $\Gamma = B_1 \cap \{X : u(X) > z\}$. Pick ξ_0 as in (2.9). If $|\Gamma| \geq \xi_0 |B_1|$, then, by (2.9) $1 = \inf_{B_{1/2}} u \geq Cz \geq Cz|\Gamma|$. If not, we apply (2.12) to obtain a collection of balls $\{B_j\}$. Since $|B_j \cap \Gamma| \geq \xi_0 |B_j|$, (2.9) shows that $\inf_{B_j/2} u \geq Cz$, and hence, (2.11) shows that $\inf_{B_j} u > \gamma z$. Thus, $\cup B_j \subset \{u > \gamma z\} \cap B_1$, and so,

$$|\{u > \gamma z\} \cap B_1| \geq |\cup B_j| \geq \{1 + 1 - \xi_0/5^n\} |\Gamma| = \rho|\Gamma|,$$

where $\rho > 1$. Then,

$$\int_{B_1} u^{p_0} = p_0 \int_0^\infty z^{p_0 - 1} |\{u > z\} \cap B_1| dz \leq C + p_0 \int_1^\infty z^{p_0-1} |\{u > z\} \cap B_1| dz$$

$$\leq C + p_0 C \int_1^\infty z^{p_0 - 2} dz + \frac{p_0}{\rho} \int_1^\infty z^{p_0 - 1} |\{u > \gamma z\} \cap B_1| dz \leq$$

$$\leq C + \frac{1}{\rho \gamma^{p_0}} \int_{B_1} u^{p_0} \qquad (\text{ if } p_0 < 1).$$

Choose now p_0 so small that $\rho \gamma^{p_0} > 1$, and (2.13) follows.

COROLLARY 2.14 (KRYLOV AND SAFONOV [35], SAFONOV [44]). *Let* $Lu = 0$, $u \geq 0$ *in* B_{2r}. *Then*

$$\sup_{B_r} u \leq C_{\lambda,n} \inf_{B_r} u$$

PROOF. Combine 2.8 and 2.13.

We now turn to the proof of 2.12: We can assume $R = 1$. Let $A = \{\text{all balls } B : B \subset B_1 \text{ and } |\Gamma \cap B| \geq \xi |B|\}$. Let $\Gamma_0 = \bigcup_{B \in A} B$, so that $\Gamma_0 \subset B_1$. Note that by the Lebesgue differentiation theorem, $|\Gamma \backslash \Gamma_0| = 0$. Let $A^0 = \{B \in A : |B \cap \Gamma| = \xi |B|\}$. We claim that $\Gamma_0 = \bigcup_{B \in A^0} B$. In fact, suppose that $B = B(X, r) \in A$, and for $0 \leq \theta \leq 1$ let $B^\theta = B((1-\theta)X, r + \theta(1-r))$, so that $B^0 = B$, $B^1 = B_1$. Let $\varphi(\theta) = |B^\theta \cap \Gamma| / |B^\theta|$, so that $\varphi(\theta)$ is continuous, $\varphi(0) \geq \xi$, $\varphi(1) < \xi$. Hence, there exists θ_0, $0 \leq \theta_0 < 1$ such that $\varphi(\theta_0) = \xi$, i.e. $B^{\theta_0} \in A^0$. Also, $B^0 = B \subset B^{\theta_0}$, and the claim follows. By the Vitali covering lemma (see [46], 1.6), we can select a subsequence of disjoint balls $\{B_k\}$, each $B_k \in A^0$, and such that $\Gamma_0 \subset \bigcup_k (B_k)^*$, where B_k^* is the ball of the same center as B_k, expanded five times. Thus, $|\Gamma_0| \leq 5^n \sum_k |B_k|$, and since the B_k's are disjoint, $|\Gamma_0 \backslash \Gamma| \geq \sum_k |(\Gamma_0 \backslash \Gamma) \cap B_k| = (1 - \xi) \sum_k |B_k|$. Hence, $\frac{|\Gamma_0|}{|\Gamma|} = 1 + \frac{|\Gamma_0 \backslash \Gamma|}{|\Gamma|} \geq 1 + \frac{|\Gamma_0 \backslash \Gamma|}{|\Gamma_0|} \geq 1 + \frac{1-\xi}{5^n}$ as desired.

We now present some well known consequences of (2.14).

COROLLARY 2.15. *Let u verify $Lu = 0$ in B_{R_0}, and let $R \le R_0$. Then, there exists $\alpha = \alpha(\lambda, n) > 0$ and $C_{\lambda,n} > 0$, such that*

$$\operatorname*{osc}_{B_R} u \le C_{\lambda,n} (R/R_0)^{\alpha} \operatorname*{osc}_{B_R} u,$$

where $\operatorname*{osc}_{B_R} u = \sup_{B_R} u - \inf_{B_R} u$

PROOF. L1 = 0. Now use 2.14 and Moser's argument ([40]).

COROLLARY 2.16. *Let $u \in C^{\infty}(B_R)$, then, for $X_0, X_1 \in \overline{B}_{R/2}$, we have*

$$|u(X_0) - u(X_1)| \le C_{\lambda,n} \left\{ \left(\frac{|X_0 - X_1|}{R} \right)^{\alpha} \|u\|_{L^{\infty}(B_R)} + R \|Lu\|_{L^n(B_R)} \right\}.$$

PROOF. Write $u = u_1 + u_2$, where $Lu_1 = 0$ in B_R, $Lu_2 = Lu$ in $B_R, u_2|_{\partial B_R} \equiv 0$. The maximum principle implies that $\|u_1\|_{L^{\infty}(B_R)} \le \|u\|_{L^{\infty}(B_R)}$. We now use (2.15), and (2.2).

THEOREM 2.17. *Let $\Omega \subset \mathbb{R}^n$ be a bounded domain, $u \in C^{\infty}(\Omega)$. Let $d(X) =$ dist$(X, \partial\Omega)$; $X_0, X_1 \in \Omega$, $|X_0 - X_1| \le d(X_0)/2$. Then, if $2|X_0 - X_1| \le \rho \le d(X_0)$, we have, with $\sigma = \frac{\alpha}{1+\alpha}$, ($\alpha$ as in (2.15)), that*

$$|u(X_0) - u(X_1)| \le C \left(\frac{|X_0 - X_1|}{\rho} \right)^{\sigma} \{ \|u\|_{L^{\infty}(\Omega)} + \rho \|Lu\|_{L^n(\Omega)} \}$$

PROOF. Let $\tau = 2^{\sigma} \rho^{1-\sigma} |X_0 - X_1|^{\sigma}$, so that $2|X_0 - X_1| \le \tau \le \rho \le d(X_0)$ and hence $B(X_0, \tau) \subset \Omega$, $X_1 \in B(X_0, \tau/2)$. Thus, by (2.16),

$$\begin{aligned} |u(X_1) - u(X_0)| &\le C \{ \tau^{-\alpha} |X_0 - X_1|^{\alpha} \|u\|_{L^{\infty}(B(X_0,\tau))} + \\ &\quad + \tau \|Lu\|_{L^n(B(X_0,\tau))} \} \le \\ &\le C \{ |X_0 - X_1|^{\alpha} 2^{-\alpha\sigma} \rho^{-\alpha + \alpha\sigma} |X_0 - X_1|^{-\alpha\sigma} \|u\|_{L^{\infty}(B(X_0,\tau))} + \\ &\quad + 2^{\sigma} \rho^{1-\sigma} |X_0 - X_1|^{\sigma} \|Lu\|_{L^n(B(X_0,\tau))} \}. \end{aligned}$$

But, $\alpha\sigma - \alpha = -\sigma$, and (2.17) follows.

REMARK 2.18. *Safonov [45] has shown that any $0 < \alpha \le 1$ can arise in (2.15)–(2.17).*

§3 Interior estimates for non-negative adjoint solutions, and some applications

In this section we will present work of P. Bauman [7], [8], and E. Fabes and D. Stroock [28] on non-negative adjoint solutions, and some applications, due to L.C. Evans [26] and F.H. Lin [37].

We start out with a doubling property of adjoint solutions.

Lemma 3.1 ([7]). *Let $L^*v \le 0$, $v \ge 0$ in B_{2r}. Then,*

$$\int_{B_r} v \le C_{\lambda,n} \int_{Br/2} v$$

Proof. For $\delta > 0$, set $\varphi_\delta(X) = [(1+\delta)^2 r^2 - |X|^2]^2$. For δ small, depending only on λ, n we have:

(a) $$L\varphi_\delta \ge Cr^2 \qquad (1+\delta)r \ge |X| \ge (1-\delta)r$$

(b) $$|L\varphi_\delta| \le Cr^2 \qquad |X| \le (1+\delta)r$$

Hence, $\int_{B_{(1+\delta)r}\setminus B_{(1-\delta)r}} v \le C \int_{B_{(1+\delta)r}\setminus B_{(1-\delta)r}} vL(\varphi_\delta/r^2)$. But $L^*v \le 0$, and φ_δ vanishes of 2nd order on $\partial B_{(1+\delta)r}$, and hence $\int_{B_{(1+\delta)r}} vL(\varphi_\delta/r^2) \le 0$. Thus,

$$\int_{B_{(1+\delta)r}\setminus B_{(1-\delta)r}} vL(\varphi_\delta/r^2) \le -\int_{B_{(1-\delta)r}} vL(\varphi_{\delta/r^2}) \le C \int_{B_{(1-\delta)r}} v,$$

and so

$$\int_{B_{(1+\delta)r}\setminus B_{(1-\delta)r}} v \le C \int_{B_{(1-\delta)r}} v.$$

A covering argument finishes the proof of the Lemma.

We now turn to a 'scale invariant' version of (2.3).

Theorem 3.2 ([28]). *Let $\Omega \subset \mathbb{R}^n$ be open. Suppose that $L^*v = 0$ in $\Omega, v \ge 0$ in $\Omega, B_{2r} \subset \Omega$. Then, there exists $C_{\lambda,n} > 0$ such that*

$$\left(⨍_{B_r} v^{n/n-1}\right)^{\frac{n-1}{n}} \le C_{\lambda,n} ⨍_{B_r} v$$

Proof. (3.2) is equivalent to

$$\left(\int_{B_r} v^{n/n-1}\right)^{\frac{n-1}{n}} \le \frac{C_{\lambda,n}}{r} \int_{B_r} v$$

Pick $f \in C_0^\infty(B_r)$, $f \ge 0$, $\int f^n = 1$ and let u solve $Lu = f$ in B_{2r}, $u|_{\partial B_{2r}} \equiv 0$. Also, let $\varphi \in C_0^\infty(B_{2r}), \varphi \equiv 1$ on B_r. Then,

$$\int vf = \int v\,Lu = \int v\,Lu\,\varphi = \int vL(u\varphi) - \int vu\,L\varphi - 2\int vA\nabla u\nabla\varphi$$
$$= -\int vu\,L\varphi - 2\int vA\nabla u\nabla\varphi = \mathrm{I} + \mathrm{II}.$$

Since $\|u\|_{L^\infty(B_{2r})} \leq Cr$ by (2.2), $|\mathrm{I}| \leq \frac{C}{r} \int\limits_{B_{2r}} v \leq \frac{C}{r} \int\limits_{B_r} v$ by (3.1).

For II, choose $\Psi \in C_0^\infty(B_{2r})$, $\Psi \equiv 1$ on $\operatorname{supp} \nabla \varphi$, and so $|\mathrm{II}| \leq 2|\int vA\nabla u \nabla \varphi \Psi| \leq \frac{\varepsilon}{r}\int vA\nabla u \nabla u \psi^2 + \frac{r}{\varepsilon}\int vA\nabla\varphi\nabla\varphi = \mathrm{II}_1 + \mathrm{II}_2$, by Cauchy-Schwarz. For II_1, we use (2.4) to get

$$
\begin{aligned}
|\mathrm{II}_1| &\leq C\frac{\varepsilon}{r}\int vu^2\{|L\psi^2| + |\nabla\psi|^2\} + \frac{C\varepsilon}{r}\int v|u|\,|f|\psi^2 \leq \\
&\leq \frac{C\varepsilon}{r}\int\limits_{B_{2r}} v + C\varepsilon\int vf \leq \frac{C\varepsilon}{r}\int\limits_{B_r} v + C\varepsilon\int vf,
\end{aligned}
$$

where we have used (2.2) and (3.1).

Clearly, $|\mathrm{II}_2| \leq \frac{C}{\varepsilon r}\int\limits_{B_{2r}} v \leq \frac{C}{\varepsilon r}\int\limits_{B_r} v$, by (3.1), and so, choosing ε so that $C\varepsilon < 1/2$, the Theorem follows. We now want to extend Theorem 3.2 to include non-negative adjoint supersolutions. To do so, we need to treat first the main case of the Green's function. We need a preliminary fact.

LEMMA 3.3. *Let $G_r(X,Y)$ be the Green's function for L in B_{3r}. Then,*

$$r^{2-n} \simeq \fint\limits_{B_r} G_r(X,Y)\,dY \qquad \text{for } X \in B_{2r}.$$

PROOF. A dilation shows that $r^{n-2}G_r(rX, rY) = G_3(X,Y)$, $|X| \leq 3$, $|Y| \leq 3$ is the Green's function of L_r in B_3, where the coefficient matrix for L_r is $A(rX)$. Thus, all that we need to show is that $\int\limits_{B_1} G_3(X,Y)dY \simeq 1, X \in B_2$. The estimate from above follows from (2.1), while the one from below follows in the same way as (2.10) but using as a barrier the function φ_δ used in (3.1), and Harnack's principle (2.14).

THEOREM 3.4. *Let Ω be a bounded domain, and $G(X,Y)$ the Green's function corresponding to L and Ω. Then, there exists $C_{\lambda,n} > 0$ such that, for any ball B_r with $B_{4r} \subset \Omega$ we have,*

$$\left(\fint\limits_{B_r} G(X,Y)^{n/n-1}dY\right)^{\frac{n-1}{n}} \leq C_{\lambda,n}\left(\fint\limits_{B_r} G(X,Y)dY\right), \forall X \in \Omega.$$

PROOF. If $X \notin B_{2r}$, the result follows from (3.2). Assume then that $X \in B_{2r}$, and let $G_r(X,Y)$ be as in (3.3). Note that the maximum principle shows that $G(X,Y) \geq G_r(X,Y)$, and that $L_Y^*(G(X,-) - G_r(X,-)) = 0$ in B_{2r}. Hence, using (3.2) we have:

$$
\begin{aligned}
&\fint\limits_{B_r} G(X,Y)^{n/n-1}\,dY \leq C\fint\limits_{B_r} [G(X,Y) - G_r(X,Y)]^{n/n-1}\,dY + \\
&+C\fint\limits_{B_r} G_r(X,Y)^{n/n-1}\,dY \leq C\left(\fint\limits_{B_r} [G(X,Y) - G_r(X,Y)]\,dY\right)^{\frac{n}{n-1}} + \\
&+C\fint\limits_{B_r} G_r(X,Y)^{\frac{n}{n-1}}dY
\end{aligned}
$$

But, (2.2) shows that $\left(\int\limits_{B_r} G_r(X,Y)^{n/n-1}dY\right)^{n-1/n} \leq Cr$, and this, combined with (3.3) gives that $\left(\int\limits_{B_r} G_r(X,Y)^{\frac{n}{n-1}}\right)^{\frac{n-1}{n}} \leq C\left(\int\limits_{B_r} G_r(X,Y)dY\right)$, and this establishes the theorem.

REMARK 3.5. *Once (3.4) and (3.2) are established, using the Riesz representation of non-negative L^* subsolutions as a non-negative L^* solution plus a Green's potential of a non-negative measure, (3.2) can be extended to apply to $v \geq 0$, $L^*v \leq 0$.*

In order to deduce some consequences of (3.4) that will be of interest to us, we need to recall some basic facts about the real variable theory of weights. Thus, let (X,d) be a compact metric space, endowed with a measure ν, so that $\nu(B(X_0,2r)) \leq C\nu(B(X_0,r))$ for $0<r<r_0$, where $B(X_0,s) = \{Z \in X : d(X_0,Z) < s\}$, and C is independent of X_0 and r. (Such a space is called 'a space of homogeneous type').

DEFINITION 3.6 ([20]). *Let μ be another non-negative measure on X. Then, $\mu \in A_\infty(d\nu)$ if, given $\varepsilon > 0$, there exists $\delta = \delta(\varepsilon) > 0$ such that if $E \subset B = B(X_0,s)$, $s < r_0$, then*

$$\frac{\nu(E)}{\nu(B)} < \delta \Rightarrow \frac{\mu(E)}{\mu(B)} < \varepsilon.$$

This is a scale invariant version of absolute continuity. Some of the main properties of this class of weights are summarized in the next result.

THEOREM 3.7 ([20]).

(i) *If $\mu \in A_\infty(d\nu)$, then μ is absolutely continuous with respect to ν.*

(ii) *A_∞ is an equivalence relationship.*

(iii) *$\mu \in A_\infty(d\nu) \iff$ there exist ε, δ, $0<\varepsilon<1$, $0<\delta<1$ such that $\frac{\nu(E)}{\nu(B)} < \delta \Rightarrow \frac{\mu(E)}{\mu(B)} < \varepsilon, \forall E \subset B = B(X_0,s), s < r_0$.*

(iv) *$\mu \in A_\infty(d\nu) \iff$ there exist $\eta > 0, \theta > 0, C > 0$ such that $\forall E \subset B = B(X_0,s), s<r_0$, we have*

$$\frac{\mu(E)}{\mu(B)} \leq C\left[\frac{\nu(E)}{\nu(B)}\right]^\theta$$

$$\frac{\nu(E)}{\nu(B)} \leq C\left[\frac{\mu(E)}{\mu(B)}\right]^\eta$$

(v) *$A_\infty(d\nu) = \bigcup\limits_{q>1} Bq(d\nu)$, where $\mu \in B_q(d\nu)$ if μ is absolutely continuous with respect to ν and $k = \frac{d\mu}{d\nu} \in L^q(d\nu)$ and verifies:*

$$\left(\int\limits_B k^q d\nu\right)^{1/q} \leq C\int\limits_B k\, d\nu$$

for all balls $B = B(X_0,s)$ $s<r_0$.

(vi) *If $\mu \in B_q(d\nu), q>1$ then there exists $\varepsilon > 0$ such that $\mu \in B_{q+\varepsilon}(d\nu)$.*

(vii) *$A_\infty(d\nu) = \bigcup\limits_{p>1} A_p(d\nu)$, where $\mu \in Ap(d\nu)$ if μ is absolutely continuous with respect to ν, and $k = \frac{d\mu}{d\nu}$ is such that $k^{\frac{-1}{p-1}} \in L^1(d\nu)$ and $\left(\int\limits_B k\, d\nu\right)\left(\int\limits_B k^{\frac{-1}{p-1}} d\nu\right)^{(p-1)} \leq C$, for all balls $B = B(X_0,s), s<r_0$.*

(viii) *If $\mu \in A_p(d\nu), p>1$, then there exists $\varepsilon > 0$ such that $\mu \in A_{p-\varepsilon}(d\nu)$.*

These conditions were introduced because of their usefulness in the study of maximal functions. Thus, we have

(ix) $\mu \in B_q(d\nu) \iff M_\mu(f)(X_0) = \sup_{\substack{X_0 \in B \\ \operatorname{diam}(B) < r_0}} \frac{1}{\mu(B)} \int_B |f| d\mu$ *verifies*

$$\|M_\mu(f)\|_{L^{q'}(d\nu)} \leq C\|f\|_{L^{q'}(d\nu)}, 1/q + 1/q' = 1.$$

Likewise, $\mu \in A_p(d\nu) \iff$

$$\|M_\nu(f)\|_{L^p(d\mu)} \leq C\|f\|_{L^p(d\mu)}.$$

COROLLARY 3.8. *Let Ω be a bounded domain, then*

(i) *There exists $C = C(\lambda, n, \operatorname{diam}\Omega)$ and $q_\lambda > \frac{n}{n-1}$ such that*

$$\sup_{x \in \Omega} \int_\Omega G(X,Y)^{q_\lambda} dY \leq C$$

(ii) *There exist τ, C, depending only on λ, n, such that, for all measurable subsets E of B_r, with $B_{2r} \subset \Omega$, we have*

$$\frac{\int_E G(X,Y)\,dY}{\int_{B_r} G(X,Y)\,dY} \geq C\left(\frac{|E|}{|B_r|}\right)^\tau \quad \forall X \in \Omega$$

(iii) *There exists p_λ, $0 < p_\lambda \leq 1$ such that, if $B_{2r} \subset \Omega$*

$$\left(\int_{B_r} G(X,Y)^{-p_\lambda}\right)^{1/p_\lambda} \leq C_\lambda \frac{|B_r|}{\int_{B_r} G(X,Y)dY}.$$

(iv) *If $Lu = 0$ in $\overline{B}_2$, $u \geq 0$, then, if $u \geq 1$ on F, $F \subset \overline{B}_1$, then $\inf_{B_{1/2}} u \geq C|F|^\tau$.*

PROOF. For (i), pick a large ball B, $\overline{\Omega} \subset B$, $\operatorname{dist}(\partial B, \overline{\Omega}) \geq 1$, and note that if $\tilde{G}$ is the Green's function for $2B$, $G \leq \tilde{G}$ by the maximum principle. Apply now (3.4) and (3.7) (vi) to $\tilde{G}$ in B. For (ii), (iii) apply (3.4) and (3.7) (iv), (vii) (with $p_\lambda = \frac{1}{p-1}$). For (iv), note that (3.7) (iv) shows that, if G is the Green's function for L in B_2, and $W_F(X) = \int_F G(X,Y)\,dY, W_F(X) \geq C|F|^\tau W_{B_1}(X)$. But, (2.10) shows that $\inf_{B_{1/2}} W_{B_1}(X) \geq C$, and the maximum principle shows that $u(X) \geq W_F(x)$ and hence (iv) follows. Note that (3.8) (iv) is a strengthening of (2.9), and can be used to give an alternate proof of (2.14) (see [28]).

REMARK 3.9. *It of course would be very interesting if one could show that one can take $p_\lambda = 1$ in (3.8) (iii). Unfortunately, this is false, as was shown by Cerutti [17].*

REMARK 3.10. *It would also be very interesting if one could take $q_\lambda = +\infty$ in (3.8) (i). In fact, this is false, even away from the pole, as was shown by Bauman in [6].*

Next, we will show how the results we have proved on adjoint solutions can be used to obtain (weak) interior L^p estimates of gradients and second derivatives of solutions. The weakness of the estimates resides in the fact that we must take p very near 0.

THEOREM 3.11 ([26]). *There exists $0 < p_0 = p_0(\lambda, n)$ such that if $u \in C^\infty(\Omega)$, $L u = f$ in Ω, and $B_{2r} \subset \Omega$, then*

$$\left(\fint_{B_r} |\nabla u|^{p_0}\right)^{1/p_0} \le C_{\lambda,n}\left\{\frac{1}{r}\left(\fint_{B_{2r}} |u|^{2n}\right)^{\frac{1}{2n}} + \left(\fint_{B_{2r}} |u|^n |f|^n\right)^{\frac{1}{2n}}\right\}.$$

PROOF. We will use (2.4), with $\Psi \equiv 1$ on B_r, $\operatorname{supp}\Psi \subset B_{\frac{3r}{2}}$, $v(y) = G_r(X,Y)$, where G_r is the Green's function for B_{2r} and X is fixed, with $|X| = \frac{3}{2}r + \frac{1}{4}r$. By (3.8) (iii),

$$\left(\fint_{B_r} G_r(X,Y)^{-p_\lambda} dY\right)^{1/p_\lambda} \le C_{\lambda,n}\frac{|B_r|}{\int_{B_r} G_r(X,Y)\,dY},$$

and, by (3.3), $\int_{B_r} G_r(X,Y)\,dY \simeq r^2$, and thus, $\left(\fint_{B_r} G_r(X,Y)^{-p_\lambda}\,dY\right)^{1/p_\lambda} \le C_{\lambda,n}$ r^{n-2}. Also, (3.4) shows that $\left(\fint_{B_{\frac{3}{2}r}} G_r(X,Y)^{\frac{n}{n-1}}\,dY\right)^{\frac{n-1}{n}} \le C_{\lambda,n} r^{2-n}$.

Using (2.4) we see that

$$\int_{B_r} G_r(X,Y) A\nabla u\,\nabla u \le \frac{C_{\lambda,n}}{r^2}\int_{B_{\frac{3}{2}r}} G_r(X,Y) u^2(Y)\,dY + $$
$$+ C_{\lambda,n}\int_{B_{\frac{3}{2}r}} G_r(X,Y)|u(Y)|\,|f(Y)|\,dY.$$

Here, using Hölder's inequality, we see that

$$\begin{aligned}\fint_{B_r} |\nabla u|^{p_\lambda} &\le \left(\fint_{B_r} |\nabla u|^{2p_\lambda} G_r(X,Y)^{p_\lambda}\right)^{\frac{1}{2}} \left(\fint_{B_r} G_r(X,Y)^{-p_\lambda}\right)^{1/2} \le \\ &\le \left(\fint_{B_r} |\nabla u|^{2p_\lambda/p_\lambda} G_r(X,Y)\right)^{p_\lambda/2} \left(\fint_{B_r} G_r(X,Y)^{-p_\lambda}\right)^{1/2} \le \\ &\le C\left(\fint_{B_r} |\nabla u|^2 G_r(X,Y)\right)^{p_\lambda/2} \cdot r^{(n-2)\,p_\lambda/2} \le \\ &\le C r^{(n-2)p_\lambda/2}\left\{\frac{1}{r^2} r^{2-n}\left(\fint_{B_{\frac{3}{2}r}} |u|^{2n}\right)^{\frac{1}{n}} + r^{2-n}\left(\fint_{B_{\frac{3}{2}r}} |u|^n |f|^n\right)^{1/n}\right\}^{p_\lambda/2},\end{aligned}$$

and so,

$$\left(\fint_{B_r} |\nabla u|^{p_\lambda}\right)^{1/p_\lambda} \leq C_{\lambda,n}\left\{\frac{1}{r}\left(\fint_{B_{\frac{3}{2}r}} u^{2n}\right)^{\frac{1}{2n}} + \left(\fint_{B_{\frac{3}{2}r}} |u|^n |f|^n\right)^{\frac{1}{2n}}\right\},$$

as desired.

REMARK 3.12. *Whether the value of p_0 in the above inequality can be precised further (and for example $p_0 \geq 1$) is an important open problem. Note however that p_λ cannot be taken to be 1 (3.9).*

We now turn to a second derivative estimate of the same type.

THEOREM 3.13 ([37]). *Let Ω be a bounded domain, $u \in C^\infty(\Omega)$, and assume that $B_{2r} \subset \Omega$. There exists $p_0 = p_0(\lambda, n) > 0$ so that*

$$\left(\fint_{B_r} |D^2 u|^{p_0}\right)^{1/p_0} \leq C_{\lambda,n}\left\{\frac{\|u\|_{L^\infty(B_{2r})}}{r^2} + \left(\fint_{B_{2r}} |Lu|^n\right)^{1/n}\right\},$$

where $D^2 u = \left(\frac{\partial^2 u}{\partial x_i \partial x_j}\right)$.

PROOF. Let us write $Lu = tr AD^2(u) = \sum_{i=1}^{k} \alpha_i \lambda_i - \sum_{j=1}^{k+1} \alpha_j \lambda_j$, where $\lambda_1, \lambda_2, \ldots, \lambda_k$, $-\lambda_{k+1}, \ldots, -\lambda_n$ are the eigenvalues of $D^2(u)$, $\lambda_l \geq 0$, $k = k(X)$ is measurable and the α_i are measurable functions verifying $\lambda^{-1} \leq \alpha_i \leq \lambda, 1 \leq i \leq n$ (see the proof of (2.1)). We will now introduce a new operator, whose coefficients will depend on u. Let $O(X)$ be an orthonormal matrix so that $OD^2u\,O^t = \operatorname{diag}\{\lambda_1, \ldots, \lambda_k, -\lambda_{k+1}, \ldots, -\lambda_n\}$, where $\operatorname{diag}\{c_1, c_2, \ldots, c_n\}$ is the matrix with entry c_i in the i'th diagonal position, and 0 elsewhere. We now define

$$A^*(X) = O^t(X)\operatorname{diag}\{\frac{\lambda^{-1}}{2}, \ldots, \frac{\lambda^{-1}}{2}, 2\lambda, \ldots, 2\lambda\}O(X),$$

so that A^* is symmetric, elliptic, measurable, with ellipticity. constant $\lambda/2$. Let L_{A^*} be the operator with matrix A^*. Then,

$$\begin{aligned}
L_{A^*}(u) &= tr(O(X)^t \operatorname{diag}\{\frac{\lambda^{-1}}{2}, \ldots, \frac{\lambda^{-1}}{2}, 2\lambda, \ldots, 2\lambda\}O(X)D^2u(X)) = \\
&= tr(\operatorname{diag}\{\frac{\lambda^{-1}}{2}, \ldots, \frac{\lambda^{-1}}{2}, 2\lambda, \ldots, 2\lambda\}\operatorname{diag}\{\lambda_1, \ldots, \lambda_k, -\lambda_{k+1}, \ldots, -\lambda_n\}) = \\
&= tr(\operatorname{diag}\{\alpha_1, \alpha_2, \ldots, \alpha_n\}\operatorname{diag}\{\lambda_1, \ldots, \lambda_k, -\lambda_{k+1}, \ldots, -\lambda_n\}) + \\
&\quad + tr(\operatorname{diag}\{\frac{\lambda^{-1}}{2} - \alpha_1, \ldots, \frac{\lambda^{-1}}{2} - \alpha_k, 2\lambda - \alpha_{k+1}, \ldots, 2\lambda - \alpha_n\} \cdot \\
&\quad \cdot \operatorname{diag}\{\lambda_1, \ldots, \lambda_k, -\lambda_{k+1}, \ldots, -\lambda_n\}) = \\
&= Lu - tr(\operatorname{diag}\{\alpha_1 - \frac{\lambda^{-1}}{2}, \ldots, \alpha_k - \frac{\lambda^{-1}}{2}, 2\lambda - \alpha_{k+1}, \ldots, 2\lambda - \alpha_n\} \cdot \\
&\quad \cdot \operatorname{diag}\{\lambda_1, \ldots, \lambda_n\}).
\end{aligned}$$

Now, $\frac{\lambda^{-1}}{2} \le \alpha_i - \frac{\lambda^{-1}}{2} \le \lambda$, and $\lambda \le 2\lambda - \alpha_j \le 2\lambda$, so that the second term above equals $\gamma(X)|D^2u(X)|$, where γ is a measurable function, such that $\frac{\lambda^{-1}}{2} \le \gamma \le 2\lambda$, and so, $L_{A^*}(u) = Lu - \gamma(X)|D^2u(X)|$. Let now A_ϵ^* be a mollification of A^*, so that $\frac{\lambda^{-1}}{2}|\xi|^2 \le \langle A_\epsilon^*(X)\xi, \xi\rangle \le 2\lambda|\xi|^2$, and so that

$$L_{A_\epsilon^*}(u) = Lu - \gamma(X)|D^2u(X)| + \epsilon_{ij}(X)\frac{\partial^2}{\partial X_i\,\partial X_j}u(X), \text{ where } \varepsilon_{ij} \to 0 \text{ in } L^n(\Omega).$$

Let now $\varphi \in C_0^\infty(B_{\frac{3}{2}r})$, $\varphi \equiv 1$ on B_r, and let G_ϵ be the Green's function for $L_{A_\epsilon}^*$ in B_{2r}, and fix $X_0 \in \partial B_{\frac{3}{2}r}$. Then,

$$\begin{aligned}
0 = u(X_0)\varphi^2(X_0) &= -\int G_\epsilon(X_0,Y)\,L_{A_\epsilon^*}(\varphi^2 u)(Y)\,dY = \\
= -\int G_\epsilon(X_0,Y)&[\varphi^2 L_{A_\epsilon^*}(u) + 2A_\epsilon^*\nabla\varphi^2\cdot\nabla u + uL_{A_\epsilon^*}\varphi^2](Y)\,dY = \\
= -\int G_\epsilon(X_0,Y)[\varphi^2 L(u) &- \varphi^2(Y)\gamma(Y)|D^2u(Y)| + \varphi^2(Y)\varepsilon_{ij}(Y)\frac{\partial^2}{\partial Y_i\,\partial Y_j}u(Y) + \\
&+2A_\epsilon^*\nabla\varphi^2\nabla u + u\,L_{A_\epsilon^*}\varphi^2]\,dY.
\end{aligned}$$

Hence,

$$\begin{aligned}
\int \varphi^2\gamma|D^2u|G_\epsilon \le \int G_\epsilon\varphi^2 L(u) &+ \int G_\epsilon\varphi^2\varepsilon_{ij}\frac{\partial^2 u}{\partial Y_i\,\partial Y_j} + \\
+2\int G_\epsilon A_\epsilon^*\nabla\varphi^2\nabla u &+ \int G_\epsilon uL_{A_\epsilon^*}\varphi^2 = \\
&= \mathrm{I} + \mathrm{II} + \mathrm{III} + \mathrm{IV}
\end{aligned}$$

$|\mathrm{I}| \le \int_{B_{2r}} G_\epsilon|Lu| \le C_{\lambda,n}\,r\left(\int_{B_{2r}}|Lu|^n\right)^{1/n}$ by (2.2);

$|\mathrm{II}| \le rC(u)\left(\int_{B_{2r}}|\varepsilon_{ij}|^n\right)^{1/n} = o(1)$, by (2.2) again.

$|\mathrm{III}| \le 4\left(\int G_\epsilon\varphi^2 A_\epsilon^*\nabla u\nabla u\right)^{1/2}\left(\int G_\epsilon A_\epsilon^*\nabla\varphi\nabla\varphi\right)^2 = 4\,\mathrm{III}_1^{1/2}\,\mathrm{III}_2^{1/2}$. To estimate III_1, we use (2.4), to see that

$$\begin{aligned}
\mathrm{III}_1 &\le C_{\lambda,n}\int G_\epsilon u^2\{|L_{A_\epsilon^*}(\varphi^2)| + |\nabla\varphi|^2\} + \\
&+ C_{\lambda,n}\int G_\epsilon|u||L_{A_\epsilon^*}(u)\varphi^2 \le \\
&\le \frac{C_{\lambda,n}}{r^2}\int_{B_{2r}} G_\epsilon u^2 + \\
&+ C_{\lambda,n}\int G_\epsilon|u|\{|Lu| + \gamma|D^2u| + |\varepsilon_{ij}||\frac{\partial^2 u}{\partial Y_i\,\partial Y_j}|\}\varphi^2 \le
\end{aligned}$$

$$\leq \frac{C_{\lambda,n}}{r^2} \int_{B_{2r}} + C_{\lambda,n} \int G_\varepsilon |u||Lu|\varphi^2 + o(1) +$$

$$+ C_{\lambda,n} \|u\|_{L^\infty(B_{2r})} \cdot \int G_\varepsilon \gamma |D^2 u| \varphi^2 \leq$$

$$\leq C_{\lambda,n} \|u\|^2_{L^\infty(B_{2r})} + C_{\lambda,n} r \left(\int_{B_{2r}} |u|^n |Lu|^n \right)^{1/n} + o(1) +$$

$$+ C_{\lambda,n} \|u\|_{L^\infty(B_{2r})} \cdot \int G_\varepsilon \gamma |D^2 u| \varphi^2$$

by (2.2) and (3.3). Also, (3.3) shows that $|\mathrm{III}_2| \leq C_{\lambda,n}$, and so

$$\mathrm{III} \leq C_{\lambda,n} r^{1/2} \left(\int_{B_{2r}} |u|^n |Lu|^n \right)^{1/2n} + o(1) + \frac{1}{\delta} \|u\|_{L^\infty(B_{2r})} +$$

$$+ \delta \int G_\varepsilon \gamma |D^2 u| \varphi^2 \},\ \text{where } \delta \text{ is to be chosen.}$$

To estimate IV, we use (2.2) once more, to get $|\mathrm{IV}| \leq C_{\lambda,n} \|u\|_{L^\infty(B_{2r})}$, and hence, choosing δ small, we have that

$$\int \varphi^2 \gamma |D^2 u| G_\varepsilon \leq C_{\lambda,n} \left\{ o(1) + \|u\|_{L^\infty(B_{2r})} + r \left(\int_{B_{2r}} |Lu|^n \right)^{\frac{1}{n}} \right\}.$$

Since $\left(\int_{B_r} G_\varepsilon(X_0, Y)^{-p_\lambda} dY \right)^{1/p_\lambda} \leq C_\lambda r^{n-2}$, by (3.8) (iii) and (3.3), Hölder's inequality finishes the proof.

REMARK 3.14. *When $n = 2$, we can take $p_0 = 2$ (see [49]). When $n > 2$, no $p_0 \geq 1$ can be used for all L's (see [42], [50], [43]). When we allow the estimate to depend also on the modulus of continuity of the a_{ij}'s, we can take $1 \leq p_0 < \infty$ ([49]).*

§4. Boundary estimates in smooth domains, 'good' solutions, normalized adjoint solutions and the size of the Green's function

Let Ω be a bounded domain in $\mathbb{R}^n$, of class C^2. Then, it is well known that, at each boundary point $Q \in \partial\Omega$, Ω verifies an interior and exterior uniform sphere condition. Also, if $d(X) = \operatorname{dist}(X, \partial\Omega)$, and $\Omega_\mu = \{X \in \overline{\Omega} : d(X) < \mu\}$, one can show that there exists $\mu > 0$ so that $d \in C^2(\Omega_\mu)$. Moreover, using principal coordinates near $\partial\Omega$ it is not hard to see that there exists numbers $\delta > 0$ and $C > 0$, which depend only on n, λ and L^∞ bounds for the principal curvatures of $\partial\Omega$ such that

$$\text{(4.1)} \qquad \begin{aligned} &\text{(i)}\ d(X) \pm c d^2(X) \geq d(X)/2 && X \in \Omega_\delta \\ &\text{(ii)}\ L(d(X) + c d^2(X)) \geq 1, && X \in \Omega_\delta \\ &\text{(iii)}\ L(d(X) - c d^2(X)) \leq -1, && X \in \Omega_\delta \end{aligned}$$

We will use (4.1) and barrier arguments to extend some of the results in §2 and §3 up to the boundary, for C^2 domains. In the next section we will go even further and do this for Lipschitz (and more general) domains.

We start out with

LEMMA 4.2. *Let Ω be a bounded C^2 domain, and let $u \in C^\infty(\Omega)$ solve $Lu = 0$ in $B(Q,r) \cap \Omega$, $Q \in \partial\Omega$, and vanish continuously on $\partial\Omega \cap B(Q,r)$. Then, there exists $C > 0$, $C = C(\lambda, n, \Omega)$ such that*

$$|u(X)| \le C\left(\frac{|X-Q|}{r}\right) M(u) \qquad \forall X \in B(Q, r/2) \cap \Omega$$

where $M(u) = \sup\{|u(X)| : X \in B(Q,r) \cap \Omega\}$.

PROOF. A dilation argument shows that we can assume $r = 1$. We can also choose $\tilde{\Omega}$, of class C^2, with $\tilde{\Omega} \subset B(Q,1)$, and such that, if $\tilde{d}(X) = \operatorname{dist}(X, \partial\tilde{\Omega})$,then $\tilde{d} > 0$ for $X \in \overline{\tilde{\Omega}} \backslash B(Q, 3/4)$, and such that, $\partial\tilde{\Omega} \cap \partial\Omega \supset B(Q,1/2) \cap \partial\Omega$. (Note that $\tilde{\Omega}$ must meet ${}^c\Omega$). Let us now choose C, δ corresponding to $\tilde{\Omega}$ as in (4.1), and choose another C^2 domain D with $D \subset \tilde{\Omega}_\delta \cap \Omega$, $\partial D \cap \partial\Omega \supset B(Q,1/2) \cap \partial\Omega$. Note that $\tilde{d}(X)|_{\partial D \cap \overline{\Omega} \backslash B(Q,3/4)} \ge a > 0$. We will now apply the maximum principle to $M(u) 2a\{\tilde{d}(X) - c\tilde{d}^2(X)\} - u(X)$ on D. On ∂D it is non-negative, while L of it is ≤ 0, and hence, it is ≥ 0 on D. Arguing in a similar manner for $-u$, we obtain the lemma.

THEOREM 4.3 [44]. *Let α be as in (2.16), Ω a bounded C^2 domain, $f \in \operatorname{Lip}(\partial\Omega)$, $u \in C^\infty(\Omega) \cap C(\overline{\Omega})$ verify $Lu = 0$ in Ω, $u|_{\partial\Omega} = f$, and $\beta = \min(\frac{\alpha}{2}, \frac{1}{4})$. Then, $\sup\limits_{X, X' \in \Omega} \frac{|u(X) - u(X')|}{|X - X'|^\beta} \le C\|f\|_{\operatorname{Lip}(\partial\Omega)}$, where C depends only on n, λ and bounds on the curvatures of $\partial\Omega$, and $\|f\|_{\operatorname{Lip}(\partial\Omega)} = \|f\|_{L^\infty(\partial\Omega)} + \sup\limits_{Q, Q' \in \partial\Omega} \frac{|f(Q) - f(Q')|}{|Q - Q'|}$.*

PROOF. Let $X_0, X_1 \in \Omega$, and assume that $|X_0 - X_1| \le 1$. Assume first that $2|X_0 - X_1| \le 2|X_0 - X_1|^{1-\theta} \le \max\{d(X_0), d(X_1)\}$, θ to be chosen. Assume also that (without loss of generality) $d(X_0) \le d(X_1)$. By (2.16), we have that

$$\begin{aligned} |u(X_0) - u(X_1)| &\le C\left(\frac{|X_0 - X_1|}{|X_0 - X_1|^{1-\theta}}\right)^\alpha \|u\|_{L^\infty(\Omega)} = \\ &= C|X_0 - X_1|^{\theta\alpha} \|f\|_{L^\infty(\partial\Omega)}. \end{aligned}$$

If this does not hold, then $2|X_0 - X_1|^{1-\theta} \ge d(X_1) \ge d(X_0)$. Choose now $Q_0, Q_1 \in \partial\Omega$ so that $|X_0 - Q_0| = d(X_0)$, $|X_1 - Q_1| = d(X_1)$, and hence,

$$\begin{aligned} |u(X_0) - u(X_1)| &\le |u(X_0) - u(Q_0)| + |u(Q_0) - u(Q_1)| + \\ &+ |u(Q_1) - u(X_1)| \le |u(X_0) - u(Q_0)| + |u(Q_1) - u(X_1)| + \\ &+ \|f\|_{\operatorname{Lip}(\partial\Omega)} |Q_1 - Q_0| \le |u(X_0) - u(Q_0)| + |u(Q_1) - u(X_1)| + \\ &+ \|f\|_{\operatorname{Lip}(\partial\Omega)} \{d(X_1) + d(X_0) + |X_1 - X_0|\} \le \\ &\le |u(X_0) - u(Q_0)| + |u(Q_1) - u(X_1)| + \\ &+ 5\|f\|_{\operatorname{Lip}(\partial\Omega)} |X_1 - X_0|^{1-\theta}. \end{aligned}$$

To bound the first two terms, let us take, for example, the second one. We write $u - f(Q_1) = u_1 + u_2$, where $Lu_i = 0$, $u_1|_{\partial\Omega} = [f - f(Q_1)]\,\chi_{B(Q_1 Ad(X_1))}$, $u_2|_{\partial\Omega} = [f - f(Q_1)]\chi_{{}^cB(Q_1, Ad(X_1))}$. Now, the maximum principle shows that $|u_1(X_1)| \leq \|f\|_{\mathrm{Lip}}(\partial\Omega)\, Ad(X_1)$. Also, (4.2) applied to u_2, and $X = X_1$, $Q = Q_1$, shows that

$$|u_2(X_1)| \leq C\left(\frac{d(X_1)}{Ad(X_1)}\right) \cdot \|f\|_{L^\infty(\partial\Omega)} = \frac{C}{A}\|f\|_{L^\infty(\partial\Omega)}.$$

If we now choose $A = d(X_1)^{1/2}$, we obtain

$$\begin{aligned}|u(X_1) - u(Q_1)| &\leq C\,\|f\|_{\mathrm{Lip}(\partial\Omega}\{d(X_1)^{1/2} + d(X_1)^{1/2}\} \leq \\ &\leq C\|f\|_{\mathrm{Lip}(\partial\Omega)}|X_1 - X_0|^{(1-\theta)/2}\end{aligned}$$

Similarly, $|u(X_0) - u(Q_0)| \leq C\|f\|_{\mathrm{Lip}(\partial\Omega)}|X_1 - X_0|^{(1-\theta)/2}$ We now choose $\theta = 1/2$, and the Lemma follows.

DEFINITION 4.4 [18]). *Let $A(X) = (a_{ij}(X)) \in L^\infty(\mathbb{R}^n)$, $A = A^t, \lambda^{-1}|\xi|^2 \leq \langle A(X)\xi, \xi\rangle \leq \lambda|\xi|^2$. Let Ω be a bounded C^2 domain, $f \in \mathrm{Lip}(\partial\Omega)$. We say that $u \in C^\beta(\overline{\Omega})$ (β as in (4.3)) is a 'good solution' to $Lu = \sum a_{ij}(X)\frac{\partial^2 u}{\partial X_i \partial X_j} = 0$ in $\Omega, u|_{\partial\Omega} = f$, if we can find a sequence of matrices $A^k \in C^\infty(\mathbb{R}^n)$, such that $a^k_{ij} \underset{k\to\infty}{\longrightarrow} a_{ij}$ (for each fixed i, j) almost everywhere, and uniformly on compact subsets of $\mathbb{R}^n \backslash \overline{E}$, where E is the set of points of discontinuity of a_{ij}, and such that, for each k, $\lambda^{-1}|\xi|^2 \leq \langle A^k(X)\xi, \xi\rangle \leq \lambda|\xi|^2$, and, such that the solutions u^k to $L^k u^k = 0$ in $\Omega, u^k|_{\partial\Omega} = f$ converge uniformly to u in $\overline{\Omega}$.*

Note that (4.3) and the Arzela-Ascoli theorem show that, given $f \in \mathrm{Lip}(\partial\Omega)$, there always exist at least one 'good solution' to $Lu = 0\, u|_{\partial\Omega} = f$.

REMARK 4.5 CONJECTURE: ([18]). *Given an $f \in \mathrm{Lip}(\partial\Omega)$ there exists a unique 'good solution' to $Lu = 0$ in Ω, $u|_{\partial\Omega} = f$.*

This conjecture holds when $n = 2$ or the matrix $A(X)$ has continuous entries ([49]), or if $A(X)$ is continuous except at a countable set of points having at most one accumulation point ([18]). See also [3], [19] and [23] for related results.

Next we show how the barrier argument used in (4.2) can also be used to extend (3.1) and (3.2) up to the boundary of Ω, thus giving global versions of (3.11) and (3.13). For normalization purposes, let us assume that the absolute value of the principal curvatures of $\partial\Omega$ are bounded by one.

LEMMA 4.6 ([37], SEE ALSO [5]). *Let $v \geq 0$, $L^*v = 0$ in $\{X \in \Omega : d(X) \leq 1\}$, Ω a bounded C^2 domain. Then,*

$$\left(\frac{1}{|B|}\int\limits_{B\cap\Omega} v^{n/n-1}\right)^{n-1/n} \leq C_\lambda \frac{1}{|B|}\int\limits_{B\cap\Omega} v$$

for all balls $B(X, r)$ with $r \leq 1/3, d(X) \leq r/3$.

PROOF. Using the argument in the proof of (3.2), (3.1) and (3.2), it is not hard to show that it suffices to prove that, if $Q_0 \in \partial\Omega$, $r \in (0, 1/3)$, $T_r(Q_0, r) \cap \Omega$, we have

$$\int\limits_{T_{2r}(Q_0)} v \leq C \int\limits_{T_r(Q_0)} v$$

It suffices (by a rescaling argument) to show this for $r = 1$. Consider now $\tilde{\Omega}_4$ a bounded C^2 domain, with $\tilde{\Omega}_4 \subset T_4(Q_0)$, $T_3(Q_0) \subset \tilde{\Omega}_4$, $T_3(Q_0) \cap \partial\Omega \subset \tilde{\Omega}_4 \cap \partial\Omega$, and with the principal curvatures of $\partial\tilde{\Omega}_4$ bounded in absolute value by $C(n)$. Let G be the Green's function for L in $\tilde{\Omega}_4$. We first claim that

$$v(Y) = \int_{\partial\tilde{\Omega}_4} \langle A(X)\nabla_X G(X,Y), N_X\rangle v(X)\, d\sigma(X), \tag{4.7}$$

where N_X is the outward unit normal to $\partial\tilde{\Omega}_4$ and $d\sigma$ is surface measure of $\partial\tilde{\Omega}_4$. In fact,

$$\begin{aligned} v(Y) = -\int_{\tilde{\Omega}_4} L_X G(X,Y) v(X) dX &= \int_{\tilde{\Omega}_4} L^* v(X) G(X,Y)\, dX - \\ &- \int_{\partial\tilde{\Omega}_4} \langle A(X)\nabla v(X), N_X\rangle G(X,Y) d\sigma(X) + \\ &+ \int_{\partial\tilde{\Omega}_4} \langle A(X)\nabla_X G(X,Y), N_X\rangle v(X)\, d\sigma(X), \end{aligned}$$

by Green's formula, and so (4.7) follows. Let now $u_i(X) = \int_{T_i(Q_0)} G(X,Y) dY$, $i = 1, 2$, so that $\int_{T_i(Q_0)} v(Y)\, dY = \int_{\partial\tilde{\Omega}_4} \langle A(X)\nabla_X u_i(X), N_X\rangle v(X) d\sigma(X)$. Observe that $Lu_i = -\chi_{T_i(Q_0)}$, and that $0 \le u_i \le C_{\lambda,n}$ in $\tilde{\Omega}_4$, $u_i|_{\partial\tilde{\Omega}_4} \equiv 0$. Also note that $\forall K \Subset \tilde{\Omega}_4$, $\inf_K u_2(X) \ge \inf_K u_1(X) \ge C_\lambda(K) > 0$, by comparison with interior balls and (2.10). Next, we will use these facts, and (4.1) applied to $\tilde{\Omega}_4$, to conclude, by the maximum principle that, for $X \in \tilde{\Omega}_4$, $d(X) \le \delta$,

$$\begin{aligned} u_1(X) &\ge C_\lambda[d(X) + Cd^2(X)] \\ u_2(X) &\le C_\lambda[d(X) - Cd^2(X)] \end{aligned}$$

Thus, for $X \in \partial\tilde{\Omega}_4$, we have

$$\langle A(X)\nabla_X u_2(X), N_X\rangle \le C_\lambda^2 \langle A(X)\nabla_X u_1(X), N_X\rangle,$$

and the lemma follows.

COROLLARY 4.8 ([37]). *Let Ω be a bounded C^2 domain, let $u \in C^\infty(\Omega)$, $Lu = f$, $u \in C(\overline{\Omega})$, $u|_{\partial\Omega} \equiv 0$. Then, there exists $p_0 = p_0(n, \lambda, \Omega)$ such that*

$$\left(\int_\Omega |\nabla u|^{p_0}\right)^{1/p_0} + \left(\int_\Omega |D^2 u|^{p_0}\right)^{1/p_0} \le C\|f\|_{L^n(\Omega)}.$$

The proof is similar to the one of (3.11), (3.13), using (4.7).

In order to continue with our study of the potential theoretic properties of our class of equations, we next study the notion of 'normalized adjoint solution' (n.a.s.), first introduced by Bauman in [7].

We will be working with bounded domains Ω, with $\overline{\Omega} \subset \frac{1}{4}B$, and we will fix $X_* \in \frac{3}{4}B \setminus \frac{1}{2}B$.

DEFINITION 4.9. *A normalized adjoint solution for L^* and Ω (n.a.s.) is any function $\tilde{v}$ of the form $\tilde{v}(Y) = v(Y)/G_B(X_*, Y)$, where $L^*v = 0$ in Ω and G_B is the Green's function for L in B.*

We start out by studying the Dirichlet problem for (n.a.s.).

THEOREM 4.10. *Let Ω be a bounded smooth domain. Then, given $\varphi \in C(\partial\Omega)$ there exists a unique (n.a.s.) $\tilde{v}$ such that $\tilde{v} \in C(\overline{\Omega})$ and $\tilde{v}|_{\partial\Omega} = \varphi$. Moreover,*

$$\tilde{v}(Y) = \int_{\partial\Omega} \varphi(Q) \frac{G_B(X_*, Q)}{G_B(X_*, Y)} \frac{\partial G}{\partial \nu_Q}(Q, Y) d\sigma(Q),$$

where G is the Green's function for L and Ω, and

$$\frac{\partial G}{\partial \nu_Q}(Q, Y) = \langle A(Q) \nabla_Q G(Q, Y), N_Q \rangle.$$

PROOF. Recall that, since the coefficients of L are smooth, and $\partial\Omega$ is smooth, given $\Psi \in C(\partial\Omega)$ there exists a unique v with $L^*v = 0$ in Ω, $v \in C(\overline{\Omega}), v|_{\partial\Omega} = \Psi$, and that $v(Y) = \int_{\partial\Omega} \Psi(Q) \frac{\partial G}{\partial \nu_Q}(Q, Y) d\sigma(Q) \quad \forall Y \in \Omega$ (see (4.7)). Thus, given $\varphi \in C(\partial\Omega)$, set $\Psi(Q) = \varphi(Q) \cdot G_B(X_*, Q)$, and set $\tilde{v}(Y) = v(Y)/G_B(X_*, Y)$. The theorem now follows.

REMARK 4.11. *Note that constants are (n.a.s.). Hence, it follows from (4.10) that $1 = \int_{\partial\Omega} \frac{G_B(X_*, Q)}{G_B(X_*, Y)} \cdot \frac{\partial G}{\partial \nu_Q}(Q, Y) d\sigma(Q)$ for each $Y \in \Omega$.*

COROLLARY 4.12. *If Ω is a bounded Lipschitz domain, and $\tilde{v}$ is a normalized adjoint solution, $\tilde{v} \in C(\overline{\Omega})$, then*

$$\sup\{\tilde{v}(Y) : Y \in \overline{\Omega}\} = \sup\{\tilde{v}(Q) : Q \in \partial\Omega\}$$
$$\inf\{\tilde{v}(Y) : Y \in \overline{\Omega}\} = \inf\{\tilde{v}(Q) : Q \in \partial\Omega\}$$

PROOF. It clearly suffices to prove the first statement. If $\partial\Omega$ is smooth, it follows from (4.10) and (4.11). If Ω is merely Lipschitz, we approximate from the inside by smooth domains.

Our next aim is to prove that non-negative (n.a.s.) verify a Harnack principle. As we saw in the proof of (4.6), the behavior of adjoint solutions is closely tied to the boundary behavior of solutions that vanish on a portion of the boundary. Thus, we first need a scale invariant version of Hopf's maximum principle.

LEMMA 4.13. *Suppose that $Lu = 0$, $u \geq 0$ in $B_{7/4}r \setminus B_{\frac{5}{4}}r$, and that u vanishes continuously on $\partial B_{7/4}r$. Then,*

$$\frac{\partial u}{\partial \nu_Q}(Q) \simeq \frac{u(Q + \frac{r}{4}N_Q)}{r},$$

for each $Q \in \partial B_{7/4}r$, with comparability constants which depend only on ellipticity and dimension.

Let us take (4.13) for granted for the moment, and use it to establish the Harnack principle for (n.a.s.).

THEOREM 4.14 ([7]). *Suppose that $\tilde{v}$ is a (n.a.s.) in $B_{2r}, \tilde{v} \geq 0$ in B_{2r}. Then, there exists $C = C(\lambda, n) > 0$ such that* $\sup\{\tilde{v}(Y) : Y \in B_r\} \leq C \inf\{\tilde{v}(Y) : Y \in B_r\}$.

PROOF. Let G_r be the Green's function for L and $B_{7/4r}$, and let $Y \in B_r, Q \in \partial B_{7/4r}$, $X_Q = Q + \frac{r}{4}\frac{(A-Q)}{|A-Q|}$, where A is the center of B_{2r}. By (4.13) we know that $\frac{G_r(X_Q,Y)}{r} \simeq \frac{\partial G_r}{\partial \nu_Q}(X,Y)$. Hence, (by 4.10)

$$\tilde{v}(Y) = \int_{\partial B_{7/4r}} \tilde{v}(Q) \frac{G_B(X_*,Q)}{G_B(X_*,Y)} \frac{\partial G_r}{\partial \nu_Q}(Q,y)\, d\sigma(Q) \simeq$$

$$\simeq \frac{1}{r} \int_{\partial B_{7/4r}} \tilde{v}(Q) \cdot \frac{G_B(X_*,Q)}{G_B(X_*,Y)} \cdot G_r(X_Q,Y)\, d\sigma(Q).$$

Fix now $Q_0 \in \partial B_{7/4r}$, so that $X_{Q_0} \in \partial B_{\frac{3}{2}r}$ is fixed, and $Y \in B_r$. Apply now (2.14) to $G_r(-,Y)$ to conclude that $G_r(X_Q,Y) \simeq G_r(X_{Q_0},Y)$ for $Q \in \partial B_{7/4r}$, with comparability constants which depend only on ellipticity and dimension. Thus, $\tilde{v}(Y) \simeq \frac{G_r(X_{Q_0},Y)}{G_B(X_*,Y)} \cdot \frac{1}{r} \int_{\partial B_{7/4}r} \tilde{v}(Q) \cdot G_B(X_*,Q)\, d\sigma(Q)$. But, (4.11) shows that $G_B(X_*,Y) = \int_{\partial B_{7/4r}} G_B(X_*,Q) \cdot \frac{\partial G_r}{\partial \nu_Q}(Q,Y)\, d\sigma(Q)$, for all $Y \in B_r$. Applying the same argument, we see that $G_B(X_*,Y) \simeq \frac{G_r(X_{Q_0},Y)}{r} \int_{\partial B_{7/4r}} G_B(X_*,Q)\, d\sigma(Q)$, and the theorem follows.

COROLLARY 4.15. *(n.a.s.) are Hölder continuous on compact subsets of Ω, with Hölder exponent that depends only on ellipticity and dimension.*

PROOF. Same as the proof of (2.15), using the fact that constants are (n.a.s.) and (4.14).

In the next section we will extend this result up to the boundary. We need to establish (4.13). A preliminary step is:

LEMMA 4.16. *Let u be as in (4.13). Fix $Q \in \partial B_{7/4r}$, $0 < s < \frac{3}{8}r$. Then, $u(X) \leq Cu(Q + \frac{r}{4}N_Q)$ for all $X \in B_{7/4r} \backslash B_{5/4r} \cap B(Q,s)$, and $C = C(\lambda,n) > 0$.*

Let us take (4.16) for granted momentarily, and use it to establish (4.13).

PROOF OF (4.13). Fix $Q \in \partial B_{7/4r}$, and assume (without loss of generality) that $u(Q + \frac{r}{4}N_Q) = 1$, and let $X_Q = Q + \frac{r}{4}N_Q$. Define $S = S(Q,r) = B(X_Q, \frac{r}{4}) \cap B(Q, r/8)$, and let $\gamma_1 = \partial S \cap \partial B(X_Q, r/4)$, $\gamma_2 = \partial S \cap \partial B(Q, r/8)$, so that dist $(\gamma_2, \partial B_{7/4r}) \geq Cr$. By (2.14), $u(X) \geq \eta u(X_Q) = \eta \quad \forall X \in \gamma_2$. For each $X \in \overline{S}$, let $R = R(X) = |X - X_Q|$, and let $h(X) = \eta[\frac{e^{-64n\lambda^2R^2/r^2} - e^{-4n\lambda^2}}{2e^{-n\lambda^2} - e^{-4n\lambda^2}}]$. It is not hard to check that $Lh > 0$ in S, and hence $L(h - u) > 0$ in S. Next, we claim that $h \leq u$ on ∂S : on $\gamma_1, R = r/4$, and so $h \equiv 0$; on $\gamma_2, u \geq \eta$, while $h \leq \eta$, since $\frac{R}{r} \geq \frac{1}{8}$. Thus, the maximum principle shows that $h \leq u$ in $\overline{S}$. Since

$$h(Q) = u(Q) = 0, \frac{\partial u}{\partial \nu_Q}(Q) \geq \frac{\partial h}{\partial \nu_Q}(Q) =$$

$$= \sum a_{ij}(Q) \frac{(X_Q - Q)j}{|Q - X_Q|} \cdot \eta \frac{e^{-64n\lambda^2R^2/r^2}}{2[e^{-n\lambda^2} - e^{-4n\lambda^2}]} 128 n \lambda^2 \frac{R}{r^2} \frac{(X_Q - Q)i}{R},$$

evaluated at $R = r/4$. This equals

$$512\eta \frac{e^{-4n\lambda^2}}{2[e^{-n\lambda^2} - e^{-4n\lambda^2}]} n \frac{\lambda^2}{r^2} \sum a_{ij}(Q)(X_Q - Q)j\,(X_Q - Q)i \geq \frac{c}{r} u(X_Q).$$

To estimate $\frac{\partial u}{\partial \nu_Q}$ from above, we use (4.2) and (4.16). The lemma follows.
All that remains is to establish (4.16).

PROOF OF (4.16). First note that we can assume, without loss of generality, that $u(X_Q) = 1$. Next, note that (4.2) implies that, if $P \in \partial B_{7/4r}$, and $B(P, 2s) \subset B(Q, (\frac{3}{8} + \frac{1}{16})r)$,

$$\text{(4.17)} \quad \begin{aligned} &\sup\{u(X) : X \in B(P, s/C_1) \cap B_{7/4r} \backslash B_{5/4r}\} \leq \\ &\frac{1}{2} \sup\{u(X) : X \in B(P, s) \cap B_{7/4r} \backslash B_{5/4r}\}, \end{aligned}$$

where C_1 depends only on λ, n. We next claim that there exists A, depending only on λ, n such that, if $h \in \mathbb{N}, u(Y) \leq A^h u(X)$ for all $X, Y \in B(Q, (\frac{3}{8} + \frac{1}{32})r)$ with $\operatorname{dist}(X, \partial B_{7/4r}) \geq r/2^h$, $\operatorname{dist}(Y, \partial B_{7/4r}) \geq r/2^h$. To establish the claim, note that there exists $M \in \mathbb{N}$, M depending only on dimension, and a chain of balls $B_0, B_1, \ldots, B_{j_0}$, with $j_0 \leq Mh$ such that

(i) $$X \in B_0, \qquad Y \in B_{j_0}$$

(ii) $$B_i \cap B_{i+1} \neq \emptyset$$

(iii) $$2B_i \subset B\left(Q, \left(\frac{3}{8} + \frac{1}{16}\right) r\right), \operatorname{diam}(B_i) \simeq \operatorname{dist}(B_i, \partial B_{7/4r})$$

(iv) $$\operatorname{diam} B_i \geq C^{-1} \min\{\operatorname{dist}(X, B_i), \operatorname{dist}(Y, B_i)\}.$$

The Harnack principle (2.14) now establishes the claim. Note that the claim implies that there exists $C_2 > 1$, depending only on C_1 such that if $Y \in B(Q, (\frac{3}{8} + \frac{1}{32})r)$ and $u(Y) \geq C_2^h$, then $\operatorname{dist}(Y, \partial B_{7/4r}) < C_1^{-h} r$. Choose now $M \geq 1$ such that $2^M \geq C_2$, let $N = M + 5, C = C_2^N$. Suppose that there exists $Y_0 \in B(Q, \frac{3}{8}r)$ such that $u(Y_0) > Cu(X_Q) = C = C_2^N$. Then, $\operatorname{dist}(Y_0, \partial B_{7/4r}) < C_1^{-N} r$. If $Q_0 \in \partial B_{7/4r}$ is the point nearest Y_0, then, $|Q - Q_0| \leq \frac{3}{8}r + C_1^{-N} r \leq (\frac{3}{8} + \frac{1}{32})r$. But, (4.17) implies that

$$\begin{aligned} \sup\{u(X) : X \in B(Q_0, C_1^{-N+M} r) \cap B_{7/4r} \backslash B_{5/4r}\} \geq 2^M \sup\{u(X) :\\ X \in B(Q_0, C_1^{-N} r) \cap B_{7/4r} \backslash B_{5/4r}\} \geq 2^M u(Y_0) \geq C_2^{N+1}. \end{aligned}$$

Hence, we can choose $Y_1 \in B(Q_0, C^{-5} r) \cap B_{7/4r} \backslash B_{5/4r}$ such that $u(Y_1) \geq C_2^{N+1}$. Therefore, $\operatorname{dist}(Y_1, \partial B_{7/4r}) \leq C_1^{-N-1} r$. We then obtain sequences of points $\{Y_k\}$, $\{Q_k\}$ such that $Y_k \in B(Q_{k-1}, C_1^{M-N-(k-1)} r) \cap B_{7/4r} \backslash B_{5/4r}$, and $B(Q_{k+1}, C_1^{M-N-(k-1)} r) \subset B(Q, (\frac{3}{8} + \frac{1}{32})r)$, and $\operatorname{dist}(Y_k, \partial B_{7/4r}) = |Y_k - Q_k| \leq C_1^{-N-k} r$, and $u(Y_k) > C_2^{N+k}$. Now, since $\operatorname{dist}(Y_k, \partial B_{7/4r}) \to 0, u(Y_k) \to \infty$, we reach a contradiction, which establishes (4.16).

We end this section with some applications (due to Bauman [8]) of (4.14) and (4.15) to estimating the size of the Green's function.

LEMMA 4.17. *Suppose that $B(X,4r) \subset B\backslash\{X_*\}$. Let G_r be the Green's function for L in $B(X,4r) = B_{4r}$. If $Y \in \partial B_r = \partial B(X,r)$, then*

$$G_r(X,Y)/G_B(X_*,Y) \simeq r^2 \Big/ \int_{B_r} G_B(X_*,Z)\,dZ \simeq r^2 \Big/ \int_{B(Y,r)} G_B(X_*,Z)\,dZ,$$

with constants of proportionality depending only on λ and n.

PROOF. A standard consequence of (3.4) and (3.7) (see [20]) is that $G_B(X_*,Z_0) \simeq \fint_{B(Y,r/2)} G_B(X_*,Z)\,dZ$ and $G_r(X,Z_0) \simeq \fint_{B(Y,r/2)} G_r(X,Z)\,dZ$ for all Z_0 in some subset of $B(Y,r/2)$ with Lebesgue measure greater than $\theta \cdot |B(Y,r/2)|$. Here θ and the proportionality constants depend only on the A_∞ constants, and hence only on λ, n. Combining this with the doubling property in (3.1), we obtain $\frac{G_r(X,Z_0)}{G_B(X_*,Z_0)} \simeq \frac{\int_{B_r} G_r(X,Z)\,dZ}{\int_{B_r} G_B(X_*,Z)\,dZ}$ for all such Z_0. (3.3) shows that $\int_{B_r} G_r(X,Z)\,dZ \simeq r^2$. Also, (4.14) shows that, for all $Z_0 \in B(Y,r/2)$, we have $\frac{G_r(X,Z_0)}{G_r(X_*,Z_0)} \simeq \frac{G_r(X,Y)}{G_r(X_*,Y)}$. Thus, using (3.1) once more, (4.17) follows.

THEOREM 4.18. *If $X, Y \in \frac{1}{4}B$, then,*

$$\frac{G_B(X,Y)}{G_B(X_*,Y)} \simeq \int_{|X-Y|}^{\operatorname{diam} B} \frac{s^2}{\int_{B(Y,s)} G_B(X_*,Z)\,dZ}\frac{ds}{s} \simeq$$

$$\simeq \int_{|X-Y|}^{\operatorname{diam} B} \frac{s^2}{\int_{B(X,s)} G_B(X_*,Z)\,dZ}\frac{ds}{s}.$$

PROOF. By (4.14) and (3.1), it is enough to prove this when $|X-Y| \le 4^{-5}\operatorname{dist}(X_*, \partial B \cup \partial\frac{1}{4}B) \simeq 4^{-5}\operatorname{diam}(B)$. Let $r = |X-Y|$, and choose N such that $4^{N+2}r \le \operatorname{dist}(X_*, \partial\frac{1}{4}B) \le 4^{N+3}r$. Let G_j be the Green's function for L in $B(X,4^j r)$ for $j = 1,\dots,N$. By (4.17) and (3.1), $G_j(X,Z)/G_B(X_*,Z) \simeq \frac{(4^j r)^2}{\int_{B(X,4^j r)} G_B(X_*,Y)\,dY}$ if $Z \in \partial B(X,4^{j-1}r)$. Let $\tilde v_j(Z) = \frac{[G_j(X,Z) - G_{j-1}(X,Z)]}{G_B(X_*,Z)}$, for all $Z \in B(X,4^{j-1}r)$. By (4.10) and (4.11), we have

$$\tilde v_j(Z) \simeq \frac{(4^j r)^2}{\int_{B(X,4^j r)} G_B(X_*,Y)\,dY} \quad \text{for } Z \in B(X,4^{j-1}r).$$

Similarly, $\tilde v(Z) = \frac{[G_B(X,Z) - G_N(X,Z)]}{G_B(X_*,Z)}$, satisfies

$$\tilde v(Z) \simeq \frac{(4^N r)^2}{\int_{B(X,4^N r)} G_B(X_*,Y)\,dY} \qquad \text{for all } Z \in B(X,4^N r).$$

Since $G_B(X,Y) = \tilde{v}(Y) + \tilde{v}_N(Y) + \tilde{v}_{N-1}(Y) + \cdots + \tilde{v}_2(Y) + \frac{G_1(X,Y)}{G_B(X_*,Y)} \simeq \sum_{j=1}^{N} \frac{(4^j r)^2}{\int_{B(X,4^j r)} G_B(X_*,Z)\,dZ}$ by the previous estimates and (4.17), (3.1) finishes the proof.

COROLLARY 4.19. *If* $X, Y \in \frac{1}{4}B, \frac{G_B(X,Y)}{G_B(X_*,Y)} \simeq \frac{G_B(Y,X)}{G_B(X_*,X)}$.

In [8] Bauman introduces an appropriate notion of capacity: if $K \Subset \frac{1}{4}B$, then

$$\operatorname{cap}_L K = \inf\{u(X_*) : u \text{ is an } L \text{ supersolution in } B, u \geq 0 \text{ in } B, u \geq 1 \text{ in} K\}.$$

If $\tilde{G}(X,Y) = G_B(X,Y)/G_B(X_*,Y)$, Bauman proves that $\operatorname{cap}_L K = \inf\{\nu(K) : \int_B \tilde{G}(X,Z)d\nu(Z)$ ≥ 1 quasi everywhere in K, $\nu \geq 0\} = \sup\{\nu(K) : \int_B \tilde{G}(X,Z)d\nu(Z) \leq 1$ quasi everywhere in K, $\nu \geq 0\}$. The connection between Green's function and capacity is given by the fact that, if $B(X,2r) \subset \frac{1}{4}B, Y \in \partial B(X,r)$, then

$$\operatorname{cap} B_r(X) \simeq \frac{1}{\tilde{G}(X,Y)}$$

As a consequence of these results, and assuming the uniqueness of 'good solutions,' which gives an unambiguous meaning to the concept of regular point, and to the limiting capacity after regularization, Bauman obtained a Wiener test: If $\Omega \subset \frac{1}{4}B$, e is a unit vector in $\mathbb{R}^n, Y \in \partial\Omega$ is a regular point for $L \iff$

(a) $$\operatorname{cap}_L(\{Y\}) > 0 \text{ or}$$

(b) $$\sum_{j=N}^{\infty} \tilde{G}(Y, Y + 2^{-j} e) \operatorname{cap}_L(R_j) = +\infty,$$

where $R_j = ^c \Omega \cap B(Y, 2^{-j}) \backslash B(Y, 2^{-j-1})$.

This applies, for instance under the conditions described in the remark after (4.5). Note that (a) can actually occur ([30], [8]) and that (b) need not coincide with the corresponding condition for the Laplacian ([39], [36], [8]).

§5 Boundary estimates for solutions and normalized adjoint solutions in non-smooth domains

In this section we will study the boundary behavior of solutions and normalized adjoint solutions in non-smooth domains. We will also introduce the concept of harmonic measure and normalized adjoint harmonic measure, and study their connections with boundary behavior. We first start out with a discussion of the general class of domains to be considered. Let Ω be a bounded domain in $\mathbb{R}^n$.

DEFINITION 5.1. *Ω is said to be of class S if there exists two numbers $\alpha\,(0 < \alpha < 1)$ and $r_0 > 0$ such that*

$$|B(X_0,r) \cap^c \Omega| \geq \alpha |B(X_0,r)|$$

for all $X_0 \in \overline{\Omega}$, $0 < r < r_0$.

DEFINITION 5.2. *Ω is said to be a non-tangentially accessible domain (N.T.A.) (see [32]) if: (i) given $Q \in \partial\Omega$, $0 < r < r_0$, there exists a point $A_r(Q) \in \Omega$ with*

$$\operatorname{dist}(A_r(Q), \partial\Omega) \simeq |A_r(Q) - Q| \simeq r$$

(ii) If $X_1, X_2 \in \Omega$,
$\operatorname{dist}(X_j, \partial\Omega) > \varepsilon, |X_1 - X_2| < 2^k \varepsilon$, *then there is a chain of Mk balls $B_1, \dots, B_{Mk}$, connecting X_1 and X_2 (i.e. $X_1 \in B_1, X_2 \in B_{Mk}, B_j \cap B_{j+1} \neq \emptyset, B_j \subset \Omega$) so that* $\operatorname{diam} B_j \simeq \operatorname{dist}(B_j, \partial\Omega)$ *and* $\operatorname{diam}(B_j) \geq C \min\{\operatorname{dist}(X_1, B_j); \operatorname{dist}(X_2, B_j)\}$.
(iii) ${}^c\Omega$ verifies (i).

If Ω is a bounded Lipschitz domain (i.e. it is locally given as the domain above the graph of a Lipschitz function) then Ω verifies an exterior and interior uniform cone condition, and it is not hard to see that it is an N.T.A. However, it is easy to construct examples of N.T.A. domains which are not locally given as the domain above a graph. The results proved below will be established for bounded Lipschitz domains. They also hold for N.T.A. domains, with similar arguments as in [32]. The results we present now are due to Fabes, Garofalo, Marín-Malave and Salsa [27]. We start out by studying boundary regularity of solutions.

LEMMA 5.3. *Let Ω be a bounded Lipschitz domain, u a solution of $Lu = 0$ in $B(Q, 2r) \cap \Omega$, with $u \equiv 0$ on $B(Q,2r) \cap \Omega$, u continuous in $\overline{B(Q,2r) \cap \Omega}$. Then, there exist $C, \overline{\alpha} > 0$, depending only on λ, n, and the Lipschitz character of $\partial\Omega$ such that, for $X \in B(Q,r) \cap \Omega$*

$$|u(X)| \leq C(|X - Q|/r)^{\overline{\alpha}} M(u),$$

where $M(u) = \sup\{|u(X)| : X \in B(Q,2r) \cap \Omega\}$.

PROOF. The Lemma is a consequence of the corresponding estimate (4.2) for smooth domains. Let $v(X)$ be the solution in $B(Q,2r) \cap \Omega$ of $Lu = 0$ with boundary values 1 on $\overline{\Omega} \cap \partial B(Q,2r)$, 0 on $\partial\Omega \cap B(Q,2r)$. By the maximum principle, $|u(X)| \leq M(u) v(X)$, and so, we only have to show that

$$v(X) \leq C(|X - Q|/r)^{\overline{\alpha}}$$

By a dilation argument, we can assume, without loss of generality, that $r = 1$. Let w be the solution of $Lw = 0$ in $B(Q,2)$, with $w \equiv 1$ on $\partial B(Q,2) \cap \overline{\Omega}, w \equiv 0$ on $\partial B(Q,2) \cap^c \Omega$. Notice that the exterior cone condition implies that $\partial B(Q,2) \cap^c \Omega$ contains $B(\tilde{Q}, C) \cap \partial B(Q,2)$ with $\tilde{Q} \in \partial B(Q,2) \cap^c \Omega$, and C depends only on the Lipschitz character of Ω. Notice that $w \leq 1$ on $B(Q,2)$ by the maximum principle, and that $v \leq w$ on $B(Q,2) \cap \Omega$. By (4.2), we see that $w \leq 1/2$ near $\tilde{Q}$. By Harnack's principle applied to $1 - w$ (2.14), on $B(Q,1)$ we have $w \leq \theta < 1$. An iteration gives the desired bound.

THEOREM 5.4. *Let* $\alpha_0 = \min(\overline{\alpha}, \alpha)$, *where* $\overline{\alpha}$ *is as in (5.3) and* α *as in (2.16). Let* $f \in \mathrm{Lip}(\partial\Omega)$, $u \in C^\infty(\Omega) \cap C(\overline{\Omega})$ *verify* $Lu = 0$ *in* Ω, $u|_{\partial\Omega} = f$, *and* $\beta = \min(\frac{\alpha_0}{2}, \frac{1}{4})$. *Then,* $\sup\limits_{X, X' \in \Omega} \frac{|u(X)-u(X')|}{|X-X'|^\beta} \leq C\|f\|_{\mathrm{Lip}(\partial\Omega)}$, *where* C *depends only on* n, λ *and the Lipschitz character of* $\partial\Omega$.

PROOF. This follows from (5.3) in the same way that (4.3) followed from (4.2).

Another corollary of (5.3) is a 'boundary Harnack principle.'

LEMMA 5.5. *Suppose that* $u \geq 0$ *solves* $Lu = 0$ *in* $B(Q, 2r) \cap \Omega$, Ω *a bounded Lipschitz domain in* $\mathbb{R}^n$, $u \equiv 0$ *on* $B(Q,2r) \cap \partial\Omega$, $u \in C(\overline{B(Q,2r) \cap \Omega})$. *Let* $A_r(Q)$ *be a point as in (5.2) (i). Then*

$$u(X) \leq Cu(A_r(Q))$$

for all $X \in B(Q, \Omega) \cap \Omega$.

PROOF. Once (5.3) is established, the proof is identical to that of (4.16), using (5.2) (ii). We now turn to the boundary behavior of normalized adjoint solutions. We start out by first establishing the analog of (5.3) for smooth domains. Recall that if Ω is a bounded smooth domain, then, given $\varphi \subset C(\partial\Omega)$ there exists a unique (n.a.s.) $\tilde{v}$ such that $\tilde{v} \in C(\overline{\Omega})$ and $\tilde{v}|_{\partial\Omega} = \varphi$. Moreover,

$$\tilde{v}(Y) = \int\limits_{\partial\Omega} \varphi(\theta) \frac{G_B(X_*, Q)}{G_B(X_*, Y)} \frac{\partial G}{\partial \nu_Q}(Q, Y)\, d\sigma(Q),$$

where G is the Green's function for L and Ω. (See (4.10)). We will denote $d\tilde{\omega}^Y = \frac{G_B(X_*,Q)}{G_B(X_*,Y)} \cdot \frac{\partial G}{\partial \nu_Q}(Q, Y) d\sigma(Q)$, and call it the normalized adjoint harmonic measure for Ω. Our first estimate on normalized adjoint harmonic measure is:

LEMMA 5.6. *Let* $\tilde{\omega}^Y_{2r}$ *denote the normalized adjoint harmonic measure for* B_{2r}. *For* $\overline{Q} \in \partial B_{2r}$, $0 < \delta < 1$, *there exists* $C = C(\lambda, n, \delta)$ *such that*

$$\inf_{Y \in B_r} \tilde{\omega}^Y_{2r}(B(\overline{Q}, \delta r) \cap \partial B_{2r}) \geq C.$$

PROOF. We can assume, without loss of generality, that $r = 1$. Recall that $\tilde{\omega}^Y_{2r}(B(\overline{Q}, \delta r) \cap \partial B_{2r}) = \int\limits_{\partial B_{2r} \cap B(\overline{Q}, \delta r)} \frac{G_B(X_*,Q)}{G_B(X_*,Y)} \cdot \frac{\partial G_{B_2}}{\partial \nu_Q}(Q, Y) d\sigma(Q)$. Fix $A \in \partial B_{3/2}$. By (4.13), and (2.14), $\forall Y \in B_1, Q \in \partial B_2$, $\frac{\partial G_{B_2}}{\partial \nu_Q}(Q, Y) \geq C_1 G_{B_2}(A, Y)$, so that, for $Y \in B_1$,

$$\int\limits_{\partial B_2 \cap B(\overline{Q}, \delta)} G_B(X_*, Q) d\sigma(Q) / G_B(X_*, Y)$$

$$\tilde{\omega}^Y_2(B(\overline{Q}, \delta) \cap \partial B_2) \geq C_1 G_{B_2}(A, Y)$$

Let now $G_3(X, Y)$ denote the Green's function for B_3. Then, by Harnack's principle for (n.a.s.) (4.14), we have $\frac{G_B(X_*,Q)}{G_3(A,Q)} \geq C_2 \frac{G_B(X_*,Y)}{G_3(A,Y)}$, for $Q \in \partial B_2, Y \in B_1$. Hence,
$\tilde{\omega}^Y_2(B(\overline{Q}, \delta) \cap \partial B_2) \geq C_1 \cdot C_2 \frac{G_{B_2}(A,Y)}{G_3(A,Y)} \cdot \int\limits_{\partial B_2 \cap B(\overline{Q}, \delta)} G_3(A, Q) d\sigma(Q)$.

Using (4.14) once more, and the argument at the beginning of the proof of (4.17), we see that

$$\frac{G_{B_2}(A,Y)}{G_3(A,Y)} \geq C_3 \frac{\int\limits_{B_1} G_{B_2}(A,Z)\,dZ}{\int\limits_{B_1} G_3(A,Z)\,dZ} \geq C_4 > 0$$

The last inequality follows from (2.2) and (2.3). All that remains to show is:

LEMMA 5.7. $\int\limits_{\partial B_2 \cap B(\overline{Q},\delta)} G_3(A,Q)d\sigma(Q) \geq C > 0.$

PROOF. We consider two smooth domains, $\Omega \subset \Omega' \Subset B_3 \backslash B_{7/4}$, with $\partial B_2 \cap B(\overline{Q},\delta) \subset \partial\Omega$, and $\partial\Omega \backslash (\partial B_2 \cap B(\overline{Q},\delta)) \subset \partial\Omega'$, $\partial B_2 \cap B(\overline{Q},\delta) \cap \partial\Omega' = \emptyset$. Pick $A' \in \Omega' \backslash \overline{\Omega}$, so that

$$\int\limits_{\partial B_2 \cap B(\overline{Q},\delta)} G_3(A,Q)\,d\sigma(Q) \geq$$

$$\geq C_1 \int\limits_{\partial B_2 \cap B(\overline{Q},\delta)} G_3(A',Q)d\sigma(Q) \geq$$

$$\geq C_1 \int\limits_{\partial B_2 \cap B(\overline{Q},\delta)} G_{\Omega'}(A',Q)d\sigma(Q),$$

where the first inequality follows from Harnack's inequality, and the second one from the maximum principle. (Here $G_{\Omega'}$ is the Green's function for Ω'). Since $G_{\Omega'}(A',Q) \equiv 0$ on $\partial\Omega \backslash \partial B_2 \cap B(\overline{Q},\delta)$, the last integral above equals $\int_{\partial\Omega} G_{\Omega'}(A',Q)\,d\sigma(Q)$. Let $B' \Subset B_2 \backslash B_{7/4} \cap \Omega$, and consider $w(X) = -\int\limits_{B'} G_\Omega(X,Y)dY$, which verifies $Lw = \chi_{B'}, w|_{\partial\Omega} \equiv 0$. Since $L_Y^* G_{\Omega'}(A',Y) = 0$ in Ω, an integration by parts gives

$$\int\limits_{B'} G_{\Omega'}(A',Y)dY = \int\limits_{\Omega} G_{\Omega'}(A',Y)\,Lw\,(Y)dY =$$

$$= \int\limits_{\partial\Omega} G_{\Omega'}(A',Q)\frac{\partial w}{\partial \nu_Q}(Q)d\sigma(Q).$$

But, (4.2) shows that $\frac{\partial w}{\partial \nu_Q}(Q) \leq C \sup\limits_{\Omega} \int\limits_{B'} G_\Omega(X,Y)dY \leq \tilde{C}$, by (2.3). Hence,

$$\int\limits_{B'} G_{\Omega'}(A',Y)dY \leq C \int\limits_{\partial B_2 \cap B(\overline{Q},\delta)} G_{\Omega'}(A',Q)d\sigma(Q)$$

Moreover, a maximum principle argument, together with (3.3) shows that $\int\limits_{B'} G_{\Omega'}(A',Y)dY \geq C$, and (5.7) follows. (5.6) has many consequences, which we now exploit.

LEMMA 5.8. *Let* $\tilde{\omega}_{2r}^{Y}$ *be as in (5.6). Then,*

$$\sup_{Y \in B_r} \tilde{\omega}_{2r}^{Y}\left(B\left(\overline{Q},\delta r\right) \cap \partial B_{2r}\right) \leq \Theta < 1, \text{ where } \Theta = \Theta(\lambda, n, \delta).$$

PROOF. Note that $\tilde{v}(Y) = \tilde{\omega}_{2r}^{Y}\left(B\left(\overline{Q},\delta r\right) \cap \partial B_{2r}\right) \leq 1$, and is a non-negative (n.a.s.). We apply (5.8) to $\tilde{\omega}_{2r}^{Y}\left({}^{c}B\left(\overline{Q},\delta r\right) \cap \partial B_{2r}\right)$ and the Harnack principle for non-negative (n.a.s.) (4.14) to conclude (5.8).

LEMMA 5.9. *Let* Ω *be a bounded Lipschitz domain,* $\tilde{v}$ *a (n.a.s.) in* $\Omega \cap B(Q,2r), Q \in \partial\Omega, \tilde{v} \in C(\overline{\Omega \cap B(Q,2r)}), \tilde{v} \equiv 0$ *on* $\partial\Omega \cap B(Q,2r)$. *Then,*

$$|\tilde{v}(Y)| \leq C\left(\frac{|Y-Q|}{r}\right)^{\overline{\alpha}} M(\tilde{v}), \text{ for } Y \in \Omega \cap B(Q,r),$$

where $M(\tilde{v}) = \sup\limits_{B(Q,2r)\cap\Omega} |\tilde{v}|$, *and* $\overline{\alpha} = \overline{\alpha}(\lambda, n, \text{Lip}$ *character of* $\Omega)$.

PROOF. The proof is the same one as that of (5.3), using (5.8) and the maximum principle (4.12).

THEOREM 5.10. *Let* $\alpha_0 = \min(\overline{\alpha},\alpha)$ *where* $\overline{\alpha}$ *is as in (5.9), and* α *the Hölder constant in (4.15). Let* $f \in \text{Lip}(\partial\Omega), \tilde{v} \in C(\overline{\Omega})$ *be a (n.a.s.) in* Ω, *with* $\tilde{v}|_{\partial\Omega} = f$, *and* $\beta = \min\left(\frac{\alpha_0}{2}, \frac{1}{4}\right)$. *Then,*

$$\sup_{Y,Y' \in \Omega} \frac{|\tilde{v}(Y) - \tilde{v}(Y')|}{|Y-Y'|^{\beta}} \leq C\|f\|_{\text{Lip}(\partial\Omega)}.$$

PROOF. Follows from (5.9) and (4.12) as in (5.4).

COROLLARY 5.11. *Let* Ω *be a bounded Lip. domain,* $f \in C(\partial\Omega)$. *Then, there exists a unique* $\tilde{v}$ *(n.a.s.) in* Ω, *with* $\tilde{v} \in C(\overline{\Omega})$, *and* $\tilde{v}|_{\partial\Omega} = f$.

PROOF. Uniqueness follows from (4.12). To prove existence we proceed in stages. The first stage is when $f \in \text{Lip}(\partial\Omega)$. We then approximate Ω from outside by a decreasing family of smooth domains $\Omega_j \downarrow \Omega$, where the Ω_j are of uniformly Lipschitz character, and $\partial\Omega_j$ is a bi-Lipschitzian image of $\partial\Omega$, uniform in j. We then solve in Ω_j, with data f_j, (obtained from f by the bi-Lipschitzian transformation), by (4.10). (5.10) gives equicontinuity of the resulting $\tilde{v}_j$ in $\overline{\Omega}$, and hence a subsequence converges uniformly in $\overline{\Omega}$. It is easy to see that the limit fulfills the desired properties.

For the next stage, given $f \in C(\partial\Omega)$, we find $f_j \in \text{Lip}(\partial\Omega)$, with $f_j \to f$ uniformly on $\partial\Omega$. The corresponding solutions $\tilde{v}_j$ converge uniformly on $\overline{\Omega}$, by the maximum principle (4.12). Their limit is the desired (n.a.s.).

DEFINTION 5.12. *Let* $f \in C(\partial\Omega), \tilde{v}$ *be as in (5.11). By the Riesz representation theorem, and the maximum principle (4.12), there exists a family of probability Borel measures* $\{\tilde{\omega}^{Y}\}$ *such that* $\tilde{v}(Y) = \int_{\partial\Omega} f d\tilde{\omega}^{Y}$. *This family of measures is called normalized adjoint L-harmonic measure, or simply normalized adjoint harmonic measure. We sometimes fix* $X_0 \in \Omega$, *and call* $\tilde{\omega} = \tilde{\omega}^{X_0}$ *the normalized adjoint harmonic measure of* $\partial\Omega$.

LEMMA 5.13. *Let Ω be a bounded Lipschitz domain in $\mathbb{R}^n, \tilde{v} \geq 0$ a (n.a.s.) in $\Omega \cap B(Q,2r), \tilde{v}$ continuous in $\overline{\Omega \cap B(Q,2r)}, \tilde{v} \equiv 0$ on $\partial\Omega \cap B(Q,2r)$. Then, if $A_r(Q)$ is a point as in (5.2) (i),*

$$\tilde{v}(X) \leq C\tilde{v}(A_r(Q)),$$

for all $X \in \Omega \cap B(Q,\Omega)$.

PROOF. Once (5.9) is established, the proof is identical to that of (4.16) using (4.14) and (5.2) (ii).

COROLLARY 5.14. *Let Ω be a bounded Lipschitz domain, and $v \geq 0$ a solution to $L^*v = 0$ in $\{Y \in \Omega : \operatorname{dist}(Y, \partial\Omega) \leq 1\}$ $v|_{\partial\Omega} \equiv 0$. Then,*

$$\left(\frac{1}{|B|}\int_{B\cap\Omega} v^{n/n-1}\right)^{n-1/n} \leq C\left(\frac{1}{|B|}\int_{B\cap\Omega} v\right),$$

where $B = B(Y_0, r)$ and $\operatorname{dist}(Y_0, \partial\Omega) \leq r/3$.

PROOF. As in the proof of (4.6), it suffices to show that, if $Q_0 \in \partial\Omega, T_r(Q_0) = B(Q_0, r) \cap \Omega$, we have

$$\int_{T_{2r}(Q_0)} v \leq C \int_{T_r(Q_0)} v \tag{5.15}$$

We first claim that

$$\int_{T_{2r}(Q_0)} G_B(X_*, Y)dY \leq C\int_{B_r} G_B(X_*, Y)dY,$$

where B_r is ball contained in $T_r(Q_0)$, of diameter comparable to r, and whose distance to Q_0 and to $\partial\Omega$ is comparable to r. This is a consequence of (3.1). Next, $\tilde{v}(Y) = v(Y)/G_B(X_*, Y)$ is a (n.a.s.), whch vanishes on $\partial\ \Omega$, and hence, (5.13) proves that $\tilde{v}(Y) \leq C\tilde{v}(A_{2r}(Q_0)), \forall Y \in T_{2r}(Q_0)$. By Harnack's principle for (n.a.s.), (4.14), $\tilde{v}(A_{2r}(Q_0)) \leq C \inf_{B_r} \tilde{v}$. Thus,

$$\int_{T_{2r}(Q_0)} v(Y)dY \leq \int_{T_{2r}(Q_0)} \tilde{v}(Y)G_B(X_*, Y)dY \leq$$
$$\leq C\inf_{B_r}\tilde{v}\int_{B_r} G_B(X_*, Y)dY \leq C\int_{B_r}\tilde{v}(Y)G_B(X_*, Y)dY =$$
$$= C\int_{B_r} v(Y)dY$$

COROLLARY 5.16. *Let Ω be a bounded Lipschitz domain, $Lu = f$ in $\Omega, u|_{\partial\Omega} \equiv 0$. Then, there exists $p_0 > 0$, $p_0 = p_0(\lambda, n,$ Lip character of $\partial\Omega)$ such that*

$$\left(\int_\Omega |\nabla u|^{p_0}\right)^{1/p_0} + \left(\int_\Omega |D^2 u|^{p_0}\right)^{1/p_0} \leq C\|f\|_{L^n(\Omega)}$$

The proof is similar to the one of (3.11), (3.13), using (3.14).

Before obtaining further corollaries of these results, we introduce one more definition:

DEFINITION 5.17. *Let Ω be a bounded Lipschitz domain. Given $f \in C(\partial\Omega)$, there exists a unique $u \in C(\bar{\Omega})$, with $Lu = 0$ in $\Omega, u|_{\partial\Omega} = f$. (This follows in exactly the same manner as (5.11)). By the Riesz representation theorem, and the maximum principle, there exists a family of probability Borel measures $\{\omega^X\}$ such that $u(X) = \int_{\partial\Omega} f d\omega^X$. This family of measures is called L-harmonic measure, or simply harmonic measure. We sometimes fix $X_0 \in \Omega$, and call $\omega = \omega^{X_0}$ the harmonic measure of $\partial\Omega$.*

We next prove a theorem that replaces the Hopf maximum principle (4.13) for Lipschitz domains.

LEMMA 5.18. *Let Ω be a bounded Lipschitz domain, $Q \in \partial\Omega$, $A_r(Q)$ a point as in (5.2) (i). Then, for $X \in \Omega \cap {}^cB(Q, 2r)$ we have:*

$$\frac{1}{r^2} \frac{G(X, A_r(Q))}{G_B(X_*, A_r(Q))} \cdot \int_{B(A_r(Q), r/2)} G_B(X_*, Z)\, dZ \simeq \omega^X(B(Q,r) \cap \Omega),$$

with comparability constants which depend only on λ, n and the Lipschitz character of $\partial\Omega$.

PROOF. Let B_r be a ball centered at $A_r(Q)$, of diameter comparable to r, and whose distance to $\partial\Omega$ and to Q is comparable to r. We first claim that, for $X \in \Omega \backslash B_r$, $\omega^X(B(Q,r) \cap \partial\Omega) \geq \frac{C}{r^2} \int_{B_{r/2}} G(X,Y)\, dY$. In fact, $\inf_{X \in B_r} \omega^X(B(Q,r) \cap \partial\Omega) \geq C$ because of (5.3) and Harnack's inequality (2.14), while (2.3) shows that $\|\frac{1}{r^2} \int_{B_{r/2}} G(X,Y) dY\|_{L^\infty(\Omega)} \leq C$. Thus, the maximum principle in $\Omega \backslash B_r$ establishes the claim. Next, we claim that, for $X \in \Omega \cap {}^cB(Q, 2r)$ we have

$$\omega^X(B(Q,r) \cap \partial\Omega) \leq \frac{C}{r^2} \int_{T_{\frac{3}{2}r}(Q)} G(X,Y)\, dY.$$

To see this, pick $\varphi \in C_0^\infty(B(Q, \frac{3}{2}r))$, $\varphi \equiv 1$ on $B(Q,r)$. Then, for $X \in \Omega \cap$

${}^cB(Q,2r)$ we have

$$\omega^X(B(Q,r)\cap\partial\Omega)\le\int_{\partial\Omega}\varphi(Q)d\omega^X(Q)=$$

$$=\int_{T_{\frac{3}{2}r}(Q)}G(X,Y)L\varphi(Y)dY\le$$

$$\le\frac{C}{r^2}\int_{T_{\frac{3}{2}r}(Q)}G(X,Y)dY$$

where the equality follows by approximating by smooth domains and integrating by parts. Next, note that (5.14) proves that $\int_{T_{\frac{3}{2}r}(Q)}G(X,Y)dY\le C\int_{B_r}G(X,Y)dY$, for $X\in\Omega\cap{}^cB(Q,2r)$. Finally, Harnack's principle for (n.a.s.) (4.14), and (3.1) show that $\int_{B_r}G(X,Y)dY\simeq\frac{G(X,A_r(Q))}{G_B(X_*,A_r(Q))}\cdot\int_{B(A_r(Q),r/2)}G_B(X_*,Z)\,dZ$, for $X\in\Omega\,{}^cB(Q,2r)$ and the lemma follows.

THEOREM 5.19. *Let Ω be a bounded Lipschitz domain, u_1,u_2 be non-negative solutions of $Lu=0$ in $B(Q,2r)\cap\Omega, Q\in\partial\Omega$, which vanish on $\partial\Omega\cap B(Q,2r)$. Let $A_r(Q)$ be a point as in (5.2) (i). Then*

$$\sup_{T_r(Q)}\frac{u_1(X)}{u_2(X)}\le C\frac{u_1(A_r(Q))}{u_2(A_r(Q))}$$

PROOF. Consider a Lipschitz domain $\Omega_{3/2r}(Q)$ which verifies: $\overline{T_r(Q)}\Subset\overline{\Omega_{3/2r}(Q)}\Subset\overline{T_{2r}(Q)}$, and such that $\partial\Omega_{3/2r}(Q)\cap\Omega=L_1\cup L_2$, where $\operatorname{dist}(L_1,\partial\Omega)\simeq r$. Note that (5.18) implies that, for all $X\in T_r(Q)$ we have $\omega^X_{\Omega_{3/2r}}(L_1\cup L_2)\le C\omega^X_{\Omega_{3/2r}}(L_1)$, by (4.14) and (3.1), where $\omega^X_{\Omega_{3/2r}}$ denotes harmonic measure for the domain $\Omega_{3/2r}(Q)$. Because of (5.5), $u_1(X)\le Cu_1(A_r(Q)$ for all $X\in L_1\cup L_2$. Hence, the maximum principle implies that $u_1(X)\le Cu_1(A_r(Q))\omega^X_{\Omega_{3/2r}}(L_1\cup L_2)$ for all $X\in\Omega_{3/2r}(Q)$. For $X\in L_1$, Harnack's principle (2.14) gives $u_2(X)\ge Cu_2(A_r(Q))$, and so, the maximum principle implies that $u_2(X)\ge Cu_2(A_r(Q))\cdot\omega^X_{\Omega_{3/2r}}(L_1)$ in $\Omega_{3/2r}(Q)$. The theorem follows.

COROLLARY 5.20. *Let Ω be a bounded Lipschitz domain, and let u_1,u_2 be as in (5.19). There exists $\beta=\beta(\lambda,n,$ Lip character of $\partial\Omega)$ such that*

$$\left|\frac{u_1(X)}{u_2(X)}-\frac{u_1(X')}{u_2(X')}\right|\le C\left(\frac{|X-X'|}{r}\right)^\beta\cdot\frac{u_1(A_r(Q))}{u_2(A_r(Q))}$$

for all $X,X'\in T_r(Q)$.

PROOF. We can assume, without loss of generality, that $u_1(A_r(Q))=u_2(A_r(Q))=1$. Fix $P\in B(Q,r)\cap\partial\Omega$. Let $M(s)=\sup\{\frac{u_1(Y)}{u_2(Y)}:Y\in T_{2s}(P)\}$, $m(s)=\inf\{\frac{u_1(Y)}{u_2(Y)}:Y\in T_{2s}(P)\}$, where $s<r/2$. Note that $M(s)-$

$u_1/u_2 = \frac{M(s)u_2 - u_1}{u_2}$ and $\frac{u_1}{u_2} - m(s) = \frac{u_1 - m(s)u_2}{u_2}$ are quotients of non-negative solutions, which vanish on $B(P, 2s) \cap \partial\Omega$. By (5.19), we obtain that $M(s/2) - m(s/2) \leq \Theta[M(s) - m(s)]$, $0 < \Theta < 1$. This gives the desired estimate on $\partial\Omega \cap B(Q, r)$. The remainder of the estimate follows from this, (5.19) and (2.15) in a routine fashion (see (4.3) for example).

REMARK 5.21. *(5.18), (5.19) and (5.20) have many important consequences. Among them is the fact that non-negative solutions have non-tangential limits a.e. with respect to ω, that ω is a 'doubling measure' (see the remarks before (3.6)), all with constants independent of the smoothness of the coefficients (see [14], [32], [7] and [27]). This has applications to prove Fatou theorems for certain non-linear equations ([38], [27]). One can also prove atomic decompositions of Hardy spaces, and duality with appropriate BMO spaces, in the same way as in [32]. Another important application is the fact that the Euclidean boundary and the Martin boundary coincide (see [32]). All of these facts hold with constants independent of the smoothness of the coefficients, and therefore, if one can show that 'good solutions' are unique (see [4.4]) they would automatically extend to 'good solutions.'*

We finish this section by proving the corresponding results for normalized adjoint solutions and normalized adjoint measures. Once the analog of (5.18) is established, the analogs of (5.19), (5.20) and (5.21) will automatically follow, with identical proofs.

LEMMA 5.22. *Let Ω be a bounded Lipschitz domain, $Q \in \partial\Omega$, $A_r(Q)$ a point as in (5.2) (i). Then, for $X \in \Omega \cap {}^cB(Q, 2r)$, we have*

$$\frac{1}{r^2} \frac{G(A_r(Q)X)}{G_B(X_*, X)} \int_{B(A_r(Q), r/2)} G_B(X_*, Z)\, dZ \simeq \tilde{\omega}^X(B(Q, r) \cap \partial\Omega),$$

with comparability constants which depend only on λ, n and the Lipschitz character of $\partial\Omega$.

PROOF. Let B_r be a ball as in the proof of (5.18). Using (5.9) and Harnack's principle (4.14) we see that $\inf_{Y \in B_r} \tilde{\omega}^Y(B(Q, r) \cap \partial\Omega) \geq C$. We next consider the (n.a.s.) $\int_{Br/2} G(X, Y)\, dX / G_B(X_*, Y)$ in $\Omega \backslash B_r$. By (4.14) and (3.4) and (3.7), together with the argument in the proof of (4.17), we see that, for $Y \in B_r \backslash B_{\frac{3}{4}r}$, this is comparable to

$$\frac{\int_{B_r \backslash B_{3/4r}} \int_{B_{r/2}} G(X, Y) dX dY}{\int_{B_r \backslash B_{3/4r}} G_B(X_*, Y)\, dY}.$$

By (3.1), the denominator is comparable to $\int_{B(A_r(Q), r/2)} G_B(X_*, Z)\, dZ$, while the numerator is bounded by r^{2+n}, by (2.3). Thus, the maximum principle (4.12) shows that, for $Y \in \Omega \backslash B_r$,

$$\tilde{\omega}^Y(B(Q,r)\cap\partial\Omega) \geq$$

$$\geq \frac{C}{r^2}\,\frac{\frac{1}{r^n}\int_{B_{r/2}} G(X,Y)dX}{G_B(X_*,Y)}\cdot\int_{B(A_r(Q),r/2)} G_B(X_*,Z)dZ \geq$$

$$\geq \frac{C}{r^2}\,\frac{G(A_r(Q),Y)}{G_B(X_*,Y)}\cdot\int_{B(A_r(Q),r/2)} G_B(X_*,Z)\,dZ,$$

where the last inequality follows from Harnack's inequality (2.14) applied to $G(-,Y)$, in $B_{r/2}$.

Let now φ be as in the proof of (5.18). Then, $\tilde{\omega}^Y(B(Q,r)\cap\partial\Omega) \leq \int_{\partial\Omega}\varphi\, d\tilde{\omega}^Y = \tilde{v}(Y)$, so that $G_B(X_*,Y)\,\tilde{v}(Y)$ is a solution for L^*, with boundary values $G_B(X_*,Q)\,\varphi(Q)$. Thus,

$$G_B(X_*,Y)\,\tilde{v}(Y) = \int_{T_{3/r}(Q)} L^*(G_B(X_*,Z)\,\varphi(Z))\,G(Z,Y)\,dZ.$$

Now, $L^*(G_B(X_*,Z)\varphi(Z)) = \sum_{i,j}\frac{\partial^2}{\partial Z_i\,\partial Z_j}(a_{ij}(Z)G_B(X_*,Z)\,\varphi(Z)) = G_B(X_*,Z)\,L\,\varphi(Z) + \sum_{i,j}\frac{\partial}{\partial Z_i}(a_{ij}(Z)G_B(X_*,Z))\frac{\partial}{\partial Z_j}\varphi(Z)$, since $L^*(G_B(X_*,Z)) = 0$. The first term's contribution is bounded by $\frac{C}{r^2}\int_{T_{3/2r}(Q)} G_B(X_*,Z)G(Z,Y)\,dZ \leq \frac{C}{r^2}G(A_r(Q),Y)\int_{B(A_r(Q),r/2)} G_B(X_*,Z)\,dZ$ for all $\Omega\backslash B(Q,2r)$ by (5.5) and (3.1), which gives the desired bound. The second term becomes, after an integration by parts

$$\sum_{i,j}\int_{T_{3/2r}(Q)} G_B(X_*,Z)a_{ij}(Z)\frac{\partial\varphi}{\partial Z_j}(Z)\cdot\frac{\partial}{\partial Z_i}G(Z,Y)\,dZ.$$

Since $Y \in \Omega\backslash B(Q,2r)$, we can apply Schwarz's inequality and (2.4), together with the argument above, to finish the proof.

THEOREM 5.23. *Let Ω be a bounded Lipschitz domain, $\tilde{v}_1, \tilde{v}_2$ be non-negative (n.a.s.) in $B(Q,2r)\cap\Omega, Q \in \partial\Omega$, which vanish on $\partial\Omega\cap B(Q,2r)$. Let $A_r(Q)$ be a point as in (5.2) (i). Then*

$$\sup_{T_r(Q)}\frac{\tilde{v}_1}{\tilde{v}_2} \leq C\frac{\tilde{v}_1(A_r(Q))}{\tilde{v}_2(A_r(Q))}.$$

Moreover, there exists $\beta = \beta(\lambda, n,$ Lip character of $\partial\Omega)$ such that $\left|\frac{\tilde{v}_1(X)}{\tilde{v}_2(X)} - \frac{\tilde{v}_1(X')}{\tilde{v}_2(X')}\right| \leq C\left(\frac{|X-X'|}{r}\right)^\beta \frac{\tilde{v}_1(A_r(Q))}{\tilde{v}_2(A_r(Q))}$ for all $X, X' \in T_r(Q)$.

The proof is identical to the one of (5.19), (5.20), using (5.22), (4.14), (5.9).

DEFINITION 5.24. *Again, (5.22) and (5.23) have the same kind of consequence for normalized adjoint solutions as the ones discussed in (5.21) for solutions.*

REMARK 5.25. *As mentioned at the beginning of the section, all the results in this section extend verbatim to N.T.A. domains (5.2). In [4], Bass and Burdzy have extended some of these results to even more general domains.*

§6 Square function estimates and some weighted Poincaré inequalities

In this section we will discuss (mostly without proofs) recent joint work with L. Escauriaza ([24]) on square function estimates for non-divergence form elliptic equations.

We first recall the well known results of classical Littlewood-Paley theory (see [46]), Chapter IV, and the references therein). Also, for an interesting description of the development of the theory, some of its generalizations, and some of its many applications, see [48].

Suppose then that $B = \{|X| < 1\}$, and that

$$\Delta u = 0 \quad \text{in } B \qquad (6.1)$$
$$u|_{\partial B} = f \in L^p(\partial B, d\sigma),\ 1 < p < \infty.$$

Let $g^2(f)(Q) = \int_0^1 (1-r)|\nabla u(rQ)|^2 dr$, and assume that $\int_{\partial B} f\, d\sigma = 0$, or equivalently, that $u(0) = 0$. Then

$$\|g(f)\|_{L^p(\partial B, d\sigma)} \simeq \|f\|_{L^p(\partial B, d\sigma)},\ 1 < p < \infty. \qquad (6.2)$$

More generally, let $\Gamma(Q)$ be a truncated circular cone with vertex Q, axis in the radial direction, and let

$$S^2(u)(Q) = \int_{\Gamma(Q)} |X - Q|^{2-n} |\nabla u(X)|^2 dX,$$

and $N(u)(Q) = \sup_{X \in \Gamma(Q)} |u(X)|$. Then, if $u(0) = 0$, we have

$$\|S(u)\|_{L^p(\partial B, d\sigma)} \simeq \|N(u)\|_{L^p(\partial B, d\sigma)},\ 0 < p < \infty. \qquad (6.3)$$

(See Calderón [16], Burkholder-Gundy and Silverstein [10], Burkholder and Gundy [9], C. Fefferman and Stein [2] and Gundy and Wheeden [31] for generalizations to A_∞ weights). (6.3) is indeed a generalization of (6.2), because, using the fact that on $\Gamma(Q)$, $|X - Q| \simeq \operatorname{dist}(X, \partial B)$, and the sub-mean value property of harmonic functions, one can show that $g(f)(Q) \le C\,S(u)(Q)$, that $\|S(u)\|_{L^p(\partial B, d\sigma)} \le C\|g(f)\|_{L^p(\partial B, d\sigma)}$ and that $N(u)(Q) \le CM(f)(Q)$, where

$$M(f)(Q) = \sup_{1>r>0} \frac{1}{\sigma(B(Q,r) \cap \partial B)} \cdot \int_{B(Q,r)\cap\partial B} |f| d\sigma$$

is the Hardy-Littlewood maximal function, which verifies the estimate

$$\|M(f)\|_{L^p(\partial B, d\sigma)} \leq C\|f\|_{L^p(\partial B, d\sigma)} \quad 1 < p \leq \infty.$$

In [47], E. Stein extended the g function results ($1 < p < \infty$) to the context of symmetric diffusion semigroups, but self-adjointness was crucial. Note also, for future reference, that $d\sigma = d\omega_\Delta^0$, by the mean value property of harmonic functions (see (5.17)).

Inequality (6.3), for solutions of (6.1) was extended to the case when B is replaced by a bounded Lipschitz domain Ω by Dahlberg [21]. The next extension, when Δ was replaced by a divergence form elliptic operator, with bounded measurable coefficients, i.e. when u is a weak solution of

$$(6.4) \qquad \frac{\partial}{\partial X_i} a_{ij}(X) \frac{\partial u}{\partial X_j} = 0 \qquad \text{in } B, \text{ and } u(0) = 0,$$

where $A(X) = (a_{ij}(X))$ is symmetric, real, and $A \in L^\infty(\mathbb{R}^n)$ and

$$(6.5) \qquad \lambda|\xi|^2 \leq \langle A(X)\xi, \xi\rangle \leq \lambda^{-1}|\xi|^2, \text{ is that}$$

$$(6.6) \qquad \|S(u)\|_{L^p(\partial B, d\omega)} \simeq \|N(u)\|_{L^p(\partial B, d\omega)} \qquad 0 < p < \infty, \text{ was}$$

proven by Dahlberg, Jerison and Kenig [22]. Extensions to Lipschitz domains and to A_∞ weights $d\mu$ (with respect to $d\omega$) were also carried out in [22]. Note also that counterexamples in [13] can be used to show that $d\omega$ in (6.6) cannot in general be replaced by $d\sigma$.

In [24], L. Escauriaza and C. Kenig have extended these results to the non-divergence form setting. Thus, once more $A \in C^\infty(\mathbb{R}^n)$, verifies (6.5) and $L = \sum_{i,j} a_{ij}(X) \frac{\partial^2}{\partial X_i \partial X_j}$. Keeping the normalizations in sections 4 and 5, we now define, for $Q \in \partial B$

$$S_L^2(u)(Q) = \int_{\Gamma(Q)} |X - Q|^2 \frac{G_B(X_*, X)}{\int_{B(X,\delta(X)/2)} G_B(X_*, Z)\, dZ} \cdot |\nabla u(X)|^2 dX,$$

where $\delta(X) = \operatorname{dist}(X, \partial B)$. Our first result is:

THEOREM 6.7. *If $Lu = 0$ in $B, u(0) = 0$, then*

$$\|S_L(u)\|_{L^p(\partial B, d\omega)} \simeq \|N(u)\|_{L^p(\partial B, d\omega)} \qquad 0 < p < \infty,$$

where $d\omega$ is the harmonic measure associated to L (see (5.17), and the comparability constants depend only on λ, n, p.

Extensions to bounded Lipschitz domains and measures $\mu \in A_\infty(d\omega)$ (see 3.7) are also given in [24].

There is a corresponding result (and corresponding extensions) for normalized adjoint solutions.

Theorem 6.8. *Let $\tilde{v}$ be a (n.a.s.), $\tilde{v}(0) = 0$. Then, for $0 < p < \infty$, $\|S_L(\tilde{v})\|_{L^p(\partial B, d\tilde{\omega})} \simeq \|N(\tilde{v})\|_{L^p(\partial B, d\tilde{\omega})}$.*

The proofs of these results are rather technical, and depend heavily on the results in the previous sections. (See [24] for the details). I would like to conclude these notes by isolating an important step in their proofs, and then formulating some new questions that it suggests.

Lemma 6.9. *Let u satisfy $Lu = 0$ on $B_{2r} \subset \frac{1}{4}B$. Then,*

$$\int_{B_r} |u(Z) - m(u)|^2 G_B(X_*, Z)\, dZ \le Cr^2 \int_{B_{2r}} |\nabla u(Z)|^2 G_B(X_*, Z)\, dZ,$$

where $m(u) = \dfrac{1}{\int_{B_r} G_B(X_*, Z)\, dZ} \displaystyle\int_{B_r} u(Z) G_B(X_*, Z)\, dZ$, *and C depends only on λ, n.*

Proof. Let $Z_0 =$ center of B_r. It suffices to show that $\sup\limits_{Z \in B_r} |u(Z) - u(Z_0)|^2 \cdot \int_{B_r} G_B(X_*, Z)\, dZ \le Cr^2 \int_{B_{2r}} |\nabla u(Z)|^2 G_B(X_*, Z)\, dZ$. Let $\omega_{2r}^{Z_0}, G_{2r}(Z_0, -)$ be respectively L-harmonic measure and Green's function for L in B_{2r}, evaluated at Z_0. Integration by parts shows that

$$\int_{\partial B_{2r}} [u(Q) - u(Z_0)]^2 d\omega_{2r}^{Z_0} = 2 \int_{B_{2r}} G_{2r}(Z_0, Y) A \nabla u \nabla u\, dY. \tag{6.10}$$

Let now k be a large integer, to be fixed later. Then, the right hand side of (6.10) is bounded by

$$\begin{aligned} &2 \int_{B_{2r}} [G_{2r}(Z_0, Y) - G_{2r/2^k}(Z_0, Y)] A \nabla u \nabla u\, dY + \\ &+ 2 \int_{B_{2r} \setminus B_{2r}/2^k} G_{2r}(Z_0, Y) A \nabla u \nabla u\, dY + \sup_{B\, 2r/2^k} |u(Z) - u(Z_0)|^2 = \\ &= \mathrm{I} + \mathrm{II} + \mathrm{III}, \end{aligned}$$

where the last term appears by applying (6.10) once more, on $B_{2r/2^k}$. By (2.15), $|\mathrm{III}| \le \Theta^k (\operatorname{osc}_{B_r} u)^2 \le 2\Theta^k \sup\limits_{B_r} |u(Z) - u(Z_0)|^2$, where $\Theta < 1$. Since, by the definition of harmonic measure (5.17), $u(Z) - u(Z_0) = \int_{\partial B_{2r}} [u(Q) - u(Z_0)] d\,\omega_{2r}^Z$, $Z \in B_r$, and $\left| \dfrac{d\omega_{2r}^Z}{d\,\omega_{2r}^{Z_0}} \right| \le C$, $Z \in B_r$ by Harnack's inequality (2.14), Schwarz's inequality gives

$$\sup_{B_r} |u(Z) - u(Z_0)|^2 \le C \int_{\partial B_{2r}} [u(Q) - u(Z_0)]^2 d\omega_{2r}^{Z_0}, \tag{6.11}$$

and so, for k large $|\mathrm{III}| \le \frac{1}{2} \int\limits_{\partial B_{2r}} [u(Q) - u(Z_0)]^2 d\omega_{2r}^{Z_0}$. Fix k now; to control I, II, observe that $\tilde{v}(Y) = \frac{G_{2r}(Z_0,Y) - G_{2r/2^k}(Z_0,Y)}{G_B(X_*,Y)}$ is a non-negative (n.a.s.) on $B_{2r/2^k}$, and so, by (4.12), its maximum is taken on $\partial B_{2r/2^k}$, where it equals the maximum of $G_{2r}(Z_0,Y)/G_B(X_*,Y)$. But, (4.14) and (3.14) shows that, for any $Y \in \partial B_{2r/2^k}$, the last expression is bounded by $\frac{C}{\int_{B_r} G_B(X_*,Z)\,dZ} \int\limits_{B_r} G_{2r}(Z_0,Z)\,dZ$. (See the proof of (4.17)). By (3.3), the numerator is bounded by Cr^2, and so, $|I| \le Cr^2 / \int\limits_{B_r} G_B(X_*,Z)\,dZ \cdot \int\limits_{B_r} G_B(X_*,Z) A\nabla u \nabla u\,dZ$.

Analogously, $G_{2r}(Z_0,Y)/G_B(X_*,Y)$ attains its maximum on $B_{2r} \backslash B_{\frac{2r}{2^k}}$ on $\partial B_{\frac{2r}{2^k}}$, where it is bounded by the same expression as above, and so, |II| has a similar bound. Thus, collecting all our bounds, we see that

$$\int\limits_{\partial B_{2r}} [u(Q) - u(Z_0)]^2 d\omega_{2r}^{Z_0} \le \frac{Cr^2}{\int_{B_r} G_B(X_*,Z)\,dZ} \int\limits_{B_{2r}} G_B(X_*,Z) A\nabla u \nabla u\,dZ.$$

using (6.11) once more, we obtain the Lemma.

CONJECTURE 6.12. *In (6.9) one can replace the right hand side by*

$$\left(Cr^p \int\limits_{B_{2r}} |\nabla u(Z)|^p G_B(X_*,Z)\,dZ \right)^{2/p} \cdot \left(\int\limits_{B_r} G_B(X_*,Z)\,dZ \right)^{1-2/p},$$

where $p = p(\lambda, n) < 2$.

This would imply, using (2.4) and standard results in the theory of weights that there exists $\varepsilon = \varepsilon(\lambda, n)$, such that for any solution of $Lu = 0$, $|\nabla u|^{2+\varepsilon} G_B(X_*, Z)$ is locally integrable. This is very much in the spirit of results due to Bass [3], in his proof that for coefficients which are homogeneous of degree zero 'good solutions' (see (4.4)) are unique. Thus, this could be instrumental in the possible proof of (4.5).

REMARK 6.13. *Another natural question is whether (6.9) holds for all C^1 functions u, and not just for solutions to $Lu = 0$. There are many works devoted to understanding when do inequalities such as (6.9) hold for all C^1 functions u. Nevertheless, none of the known sufficient conditions apply, since they all require that $1/G(X_*,Z) \in L^1_{\mathrm{loc}}$, which is ruled out by the counterexamples in [17]. This question also seems to be closely related to (4.5).*

References

1. A.D. Alexandrov, *Uniqueness condition and estimates of the solution of Dirichlet's problem*, Vestmik Leningrad **13** (1963), no. 3, 5-39.
2. I. Bakelman, *Theory of quasilinear elliptic equations*, Siberian Math J. **2** (1961), 179-186.
3. R. Bass, *The Dirichlet problem for radially homogeneous elliptic operators*, Trans. AMS **320** (1990), 593-674.
4. R. Bass and K. Burdzy, *The boundary Harnack principle for non-divergence form elliptic operators*, preprint.

5. B. Barcelo, L. Escauriaza and E. Fabes, *Gradient estimates at the boundary for solutions to nondivergence elliptic equations*, Contemp. Math. **107** (1990), 1-12.
6. P. Bauman, *Equivalence of the Green's functions for diffusion operators in* $\mathbb{R}^n$*: a counterexample*, Proc. AMS **91** (1984), 64-68.
7. P. Bauman, *Positive solutions of elliptic equations in nondivergence form and their adjoints*, Ark. Mat. **22** (1984), 153-173.
8. P. Bauman, *A Wiener test for nondivergence structure second order elliptic equations*, Indiana U. Math J. **4** (1985), 825-844.
9. D. Burkholder and R. Gundy, *Distribution function inequalities for the area integral*, Studia Math. **44** (1972), 527-544.
10. D. Burkholder, R. Gundy and M. Silverstein, *A maximal function characterization of the class* H^p, Trans. AMS **157** (1971), 137-153.
11. L. Caffarelli, *Interior a priori estimates for solutions of fully non-linear equations*, Annals of Math. **130** (1989), 189-213.
12. L. Caffarelli, *Interior* W^{2p} *estimates for solutions of Monge-Ampére equations*, Annals of Math. **131** (1990), 135-150.
13. L. Caffarelli, E. Fabes and C. Kenig, *Completely singular elliptic-harmonic measures*, Indiana U. Math. J. **30** (1981), 917-924.
14. L. Caffarelli, E. Fabes, S. Mortola and S. Salsa, *Boundary behavior of nonnegative solutions of elliptic operators in divergence form*, Indiana U. Math. J. **30** (1981), 621-640.
15. L. Caffarelli, L. Nirenberg and J. Spruck, *The Dirichlet problem for nonlinear second order elliptic equations*, Comm. Pure Appl. Math. **37** (1984), 369-402.
16. A. Calderón Commutators of singular integral operators, Proc. Nat. Acad. Sci., USA **53** (1965), 1092-1099.
17. C. Cerutti, Proc. AMS, to appear.
18. C. Cerutti, L. Escauriaza and E. Fabes, *Uniqueness in the Dirichlet problem for some elliptic operators with discontinuous coefficients*, Annali di Matematica Pura ed Aplicata, to appear.
19. C. Cerutti, L. Escauriaza and E. Fabes, *Uniqueness for some diffusion with discontinuous coefficients*, Annals of Probability, to appear.
20. R. Coifman and C. Fefferman, *Weighted norm inequalities for maximal functions and singular integrals*, Studia Math. **51** (1974), 241-250.
21. B. Dahlberg, *Weighted norm inequalities for the Lusim area integral and non-tangential maximal function for functions harmonic in a Lipschitz domain*, Studia Math. **67** (1980), 297-314.
22. B. Dahlberg, D. Jerison and C. Kenig, *Area integral estimates for elliptic differential operators with non-smooth coefficients*, Ark. Mat. **22** (1984), 97-108.
23. L. Escauriaza, *Uniqueness in the Dirichlet problem for time independent elliptic operators*, IMA Volumes in Math. and its Appl. **42** (1991), 115-127.
24. L. Escauriaza and C. Kenig, *Area integral estimates for solutions and normalized adjoint solutions to nondivergence form elliptic equations*, Ark. Mat., to appear.
25. L.C. Evans, *Classical solutions of fully nonlinear convex second order elliptic equations*, Comm. Pure Appl. Math. **35** (1982), 333-363.
26. L.C. Evans, *Some estimates for nondivergence structure second order elliptic equations*, Trans. AMS **287** (1985), 701-712.
27. E. Fabes, N. Garofalo, S. Marin-Malave and S. Salsa, *Fatou theorems for some nonlinear elliptic equations*, Revista Mat. Ibero-Americana **4** (1988), 227-251.
28. E. Fabes and D. Stroock, *The* L^p *integrability of Green's functions and fundamental solutions for elliptic and parabolic equations*, Duke Math. J. **51** (1984), 977-1016.
29. C. Fefferman and E. Stein, H^p *spaces of several variables*, Acta. Math. **129** (1972), 137-193.
30. D. Gilbarg and J. Serrin, *On isolated singularities of solutions of second order elliptic differential equations*, J. D' Analyse Math. **4** (1955-1956), 309-340.
31. R. Gundy and R. Wheeden, *Weighted integral inequalities for the non-tangential maximal function*, Lusin area integral and Walsh-Paley series, Studia Math. **49** (1974), 107-124.
32. D. Jerison and C. Kenig, *Boundary behavior of harmonic functions in non-tangentially accessible domains*, Adv. in Math. **46** (1982), 80-147.
33. N. Krylov, *Some new results in the theory of nonlinear elliptic and parabolic equations*, Proc. of the ICM, Berkeley, Ca, 1986, pp. 1101-1109.

34. N. Krylov and M. Safonov, *An estimate of the probability that a diffusion process hits a set of positive measure*, Soviet Math. Dokl. **20** (1979), 253-255.
35. N. Krylov and M. Safonov, *A certain property of solutions of parabolic equations with measurable coefficients*, Math. USSR Izv. **16** (1981), 151-164.
36. E. Landis, *s-capacity and the behavior of a solution of a second order elliptic equation with discontinuous coefficients in the neighborhood of a boundary point*, Soviet Math. Dokl. **9** (1968), 582-586.
37. F.L. Lin, *Second derivative L^p estimates for elliptic equations of nondivergent type*, Proc. AMS **96** (1986), 447-451.
38. J. Manfredi and A. Weitsman, *On the Fatou theorem for p-harmonic functions*, Comm. in PDE **13** (1988), 651-688.
39. K. Miller, *Nonequivalence of regular boundary points for the Laplace and nondivergence equations even with continuous coefficients*, Ann. Scuola Norm. Sup. Pisa **24** (1970), 159-163.
40. J. Moser, *On Harnack's theorem for elliptic differential equations*, Comm. in Pure Appl. Math. **13** (1960), 457-468.
41. C. Pucci, *Limitazioni per soluzioni di equazioni ellittiche*, Ann. Mat. Pura Appl. **74** (1966), 15-30.
42. C. Pucci, *Operatori ellittichi estremanti*, Ann. Mat. Pura Appl. **72** (1966), 141-170.
43. C. Pucci and G. Talenti, *Elliptic (second-order) partial differential equations with measurable coefficients and approximating integral equations*, Adv. in Math. **19** (1976), 48-105.
44. M. Safonov, *Harnack's inequality for elliptic equations and the Hölder property of their solutions*, J. Soviet Math. (1983), 851-863.
45. M. Safonov, *Unimprovability of estimates of Hölder constant for solutions of linear elliptic equations with measurable coefficients*, Math. USSR **60** (1988), 269-281.
46. E. Stein, *Singular Integrals and Differentiability Properties of Functions*, Princeton Univ. Press, Princeton, NJ, 1970.
47. E. Stein, *Topics in Harmonic Analysis Related to the Littlewood-Paley Theory*, Annals of Math. Studies #63 (1970), Princeton Univ. Press, Princeton, NJ.
48. E. Stein, *The development of square functions in the work of A. Zygmund*, Conference on Harmonic Analysis in honor of A. Zygmund, Wadsworth International Group, International Mathematical Series (1981), 2-30.
49. D. Stroock and S. Varadhan, *Multidimensional Diffusion Processes*, Springer Verlag, Heidelberg, New York, 1979.
50. N. N. Uraltseva, *Impossibility of W_q^2 bounds for multidimensional elliptic equations with discontinuous coefficients*, Zap. Naveň. Sem. Leningrad Ozdel Mat. Inst. Steklov (LOMi) **5** (1967).

General Theory of Dirichlet Forms and Applications

by

Michael Röckner
Institut für Angewandte Mathematik
Universität Bonn
Wegelerstraße 6
5300 Bonn 1
Germany

Contents

Chapter 0

Introduction

The purpose of these lectures is to give a "compact" introduction to the theory of (not necessarily symmetric) Dirichlet forms on general state spaces including both the analytic and probabilistic aspects of the theory. They are based on joint work with Z.M. Ma [MR 92], S. Albeverio [AR 91] and B.K. Driver [DR 92] whom I would like to thank at this point for agreeing that I present some of our joint results here. These and M. Fukushima's lectures are meant to complement each other. In Chapter I we start with the analytic part of the theory of Dirichlet forms. In Section I.1 we recall the relationship between strongly continuous contraction semigroups $(T_t)_{t>0}$, strongly continuous contraction resolvents $(G_\alpha)_{\alpha>0}$ and their generators $(L, D(L))$ on a Banach space. In Section I.2 we consider coercive closed forms $(\mathcal{E}, D(\mathcal{E}))$ on a Hilbert space $\mathcal{H}$ and establish one-to-one correspondences with semigroups, resolvents and their generators as above. In Section I.3 we consider the case where $\mathcal{H}$ is some L^2-space over an arbitrary measure space $(E; \mathcal{B}; m)$ and investigate the relation between "contraction properties" of the four corresponding objects $(\mathcal{E}, D(\mathcal{E}))$, $(G_\alpha)_{\alpha>0}$, $(T_t)_{t>0}$ and $(L, D(L))$ on the "state space" E.
Apart from Section II.1, where we briefly recall the notion of closability, Chapter II is devoted to examples of Dirichlet forms, in particular including cases where the state space E is infinite dimensional. We keep close to situations typical for applications starting from some linear operator (Section II.2), some bilinear form on a finite or infinite dimensional state space E (Sections II.3,4) or some semigroup of kernels (Section II.5). In each case we show under what conditions and how to obtain the corresponding Dirichlet form.
Chapter III on the potential theory of Dirichlet forms can be considered as a link between the analytic and probabilistic component of the theory. We, however, only study this potential theory to an extend necessary to understand the probabilistic results in Chapter IV, that is, we only treat exceptional sets, capacities (Section IV.1) and quasi-continuity (Section IV.2).
In Chapter IV we first recall some basics on Markov processes (Section IV.1) and then present the fundamental theorem on the existence of Markov processes associated with Dirichlet forms (Section IV.2). More precisely, we give an analytic characterization of the class of all Dirichlet forms (called "quasi-regular") on general state spaces which

are associated with (pairs of) right continuous strong Markov processes and discuss some consequences. For simplicity, we assume here that the state space E is a Borel subset of a Polish space. Subsequently, in Section IV.3 we prove that the examples in Chapter II, Sections 2-4, are quasi-regular. So far, most of the results on Dirichlet forms were obtained in the classical framework (cf. [F 80], [Si 74]), i.e., for "regular" Dirichlet forms on locally compact separable metric spaces and associated Hunt processes (cf. M. Fukushima's lectures). Therefore, we include a description of a general regularization method in Section IV.4 which makes it possible to transfer all results known in the classical framework to quasi-regular Dirichlet forms and their associated right-continuous strong Markov processes, i.e., to the most general class of Dirichlet forms having a probabilistic counterpart.

Chapter V and VI contain further applications. In Chapter V we show how to use Dirichlet forms to construct weak solutions for stochastic differential equations on infinite dimensional state spaces. In Chapter VI we construct a diffusion on the path (and loop space) W of a compact Riemannian manifold (M, g) using a natural Dirichlet form on W intrinsically given by (M, g).

Finally we would like to mention that in these lectures the emphasis is on the description of applications of the theory of Dirichlet forms. Though reasonably self-contained, these notes only give an outline of the theoretical part of the material summarized above. In particular, not all proofs are included and the list of references is kept fairly short, just sufficient for our purposes . Therefore, the reader is advised to consult e.g. [MR 92] for more details and references (at least at a later stage). Concerning the history of the theory we should mention the pioneering papers of A. Beurling and J. Deny [BeDe 58, 59] and also Deny's C.I.M.E. - lectures [De 70]. For more detailed historical information we refer to [F 80].

The financial support of the Sonderforschungsbereich 256 in Bonn for this work is gratefully acknowledged.

Chapter I

Underlying L^2-Theory and Contraction Properties

For the proofs not given here we refer e.g. to [MR 92].

1 Resolvents, semigroups, generators

In this section we recall some basic facts on semigroups, resolvents and generators. We fix a Banach space $(B, \| \ \|)$ over $\mathbb{K} := \mathbb{R}$ or $\mathbb{C}$. We call a pair $(L, D(L))$ a *linear operator* on B if $D(L)$ is a linear subspace of B and $L : D(L) \to B$ a $\mathbb{K}$-linear map. We shall sometimes write L instead of $(L, D(L))$. If L is one-to-one, L^{-1} is defined by $D(L^{-1}) = L(D(L))$ and $L^{-1}u = v$, where $v \in B$ such that $Lv = u$. Given two linear operators L_1, L_2 on B and α, $\beta \in \mathbb{K}$, the linear operator $\alpha L_1 + \beta L_2$ on B is defined by $D(\alpha L_1 + \beta L_2) = D(L_1) \cap D(L_2)$ and $(\alpha L_1 + \beta L_2)u = \alpha L_1 u + \beta L_2 u$. $L_2 L_1$ is defined as $D(L_2 L_1) = \{u \in D(L_1) | L_1 u \in D(L_2)\}$ and $L_2 L_1 u = L_2(L_1 u)$. We write $L_1 = L_2$ if $D(L_1) = D(L_2)$ and $L_1 u = L_2 u$ for all $u \in D(L_1) = D(L_2)$. We denote the identity operator on B by Id_B and abbreviate $\alpha \, \mathrm{Id}_B$ by α if $\alpha \in \mathbb{K}$. Recall that a linear operator L on B is *continuous* (or *bounded*) if

$$\|L\| := \sup\{\|Lu\| \, | \, u \in B, \ \|u\| \le 1\} < \infty .$$

Definition 1.1. Let L be a linear operator on B. The *resolvent set* $\rho(L)$ of L is defined to be the set of all $\alpha \in \mathbb{K}$ such that $(\alpha - L) : D(L) \to B$ is one-to-one and for its inverse $(\alpha - L)^{-1}$ we have

(a) $D((\alpha - L)^{-1}) = B$

(b) $(\alpha - L)^{-1}$ is continuous on B.

$\sigma(L) := \mathbb{K} \backslash \rho(L)$ is called the *spectrum* of L and $\{(\alpha - L)^{-1} | \alpha \in \rho(L)\}$ the *resolvent* of L.

Definition 1.2. A family $(G_\alpha)_{\alpha > 0}$ of linear operators on B with $D(G_\alpha) = B$ for all $\alpha \in]0, \infty[$ is called a *strongly continuous contraction resolvent* (on B) if

(i) $\lim_{\alpha\to\infty} \alpha G_\alpha u = u$ for all $u \in B$ *(strong continuity).*

(ii) αG_α is a contraction on B for all $\alpha > 0$.

(iii) $G_\alpha - G_\beta = (\beta - \alpha) G_\alpha G_\beta$ for all $\alpha, \beta > 0$ *(first resolvent equation).*

Proposition 1.3. *Let $(G_\alpha)_{\alpha>0}$ be a strongly continuous contraction resolvent on B. Then there exists exactly one linear operator $(L, D(L))$ on B such that $]0,\infty[\subset \rho(L)$ and $G_\alpha = (\alpha - L)^{-1}$ for all $\alpha > 0$. This operator is closed and densely defined.*

Remark 1.4. L in 1.3 is called the *generator* of $(G_\alpha)_{\alpha>0}$.

Definition 1.5. A family $(T_t)_{t>0}$ of linear operators on B with $D(T_t) = B$ for all $t > 0$ is called a *strongly continuous contraction semigroup* (on B) if

(i) $\lim_{t\to 0} T_t u = u$ for all $u \in B$ *(strong continuity).*

(ii) T_t is a contraction on B for all $t > 0$.

(iii) $T_t T_s = T_{t+s}$ for all $t, s > 0$ *(semigroup property).*

Definition 1.6. Given a strongly continuous contraction semigroup $(T_t)_{t>0}$ on B, the linear operator $(L, D(L))$ on B defined by

$$\begin{aligned} D(L) &:= \{u \in B \mid \lim_{t\downarrow 0} \frac{1}{t}(T_t u - u) \text{ exists } \} \\ Lu &:= \lim_{t\downarrow 0} \frac{1}{t}(T_t u - u), \ u \in D(L), \end{aligned}$$

is called the *(infinitesimal) generator* of $(T_t)_{t>0}$.

The following diagram describes the unique correspondence between strongly continuous contraction semigroups, strongly continuous contraction resolvents and certain (in general unbounded) linear operators on B. The missing link we did not mention so far is filled in by the well-known Hille-Yosida theorem (cf. e.g. [ReS 75]). Note that if B is a real Hilbert space with inner product $(\ ,\ )$, then (i), (ii) in the lower box of the diagram can be replaced by

(i)′ $(Lu, u) \le 0$ for all $u \in D(L)$ *(negative definite)*

(ii)′ $(\alpha - L)(D(L)) = B$ for some $\alpha > 0$

provided L is closed (cf. [ReS 75, Theorem X.48]). Furthermore, then also

(1.1) $$(G_\alpha f, f) = (G_\alpha f, -L\, G_\alpha f) + \alpha (G_\alpha f, G_\alpha f) \ge 0 \text{ for all } f \in B,\ \alpha > 0 .$$

Diagram 1

$B =$ Banach space over $\mathbb{K} = \mathbb{R}, \mathbb{C}$

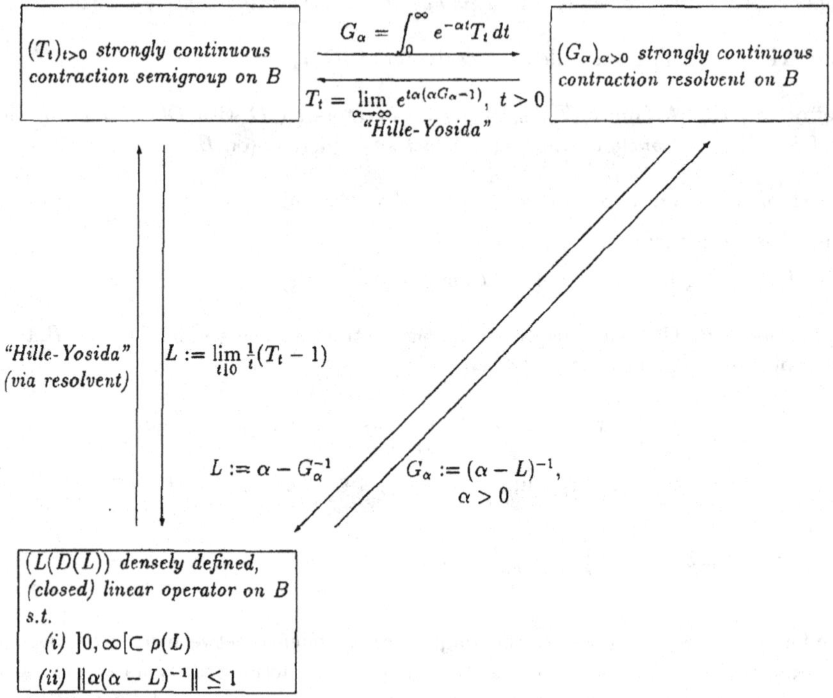

To conclude this section we briefly discuss the "symmetric case": assume B is a real Hilbert space (i.e., its norm comes from an inner product $(\ ,\ )$) and L is a *self-adjoint, negative definite* operator (i.e., L coincides with its adjoint $\hat{L}$ on B and $(Lu, u) \leq 0$ for all $u \in D(L)$). Then (i) and (ii) in the above diagram always hold and the above correspondences remain true if one adds that both, all T_t, $t > 0$, and all G_α, $\alpha > 0$, are self-adjoint. In this case $T_t = e^{tL}$, $t > 0$. e^{tL} is defined by the spectral theory for self-adjoint operators on a Hilbert space.

2 Coercive bilinear forms

In this section we want to discuss the connection between coercive bilinear forms and the objects in the preceding section in the case where B is replaced by some real Hilbert space $\mathcal{H}$ with inner product $(\ ,\ )$ and norm $\|\ \| := (\ ,\ )^{1/2}$ which we fix in this section.

Let D be a linear subspace of $\mathcal{H}$ and $\mathcal{E} : D \times D \to \mathbb{R}$ a bilinear map. We define its *symmetric part* and *antisymmetric part* $(\tilde{\mathcal{E}}, D), (\check{\mathcal{E}}, D)$ respectively by

$$(2.1) \qquad \tilde{\mathcal{E}}(u,v) := \frac{1}{2}(\mathcal{E}(u,v) + \mathcal{E}(v,u))\ ;\ \ \check{\mathcal{E}}(u,v) := \frac{1}{2}(\mathcal{E}(u,v) - \mathcal{E}(v,u))\ ,$$

$u, v \in D$. Clearly, $\mathcal{E} = \tilde{\mathcal{E}} + \check{\mathcal{E}}$. For $\alpha \geq 0$ we set

$$(2.2) \qquad \mathcal{E}_\alpha(u,v) := \mathcal{E}(u,v) + \alpha(u,v)\ ;\ \ u, v \in D\ .$$

Assume $(\mathcal{E}, D)$ is *positive definite* (i.e., $\mathcal{E}(u,u) \geq 0$ for all $u \in D$). Then $(\mathcal{E}, D)$ is said to satisfy the *weak sector condition* if

$$(2.3) \qquad \begin{array}{l} \text{there exists a constant } K > 0 \text{ (called } \textit{continuity constant}\text{)} \\ \qquad \text{such that } |\mathcal{E}_1(u,v)| \leq K\ \mathcal{E}_1(u,u)^{1/2}\mathcal{E}_1(v,v)^{1/2} \text{ for all } u, v \in D\ . \end{array}$$

Clearly, (2.3) just says that $\mathcal{E}_1$ is continuous w.r.t to the norm $\tilde{\mathcal{E}}_1^{1/2}$ on D.

Exercise 2.1. Let $(\mathcal{E}, D)$ be as above, $(\mathcal{E}, D)$ positive definite.

(i) Prove that the following assertions are equivalent:

(a) $(\mathcal{E}, D)$ satisfies the weak sector condition.

(b) For every $\alpha > 0$ there exists $K_\alpha \in]0, \infty[$ such that

$$|\mathcal{E}_\alpha(u,v)| \leq K_\alpha\ \mathcal{E}_\alpha(u,u)^{1/2}\mathcal{E}_\alpha(v,v)^{1/2} \text{ for all } u, v \in D\ .$$

(c) For every $\alpha > 0$ there exists $K'_\alpha \in]0, \infty[$ such that

$$|\check{\mathcal{E}}(u,v)| \leq K'_\alpha\ \mathcal{E}_\alpha(u,u)^{1/2}\mathcal{E}_\alpha(v,v)^{1/2} \text{ for all } u, v \in D\ .$$

(ii) Show that K_α in (ii)(b) can be chosen independently of $\alpha > 0$ if and only if $(\mathcal{E}, D)$ satisfies the *(strong) sector condition*, i.e.,

(2.4) there exists $K \in]0,\infty[$ such that

$$|\mathcal{E}(u,v)| \leq K\ \mathcal{E}(u,u)^{1/2}\mathcal{E}(v,v)^{1/2} \text{ for all } u,v \in D\ .$$

Correspondingly, a positive definite linear operator $(L, D(L))$ on $\mathcal{H}$ is said to satisfy the *(strong) sector condition* if

(2.5) there exists a constant $K > 0$ such that

$$|(Lu,v)| \leq K(Lu,u)^{1/2}(Lv,v)^{1/2} \text{ for all } u,v \in D(L)\ .$$

Note that if L is *strictly positive definite* (i.e., there exists $c \in]0,\infty[$ such that $(Lu,u) \geq c(u,u)$ for all $u \in D(L)$) and bounded, then L satisfies (2.5).

Remark 2.2. The reason for calling (2.4) or (2.5) sector condition is given in 2.13, 2.14 below. If $\mathcal{E} = \tilde{\mathcal{E}}$ then (2.4) holds with $K = 1$. This is just Cauchy-Schwarz's inequality.

Definition 2.3. A pair $(\mathcal{E}, D(\mathcal{E}))$ is called a *symmetric closed form* (on $\mathcal{H}$) if $D(\mathcal{E})$ is a dense linear subspace of $\mathcal{H}$ and $\mathcal{E} : D(\mathcal{E}) \times D(\mathcal{E}) \to \mathbb{R}$ is a positive definite bilinear form which is *symmetric* (i.e., $\mathcal{E} = \tilde{\mathcal{E}}$) and *closed* on $\mathcal{H}$ (i.e., $D(\mathcal{E})$ is complete w.r.t. the norm $\mathcal{E}_1^{1/2}$).

Definition 2.4. A pair $(\mathcal{E}, D(\mathcal{E}))$ is called a *coercive closed form* (on $\mathcal{H}$) if $D(\mathcal{E})$ is a (dense) linear subspace of $\mathcal{H}$ and $\mathcal{E} : D(\mathcal{E}) \times D(\mathcal{E}) \to \mathbb{R}$ is a bilinear form such that the following two conditions hold.

(i) Its symmetric part $(\tilde{\mathcal{E}}, D(\mathcal{E}))$ is a symmetric closed form on $\mathcal{H}$.

(ii) $(\mathcal{E}, D(\mathcal{E}))$ satisfies the weak sector condition (2.3).

Below we shall use the following result due to G. Stampacchia (cf. [St 64]) in an essential way. From now on (unless otherwise stated) we consider $D(\mathcal{E})$ to be equipped with one of the equivalent norms $\tilde{\mathcal{E}}_\alpha^{1/2}$, $\alpha > 0$.

Theorem 2.5. *Let $(\mathcal{E}, D(\mathcal{E}))$ be a coercive closed form on $\mathcal{H}$ and let C be a non-empty closed convex subset of $D(\mathcal{E})$. Let J be a continuous linear functional on $D(\mathcal{E})$ and $\alpha > 0$. Then there exists a unique $v \in C$ such that*

(2.6)
$$\mathcal{E}_\alpha(v, w-v) \geq J(w-v) \quad \textit{for all} \quad w \in C\ .$$

The following two theorems give the connection between closed forms and strongly continuous contraction resolvents resp. semigroups on $\mathcal{H}$ and their generators (studied in the preceding section).

Theorem 2.6. *Let $(\mathcal{E}, D(\mathcal{E}))$ be a coercive closed form on $\mathcal{H}$ with continuity constant K. Then there exist unique strongly continuous contraction resolvents $(G_\alpha)_{\alpha>0}$, $(\hat{G}_\alpha)_{\alpha>0}$ on $\mathcal{H}$ such that*

$$G_\alpha(\mathcal{H}),\ \hat{G}_\alpha(\mathcal{H}) \subset D(\mathcal{E}) \quad \text{and} \quad \mathcal{E}_\alpha(G_\alpha f, u) = (f, u) = \mathcal{E}_\alpha(u, \hat{G}_\alpha f) \tag{2.7}$$
$$\text{for all } f \in \mathcal{H},\ u \in D(\mathcal{E}),\ \alpha > 0\ .$$

In particular,

$$(G_\alpha f, g) = (f, \hat{G}_\alpha g) \quad \text{for all} \quad f, g \in \mathcal{H}\ , \tag{2.8}$$

i.e., $\hat{G}_\alpha$ is the adjoint of G_α for all $\alpha > 0$. And if $(T_t)_{t>0}$, $(\hat{T}_t)_{t>0}$ denote the strongly continuous contraction semigroups corresponding to $(G_\alpha)_{\alpha>0}$, $(\hat{G}_\alpha)_{\alpha>0}$ respectively, then

$$(T_t f, g) = (f, \hat{T}_t g) \quad \text{for all} \quad f, g \in \mathcal{H}\ ,\ t > 0\ . \tag{2.9}$$

Proof. Consequence of 2.5. □

Remark 2.7. (i) Note that since $\|G_\alpha\| \leq \alpha^{-1}$, it follows from (2.7) that each G_α is a continuous linear operator from $\mathcal{H}$ to $D(\mathcal{E})$.
(ii) $(G_\alpha)_{\alpha>0}$, $(T_t)_{t>0}$ are called *resolvent* resp. *semigroup associated with* $(\mathcal{E}, D(\mathcal{E}))$ and $(\hat{G}_\alpha)_{\alpha>0}$, $(\hat{T}_t)_{t>0}$ are called *coresolvent* resp. *cosemigroup associated with* $(\mathcal{E}, D(\mathcal{E}))$. Below we shall prove a complete one-to-one correspondence between $(G_\alpha)_{\alpha>0}$ and $(\mathcal{E}, D(\mathcal{E}))$ (and hence between $(\hat{G}_\alpha)_{\alpha>0}$ and $(\mathcal{E}, D(\mathcal{E}))$). Therefore, we shall call the corresponding generators $L, \hat{L}$ of $(G_\alpha)_{\alpha>0}$, $(\hat{G}_\alpha)_{\alpha>0}$ respectively, the *generator* and *cogenerator* of $(\mathcal{E}, D(\mathcal{E}))$.

Now we turn to the question whether a resolvent $(G_\alpha)_{\alpha>0}$ (with adjoint $(\hat{G}_\alpha)_{\alpha>0}$) as above, uniquely determines the form $(\mathcal{E}, D(\mathcal{E}))$ via (2.7), and under which assumptions there always exists a coercive closed form $(\mathcal{E}, D(\mathcal{E}))$ on $\mathcal{H}$ satisfying (2.7) if $(G_\alpha)_{\alpha>0}$ is a given strongly continuous contraction resolvent on $\mathcal{H}$. We begin with the following fact whose proof is obvious.

Proposition 2.8. *Let $(\mathcal{E}, D(\mathcal{E}))$, $(G_\alpha)_{\alpha>0}$ be as in 2.6 and let $(L, D(L))$ be the generator of $(G_\alpha)_{\alpha>0}$. Then*

$$D(L) \subset D(\mathcal{E}) \quad \text{and} \quad \mathcal{E}(u, v) = (-Lu, v) \quad \text{for all } u \in D(L)\ ,\ v \in D(\mathcal{E})\ . \tag{2.10}$$

In particular, $1 - L$ satisfies the sector condition (2.5).
Corresponding statements hold for the generator $\hat{L}$ of $(\hat{G}_\alpha)_{\alpha>0}$.

Theorem 2.9. *Let $(\mathcal{E}, D(\mathcal{E}))$, $(G_\alpha)_{\alpha>0}$ be as in 2.6 and let $(L, D(L))$ be the generator of $(G_\alpha)_{\alpha>0}$.*

(i) Let $u \in \mathcal{H}$. Then $u \in D(\mathcal{E})$ if and only if $\sup_{\beta>0} \mathcal{E}^{(\beta)}(u, u) < \infty$.

(ii) $D(L)$ is dense in $D(\mathcal{E})$ and moreover, for all $u \in D(\mathcal{E})$

$$\lim_{\beta\to\infty} \mathcal{E}_1(\beta G_\beta u - u, \beta G_\beta u - u) = 0\ .$$

In particular, the closed form $(\mathcal{E}, D(\mathcal{E}))$ is uniquely determined by $(G_\alpha)_{\alpha>0}$ via (2.7).

(iii) $\lim_{\beta\to\infty} \mathcal{E}^{(\beta)}(u,v) = \mathcal{E}(u,v)$ *for all* $u, v \in D(\mathcal{E})$.

The proof of 2.9 is not difficult but a little technical. So, we omit it here and rather give a proof for the converse of 2.6 (cf. 2.11 below). We first need the following

Lemma 2.10. *Let* $(G_\alpha)_{\alpha>0}$ *be a strongly continuous contraction resolvent on* $\mathcal{H}$ *with generator* L. *Then the following are equivalent:*

(i) $(1-L)$ *satisfies the sector condition (2.5).*

(ii) G_α *satisfies (2.5) for one (resp. all)* $\alpha > 0$.

Proof. Exercise. □

Theorem 2.11. *Let* $(G_\alpha)_{\alpha>0}$ *be a strongly continuous contraction resolvent on* $\mathcal{H}$ *such that each* G_α *satisfies the sector condition (2.5) and let* L *be its generator. Define*

$$\mathcal{E}(u,v) := (-Lu, v)\ ;\quad u, v \in D(L)\ .$$

Let $D(\mathcal{E})$ *be the completion of* $D(L)$ *w.r.t.* $\tilde{\mathcal{E}}_1^{1/2}$ *and denote the unique bilinear extension of* $\mathcal{E}$ *to* $D(\mathcal{E})$ *which is continuous w.r.t.* $\tilde{\mathcal{E}}_1^{1/2}$ *again by* $\mathcal{E}$. *Then* $(\mathcal{E}, D(\mathcal{E}))$ *is a coercive closed form on* $\mathcal{H}$ *such that*

$$\mathcal{E}(u,v) = (-Lu, v) \text{ for all } u \in D(L)\ ,\ v \in D(\mathcal{E})\ . \tag{2.11}$$

Furthermore, $(G_\alpha)_{\alpha>0}$ *and* $(\mathcal{E}, D(\mathcal{E}))$ *are related by (2.7).*

Proof. If $(u_n)_{n\in\mathbb{N}}$ is a Cauchy sequence in $D(L)$ w.r.t. $\tilde{\mathcal{E}}_1^{1/2}$ such that $u_n \xrightarrow[n\to\infty]{} 0$ in $\mathcal{H}$, then by 2.10

$$\begin{aligned} 0 \le \mathcal{E}_1(u_n,u_n) &= \mathcal{E}_1(u_n-u_m,u_n-u_m) + \mathcal{E}_1(u_n-u_m,u_m) + ((1-L)u_m,u_n) \\ &\le \mathcal{E}_1(u_n-u_m,u_n-u_m) + ((1-L)u_m,u_n) \\ &\quad + K\ \mathcal{E}_1(u_n-u_m,u_n-u_m)^{1/2}\mathcal{E}_1(u_m,u_m)^{1/2} \end{aligned}$$

which can be made arbitrarily small for m, n large. Hence the map mapping every $\tilde{\mathcal{E}}_1^{1/2}$-Cauchy sequence in $D(L)$ to the corresponding Cauchy sequence i.e., limit in $\mathcal{H}$, gives rise to an inclusion of the completion $D(\mathcal{E})$ of $D(L)$ w.r.t. $\tilde{\mathcal{E}}_1^{1/2}$ into $\mathcal{H}$. The unique bilinear extension of $\mathcal{E}$ satisfying

$$|\mathcal{E}_1(u,v)| \le K\ \mathcal{E}_1(u,u)^{1/2}\mathcal{E}_1(v,v)^{1/2} \text{ for all } u, v \in D(\mathcal{E})$$

with domain $D(\mathcal{E})$ is then a closed form on $\mathcal{H}$. We then have that for every $\alpha > 0$, $G_\alpha(\mathcal{H}) = D(L) \subset D(\mathcal{E})$ and for all $u \in \mathcal{H}$, $v \in D(\mathcal{E})$

$$\mathcal{E}_\alpha(G_\alpha u, v) = (-L\ G_\alpha u, v) + \alpha(G_\alpha u, v) = (u, v)\ .$$

(2.11) follows now by (2.10). □

Finally, one can obtain L directly from $(\mathcal{E}, D(\mathcal{E}))$.

Proposition 2.12. *Let $(\mathcal{E}, D(\mathcal{E}))$ be a coercive closed form on $\mathcal{H}$. Define*

$$(2.12) \qquad D(L) := \{u \in D(\mathcal{E}) \mid v \mapsto \mathcal{E}(u,v) \text{ is continuous w.r.t. } (\ ,\ )^{1/2} \text{ on } D(\mathcal{E})\}.$$

For $u \in D(L)$, let Lu denote the unique element in $\mathcal{H}$ such that $(-Lu, v) = \mathcal{E}(u,v)$ for all $v \in D(\mathcal{E})$. Then L is the generator of the strongly continuous contraction resolvent $(G_\alpha)_{\alpha>0}$ which is related to $(\mathcal{E}, D(\mathcal{E}))$ by (2.7). In particular, $1-L$ satisfies the sector condition (2.5) and we have a one-to-one correspondence between such L and coercive closed forms $(\mathcal{E}, D(\mathcal{E}))$ specified by (2.10).

Proof. Let $(G_\alpha)_{\alpha>0}$ with generator $\tilde{L}$ be the strongly continuous contraction resolvent on $\mathcal{H}$ associated with $(\mathcal{E}, D(\mathcal{E}))$ by 2.6. Then by (2.10) it follows that $D(\tilde{L}) \subset D(L)$ and since $D(\mathcal{E})$ is dense in $\mathcal{H}$, that $\tilde{L} = L$ on $D(\tilde{L})$. But for $\alpha > 0$, $(\alpha - L)(D(L)) \supset (\alpha - L)(D(\tilde{L})) = (\alpha - \tilde{L})(D(\tilde{L})) = \mathcal{H}$ and $\alpha - L$ is one-to-one since $(\alpha - L)u = 0$ for some $u \in D(L)$ implies $((\alpha - L)u, u) = \mathcal{E}_\alpha(u,u) = 0$ and hence $u = 0$. Consequently, $D(L) = D(\tilde{L})$. □

So far we have not included the semigroup $(T_t)_{t>0}$ in our considerations of this section. In order to describe the additional property of $(T_t)_{t>0}$ which corresponds to the sector condition of $1 - L$ (where L is the generator) resp. to the sector condition of the resolvent operators G_α, we need to complexify.

Let $(\mathcal{H}_\mathbb{C}, (\ ,\ ))$ be the *complexification* of $(\mathcal{H}, (\ ,\ ))$, i.e., $\mathcal{H}_\mathbb{C} = \mathcal{H} \times \mathcal{H}$ with addition given by $[f_1, g_1] + [f_2, g_2] = [f_1 + f_2, g_1 + g_2]$, scalar multiplication given by $(a + ib)[f, g] = [af - bg, ag + bf]$, and inner product given by $([f_1, g_1], [f_2, g_2]) := (f_1, f_2) + i(f_1, g_2) - i(g_1, f_2) + (g_1, g_2)$. Again $\|\ \| := (\ ,\ )^{1/2}$ denotes the corresponding norm on $\mathcal{H}_\mathbb{C}$. Any linear operator $(L, D(L))$ on $\mathcal{H}$ can be "extended" to an operator $(L^\mathbb{C}, D(L^\mathbb{C}))$ on $\mathcal{H}_\mathbb{C}$ by

$$L^\mathbb{C}([u,v]) = [Lu, Lv] \text{ for } [u,v] \in D(L^\mathbb{C}) := \{[u,v] \in \mathcal{H}_\mathbb{C} \mid u, v \in D(L)\}\ .$$

Define for $K \in]0, \infty[$

$$S(K) := \{z \in \mathbb{C} \mid |\mathrm{Im} z| \le K \mathrm{Re} z\}\ .$$

The following is easily verified.

Proposition 2.13. *Let $(L, D(L))$ be a positive definite linear operator on $\mathcal{H}$. Then L satisfies the sector condition (2.5) if and only if*

$$(2.13) \qquad \{(L^\mathbb{C}([u,v]), [u,v]) \mid [u,v] \in D(L^\mathbb{C})\} \subset S(K) \text{ for some } K \in]0, \infty[\ .$$

Remark 2.14. The set on the left hand side of (2.13) is called the *numerical range* of $L^\mathbb{C}$. So, (2.13) means that the numerical range of $L^\mathbb{C}$ is contained in the *sector* $S(K)$. Hence 2.13 justifies the terminology sector condition for (2.5).

Definition 2.15. Let $K \in]0, \infty[$. A family $(T_z)_{z \in S(K)}$ of bounded linear operators on $\mathcal{H}_\mathbb{C}$ is called a *holomorphic contraction semigroup* on $S(K)$ if

(i) $T_0 = id_{\mathcal{H}_\mathbb{C}}$

(ii) $T_{z_1} T_{z_2} = T_{z_1 + z_2}$ for all $z_1, z_2 \in S(K)$.

(iii) $T_z f \to f$ in $\mathcal{H}_{\mathbb{C}}$ as $z \to 0$ in $S(\tilde{K})$ for all $f \in \mathcal{H}_{\mathbb{C}}$ and all $\tilde{K} < K$.

(iv) $\|T_z\| \leq 1$ for all $z \in S(K)$.

(v) $z \mapsto (T_z f, g)$ is analytic on the interior of $S(K)$ for all $f, g \in \mathcal{H}_{\mathbb{C}}$.

We need the following modification of the Hille-Yosida theorem.

Theorem 2.16. *For a linear operator $(L, D(L))$ on $\mathcal{H}_{\mathbb{C}}$ and $K \in]0,\infty[$ the following are equivalent:*

(i) L is the generator of a holomorphic contraction semigroup $(T_z)_{z\in S(K^{-1})}$ on $S(K^{-1})$, i.e., for all $\tilde{K} > K$, $Lu = \lim\limits_{\substack{z\to 0\\ z\in S(\tilde{K}^{-1})}} \frac{1}{z}(T_z u - u)$ for $u \in D(L)$ where

$$D(L) := \{v \in \mathcal{H}_{\mathbb{C}} \mid \lim_{\substack{z\to 0\\ z\in S(\tilde{K}^{-1})}} \frac{1}{z}(T_z v - v) \text{ exists.}\} \, .$$

(ii) L is m-dissipative of type $S(K^{-1})$, i.e., $1 \in \rho(L)$ and for all $\lambda \in \mathbb{C} \setminus (-S(K))$ and all $u \in D(L)$

$$\|(\lambda - L)u\| \geq \operatorname{dis}(\lambda, -S(K))\|u\|$$

where $\operatorname{dis}(\lambda, -S(K))$ denotes the distance from λ to $-S(K)$ in $\mathbb{C}$.

In this case for all $t > 0$, $m \in \mathbb{N}$, and $f \in \mathcal{H}$, one has $T_t f \in D(L^m)$ and $\|L^m T_t f\| \leq C\|f\|/t^m$ for some $C \in]0,\infty[$ independent of f and t.

For the proof of 2.16 we refer to [Go 85, Sect. 5], [C 75, Theorem 1.3 and Proposition 1.4], and [ReS 75, Corollary 2 of Theorem X.52]. Now we are prepared to identify the property of $(T_t)_{t>0}$ corresponding to the weak sector condition.

Corollary 2.17. *Let $(L, D(L))$ be the generator of a strongly continuous contraction semigroup on $\mathcal{H}$. Then the following are equivalent:*

(i) $1 - L$ satisfies the sector condition (2.5).

(ii) $(e^{-t}T_t^{\mathbb{C}})_{t>0}$ is the restriction of a holomorphic contraction semigroup on the sector $S(K^{-1})$ for some $K \in]0,\infty[$.

Proof. $(i) \Rightarrow (ii)$: Suppose (i) with constant K. Let $L' := (L-1)^{\mathbb{C}}$. Clearly, $1 \in \rho(L')$ and for all $u \in D(L') \setminus \{0\}$ and $\lambda \in \mathbb{C} \setminus (-S(K))$

$$\begin{aligned}
\|(\lambda - L')u\| &\geq \|u\|^{-1}|((\lambda - L')u, u)| \\
&= |\lambda + \|u\|^{-2}(-L'u, u)|\|u\| \\
&\geq \operatorname{dis}(\lambda, -S(K))\|u\|
\end{aligned}$$

by 2.13. Hence L' is m-dissipative of type $S(K^{-1})$ and thus generates a holomorphic contraction semigroup $(T'_z)_{z\in S(K^{-1})}$ on $S(K^{-1})$ by 2.16. Clearly, $(L-1)^{\mathbb{C}}$ is the generator of both $(T'_t)_{t>0}$ and $(e^{-t}T_t^{\mathbb{C}})_{t>0}$ and (ii) is proved.

$(ii) \Rightarrow (i)$: Let $(T'_z)_{z\in S(K^{-1})}$ be the holomorphic contraction semigroup on $S(K^{-1})$ extending $(e^{-t}T_t^{\mathbb{C}})_{t>0}$. Then $L' := (L-1)^{\mathbb{C}}$ is the generator of $(T'_z)_{z\in S(K^{-1})}$ (cf. [Go

85, Remark 5.7]), thus L' is m-dissipative of type $S(K^{-1})$ by 2.16. Hence for each $\alpha \in S(K^{-1})$ and each $u \in D(L') = D(L^{\mathbb{C}})$

$$\|(\frac{1}{\alpha t} - L')u\| \geq \frac{1}{\alpha t}\|u\| \quad \text{for all } t > 0 ,$$

i.e., $\|(1 - \alpha t L')u\| \geq \|u\|$ for all $t > 0$. Consequently,

$$\begin{aligned} 0 &\leq \frac{d}{dt}\Big|_{t=0} \sqrt{\|u\|^2 + |\alpha|^2 t^2 \|L'u\|^2 + 2t\mathrm{Re}(-\alpha L'u, u)} \\ &= \|u\|^{-1} Re(-\alpha L'u, u) . \end{aligned}$$

This implies that

$$\mathrm{Re}\alpha\,\mathrm{Re}(-L'u, u) \geq -\mathrm{Im}\alpha\,\mathrm{Im}(-L'u, u) .$$

Choosing $\alpha \in S(K^{-1})\backslash\{0\}$ such that $\mathrm{Im}\alpha = \pm K^{-1}\mathrm{Re}\alpha$, we obtain that

$$|\mathrm{Im}(-L'u, u)| \leq K\mathrm{Re}(-L'u, u)$$

i.e., $(-L'u, u) \in S(K)$. Now (i) follows by 2.13. □

We summarize what we have achieved so far in the Diagram 2 (extending Diagram 1 on p. 5) on p. 13. Now let us discuss the special symmetric case. All facts appearing below which follow from spectral theory on $\mathcal{H}$ will not be used in the sequel, but are included for completeness. So, if $\mathcal{E}(u, v) = \mathcal{E}(v, u)$ for all $u, v \in D(\mathcal{E})$, then the corresponding property of the generator L is symmetry and hence self-adjointness, since $(\alpha - L)(D(L)) = \mathcal{H}$ (cf. [ReS 72, Theorem VIII.3]). In fact, in this case properties (i)-(iii) in the above diagram can be replaced by the assumptions that L is negative definite and self-adjoint on $\mathcal{H}$ (cf. 1.13 and [ReS 72, Theorem VIII.3]). The corresponding property for $(G_\alpha)_{\alpha>0}$ is again that each G_α is symmetric. The most important observation, however, is that all sector conditions automatically hold by Cauchy-Schwarz's inequality.
By spectral theory one easily shows that $D(\sqrt{-L}) = D(\mathcal{E})$ and that

$$\mathcal{E}(u, v) = (\sqrt{-L}u, \sqrt{-L}v) \; ; \quad u, v \in D(\sqrt{-L}) .$$

Define for $t > 0$

$$^{(t)}\mathcal{E}(u, v) = \frac{1}{t}(u - T_t u, v) \; ; \quad u, v \in \mathcal{H} .$$

Then $u \in D(\mathcal{E})$ if and only if $\sup_{t>0} {}^{(t)}\mathcal{E}(u, u) < \infty$ and ${}^{(t)}\mathcal{E}(u, u) \uparrow \mathcal{E}(u, u)$ as $t \to 0$ in this case. We refer to [F 80, §1.3] for details.
The following lemma is a consequence of the well-known Banach-Saks resp. Banach-Alaoglu theorem (cf. e.g. [MR 92, Appendix]). It will be used many times throughout these lectures.

Diagram 2

$\mathcal{H}$ = Hilbert space over $\mathbb{R}$ with inner product $(\,,\,)$ and norm $\|\;\| := (\,,\,)^{1/2}$.

$(T_t)_{t>0}$ *strongly continuous contraction semigroup on* $\mathcal{H}$

$$G_\alpha = \int_0^\infty e^{-\alpha t} T_t \, dt$$

$$T_t = \lim_{\alpha \to \infty} e^{t\alpha(\alpha G_\alpha - 1)}, \; t > 0$$

"Hille-Yosida"

$(G_\alpha)_{\alpha>0}$ *strongly continuous contraction resolvent on* $\mathcal{H}$

$(e^{-t}T_t^{\mathbb{C}})_{t>0}$ *is the restriction of a holomorphic contraction semigroup on* $\mathcal{H}_{\mathbb{C}}$

$|(G_1 u, v)| \leq$
$\quad const \cdot (G_1 u, u)^{1/2}(G_1 v, v)^{1/2}$

"Hille-Yosida"
(via resolvent)

$L := \lim_{t \downarrow 0} \frac{1}{t}(T_t - 1)$

$\mathcal{E}(u,v) =$
$\lim_{\beta \to \infty} \beta(u - \beta G_\beta u, v)$

$\mathcal{E}_\alpha(G_\alpha u, v)$
$= (u, v),$
$u \in \mathcal{H},$
$v \in D(\mathcal{E})$

$L := \alpha - G_\alpha^{-1}$

$G_\alpha := (\alpha - L)^{-1},$
$\quad \alpha > 0$

$(L(D(L))$ *densely defined, (closed) linear operator on* $\mathcal{H}$
s.t.
  (i) $]0,\infty[\subset \rho(L)$
  (ii) $\|\alpha(\alpha - L)^{-1}\| \leq 1$
  (iii) $|((1-L)u, v)| \leq const \cdot$
    $((1-L)u, u)^{1/2} \cdot$
    $((1-L)v, v)^{1/2}$

$D(L) := \{u \in D(\mathcal{E}) \,|\, \exists Lu$
$\in \mathcal{H} \, s.t. \; \mathcal{E}(u,v) = (-Lu, v)$
$\quad \forall v \in D(\mathcal{E})\}$

$\mathcal{E}(u,v) := (-Lu, v),$
$\quad u, v \in D(L)$
$\quad \&\; completition$

$(\mathcal{E}, \mathcal{D}(\mathcal{E}))$ *coercive closed form on* $\mathcal{H}$

Lemma 2.18. *Let $(\mathcal{E}, D(\mathcal{E}))$ be a coercive closed form and $u_n \in D(\mathcal{E})$, $n \in \mathbb{N}$, such that*

$$\sup_{n\in\mathbb{N}} \mathcal{E}(u_n, u_n) < \infty .$$

If $u \in \mathcal{H}$ such that $u_n \to u$ in $\mathcal{H}$ as $n \to \infty$, then $u \in D(\mathcal{E})$ and $u_n \to u$ weakly in the Hilbert space $(D(\mathcal{E}), \tilde{\mathcal{E}}_1)$ and there exists a subsequence $(u_{n_k})_{k\in\mathbb{N}}$ of $(u_n)_{n\in\mathbb{N}}$ such that its Cesaro mean $w_n := \frac{1}{n}\sum_{k=1}^{n} u_{n_k} \to u$ in $D(\mathcal{E})$ as $n \to \infty$. Moreover,

$$\mathcal{E}(u,u) \le \liminf_{n\to\infty} \mathcal{E}(u_n, u_n) .$$

Proof. Since

$$\sup_{n\in\mathbb{N}} \mathcal{E}_1(u_n, u_n) < \infty ,$$

by the Banach-Alaoglu theorem there exists $v \in D(\mathcal{E})$ such that $u_{n_k} \to v$ weakly in $(D(\mathcal{E}), \tilde{\mathcal{E}}_1)$ as $n \to \infty$, for some subsequence $(n_k)_{k\in\mathbb{N}}$ of $(n)_{n\in\mathbb{N}}$. By the Banach-Saks theorem the Cesaro mean $(w_n)_{n\in\mathbb{N}}$ of a subsequence of $(u_{n_k})_{k\in\mathbb{N}}$ converges to v in $D(\mathcal{E})$, hence in $\mathcal{H}$. Since $w_n \to u$ in $\mathcal{H}$ as $n \to \infty$, $u = v$. Since this reasoning holds for every subsequence, $u_n \to u$ weakly in $(D(\mathcal{E}), \tilde{\mathcal{E}}_1)$ as $n \to \infty$. Furthermore,

$$\mathcal{E}(u,u) = \lim_{n\to\infty} \tilde{\mathcal{E}}(u, u_n) \le \liminf_{n\to\infty} (\tilde{\mathcal{E}}(u,u)^{1/2} \tilde{\mathcal{E}}(u_n, u_n)^{1/2})$$

and consequently,

$$\mathcal{E}(u,u)^{1/2} \le \liminf_{n\to\infty} \mathcal{E}(u_n, u_n)^{1/2}$$

and the last part of the assertion follows. □

3 Contraction properties

In this section we replace $\mathcal{H}$ by the concrete Hilbert space $L^2(E;m) := L^2(E;\mathcal{B};m)$ with usual inner product $(\ ,\ )$ where $(E;\mathcal{B};m)$ is a measure space. As usual we set for $u, v : E \to \mathbb{R}$

$$u \vee v := \sup(u,v),\ u \wedge v := \inf(u,v),\ u^+ := u \vee 0,\ u^- := -(u \wedge 0) .$$

We write $f \le g$ or $f < g$ for $f, g \in L^2(E;m)$ (or any m-classes f, g of functions on E) if the inequality holds m-a.e. for corresponding representatives.

Definition 3.1. (i) Let G be a bounded linear operator on $L^2(E;m)$ with $D(G) = L^2(E;m)$. G is called *sub-Markovian* if for all $f \in L^2(E;m)$, $0 \le f \le 1$ implies $0 \le Gf \le 1$. A strongly continuous contraction resolvent $(G_\alpha)_{\alpha>0}$ resp. semigroup $(T_t)_{t>0}$ is called *sub-Markovian* if all αG_α, $\alpha > 0$, resp. T_t, $t > 0$, are sub-Markovian.
(ii) A closed densely defined linear operator L on $L^2(E;m)$ is called *Dirichlet operator* if $(Lu, (u-1)^+) \le 0$ for all $u \in D(L)$.

Exercise 3.2. Let G be sub-Markovian. Prove that $Gf \ge 0$ for all $f \in L^2(E;m)$ with $f \ge 0$.

Proposition 3.3. *Let $(G_\alpha)_{\alpha>0}$ be a strongly continuous contraction resolvent on $L^2(E;m)$ with corresponding generator L and semigroup $(T_t)_{t>0}$. Then the following are equivalent.*

(i) $(G_\alpha)_{\alpha>0}$ *is sub-Markovian.*

(ii) $(T_t)_{t>0}$ *is sub-Markovian.*

(iii) L *is a Dirichlet operator.*

Proof. (cf. [BH 86]) *(i)* ⇒ *(ii)*: Since for all $u \in D(L)$, $t > 0$,

$$T_t u = \lim_{\alpha \to \infty} \exp(t\alpha(\alpha G_\alpha - 1))u$$

we have that $0 \le T_t u \le 1$ if $u \in D(L)$ with $0 \le u \le 1$. Since for $f \in L^2(E;m)$, $\beta G_\beta f \underset{\beta\to\infty}{\longrightarrow} f$ in $L^2(E;m)$ and $\beta G_\beta f \in D(L)$ and since βG_β is sub-Markovian, (ii) now easily follows.

(ii) ⇒ *(iii)*: Let G be any sub-Markovian contraction operator on $L^2(E;m)$, then for all $f \in L^2(E;m)$, since $f = (f-1)^+ + f \wedge 1$ and $G(f \wedge 1) \le G(|f| \wedge 1) \le 1$,

$$\begin{aligned}(Gf,(f-1)^+) &= (G(f-1)^+,(f-1)^+) + (G(f\wedge 1),(f-1)^+)\\ &\le ((f-1),(f-1)^+) + \int (f-1)^+ dm\\ &= (f,(f-1)^+)\,.\end{aligned}$$

Hence, for all $u \in D(L)$

$$(Lu,(u-1)^+) = \lim_{t\downarrow 0} \frac{1}{t}(T_t u - u,(u-1)^+) \le 0\,.$$

(iii) ⇒ *(i)*: Let $f \in L^2(E;m)$ and $v := \alpha G_\alpha f$. If $f \le 1$, then

$$\begin{aligned}\alpha(v,(v-1)^+) &= (\alpha v - Lv,(v-1)^+) + (Lv,(v-1)^+)\\ &\le \alpha(f,(v-1)^+) \le \alpha \int (v-1)^+ dm\,.\end{aligned}$$

Hence

$$\int ((v-1)^+)^2 dm \le 0$$

i.e., $v \le 1$. If $f \ge 0$, then $-nf \le 1$ hence $-nv \le 1$ for all $n \in \mathbb{N}$. Consequently, $v \ge 0$.
□

Theorem 3.4. *Suppose* $(\mathcal{E}, D(\mathcal{E}))$ *is a coercive closed form on* $L^2(E;m)$ *with continuity constant* K *and corresponding resolvent* $(G_\alpha)_{\alpha>0}$. *Then the following are equivalent:*

(i) For all $u \in D(\mathcal{E})$ *and* $\alpha \ge 0$, $u \wedge \alpha \in D(\mathcal{E})$ *and* $\mathcal{E}(u\wedge\alpha, u - u\wedge\alpha) \ge 0$

(ii) For all $u \in D(\mathcal{E})$, $u^+ \wedge 1 \in D(\mathcal{E})$ *and* $\mathcal{E}(u^+\wedge 1, u - u^+\wedge 1) \ge 0$.

(iii) For all $u \in D(\mathcal{E})$, $u^+ \wedge 1 \in D(\mathcal{E})$ *and* $\mathcal{E}(u + u^+\wedge 1, u - u^+\wedge 1) \ge 0$.

(iv) $(G_\alpha)_{\alpha>0}$ *is sub-Markovian.*

If $(G_\alpha)_{\alpha>0}$ *is replaced by its adjoint* $(\hat{G}_\alpha)_{\alpha>0}$ *the analogous equivalences hold with the two respective entries of* $\mathcal{E}(\cdot,\cdot)$ *interchanged.*

Proof. [MR 92, II.4.4] □

Definition 3.5. A coercive closed form $(\mathcal{E}, D(\mathcal{E}))$ on $L^2(E;m)$ is called a *Dirichlet form* if for all $u \in D(\mathcal{E})$, one has that

$$
\begin{aligned}
&u^+ \wedge 1 \in D(\mathcal{E}) \quad \text{and} \quad \mathcal{E}(u + u^+ \wedge 1, u - u^+ \wedge 1) \geq 0 \\
&\qquad\qquad\qquad\quad \text{and} \quad \mathcal{E}(u - u^+ \wedge 1, u + u^+ \wedge 1) \geq 0 \,.
\end{aligned} \tag{3.1}
$$

If $(\mathcal{E}, D(\mathcal{E}))$ is in addition symmetric in which case (3.1) is equivalent with

$$
\mathcal{E}(u^+ \wedge 1, u^+ \wedge 1) \leq \mathcal{E}(u,u) \,, \tag{3.2}
$$

it is called a *symmetric Dirichlet form.*

Remark 3.6. (i) If $(\mathcal{E}, D(\mathcal{E}))$ is a Dirichlet form, the pair $(\mathcal{E}, D(\mathcal{E}))$ is also sometimes called a *Dirichlet space.*
(ii) A coercive closed form satisfying only one of the two inequalities in (3.1) is sometimes called a $\frac{1}{2}$*-Dirichlet form* as e.g. in the final diagram of this chapter (cf. p. 18 below).

$u^+ \wedge 1$ is called the *unit contraction* of u. The following shows that there is a "smoothed" version of (3.1).

Proposition 3.7. *Let $(\mathcal{E}, D(\mathcal{E}))$ be a coercive closed form on $L^2(E;m)$.*

(i) Let $u \in D(\mathcal{E})$ and assume that

$$
\begin{aligned}
&\textit{for every } \epsilon > 0 \textit{ there exists } \varphi_\epsilon : \mathbb{R} \to [-\epsilon, 1+\epsilon] \textit{ such that} \\
&\varphi_\epsilon(t) = t \textit{ for all } t \in [0,1],\ 0 \leq \varphi_\epsilon(t_2) - \varphi_\epsilon(t_1) \leq t_2 - t_1 \textit{ if } t_1 \leq t_2, \\
&\varphi_\epsilon \circ u \in D(\mathcal{E}), \textit{ and } \liminf_{\epsilon \to 0} \mathcal{E}(u \pm \varphi_\epsilon \circ u, u \mp \varphi_\epsilon \circ u) \geq 0.
\end{aligned} \tag{3.3}
$$

Then (3.1) holds.

(ii) $(\mathcal{E}, D(\mathcal{E}))$ is a Dirichlet form if and only if (3.3) holds for all $u \in D(\mathcal{E})$.

Proof. (i): Observe that by adding the two inequalities in (3.3) we obtain that

$$
\limsup_{\epsilon \to 0} \mathcal{E}(\varphi_\epsilon \circ u, \varphi_\epsilon \circ u) \leq \mathcal{E}(u,u) \,.
$$

Since clearly, $\varphi_\epsilon \circ u \underset{\epsilon \to 0}{\longrightarrow} u^+ \wedge 1$ in $L^2(E;m)$, this implies that $\varphi_{\epsilon_n} \circ u \underset{n \to \infty}{\longrightarrow} u^+ \wedge 1$ weakly in $(D(\mathcal{E}), \tilde{\mathcal{E}}_1)$ for some subsequence $\epsilon_n \downarrow 0$ and $\mathcal{E}(u^+ \wedge 1, u^+ \wedge 1) \leq \liminf_{n \to \infty} \mathcal{E}(\varphi_{\epsilon_n} \circ u, \varphi_{\epsilon_n} \circ u)$ (by 2.18). Hence by (3.3)

$$
\begin{aligned}
\mathcal{E}(u \pm u^+ \wedge 1, u \mp u^+ \wedge 1) &\geq \mathcal{E}(u,u) \mp \lim_{n \to \infty} \mathcal{E}(u, \varphi_{\epsilon_n} \circ u) \\
&\quad \pm \lim_{n \to \infty} \mathcal{E}(\varphi_{\epsilon_n} \circ u, u) - \liminf_{n \to \infty} \mathcal{E}(\varphi_{\epsilon_n} \circ u, \varphi_{\epsilon_n} \circ u) \\
&= \limsup_{n \to \infty} \mathcal{E}(u \pm \varphi_{\epsilon_n} \circ u, u \mp \varphi_{\epsilon_n} \circ u) \\
&\geq 0 \,.
\end{aligned}
$$

(ii): It is obvious that the condition is necessary since we can take $\varphi_\epsilon(t) := (t \vee 0) \wedge 1$ for all $\epsilon > 0$. The sufficiency follows by (i). □

Proposition 3.8. *A coercive closed form on $L^2(E;m)$ is a Dirichlet form if and only if (3.1) or (3.3) hold for all u in a dense subset of $D(\mathcal{E})$.*

Proposition 3.9. *Let $(\mathcal{E}, D(\mathcal{E}))$ be a Dirichlet form on $L^2(E;m)$. Let $u_1, \ldots, u_n \in D(\mathcal{E})$ and $u \in L^2(E;m)$ such that (for some m-versions) $|u(x)| \leq \sum_{k=1}^{n} |u_k(x)|$ and $|u(x) - u(y)| \leq \sum_{k=1}^{n} |u_k(x) - u_k(y)|$ for all $x, y \in E$, then $u \in D(\mathcal{E})$ and $\mathcal{E}(u,u)^{1/2} \leq \sum_{k=1}^{n} \mathcal{E}(u_k, u_k)^{1/2}$. In particular, $D(\mathcal{E})$ is stable under taking $|\cdot|$, $\wedge$, $\vee$.*

Remark 3.10. (i) If $n = 1$ in 3.9, u is called a *normal contraction* of u_1.
(ii) Note that if $1 \in D(\mathcal{E})$ such that $\mathcal{E}(u, 1) = 0$ for all $u \in D(\mathcal{E})$, then 3.9 remains true for $u \in L^2(E;m)$ and $u_1, \ldots, u_n \in D(\mathcal{E})$ such that (for some m-versions) $|u(x) - u(y)| \leq \sum_{k=1}^{n} |u_k(x) - u_k(y)|$ for all $x, y \in E$.

We close this section with a result which will become very useful later.

Proposition 3.11. *Let $(\mathcal{E}, D(\mathcal{E}))$ be a coercive closed form on $L^2(E;m)$ and let $u \in D(\mathcal{E})$. Then:*

(i) $(u \wedge n) \vee (-n) \xrightarrow[n\to\infty]{} u$ *in* $D(\mathcal{E})$,

(ii) $(u \wedge \varepsilon) \vee (-\varepsilon) \xrightarrow[\varepsilon\to 0]{} 0$ *in* $D(\mathcal{E})$.

Proof. Exercise (Hint: Use 2.18). □

The following diagram summarizes the main achievements of this chapter and completes Diagram 2 on p. 13 for $\mathcal{H} := L^2(E;m)$.

Diagram 3

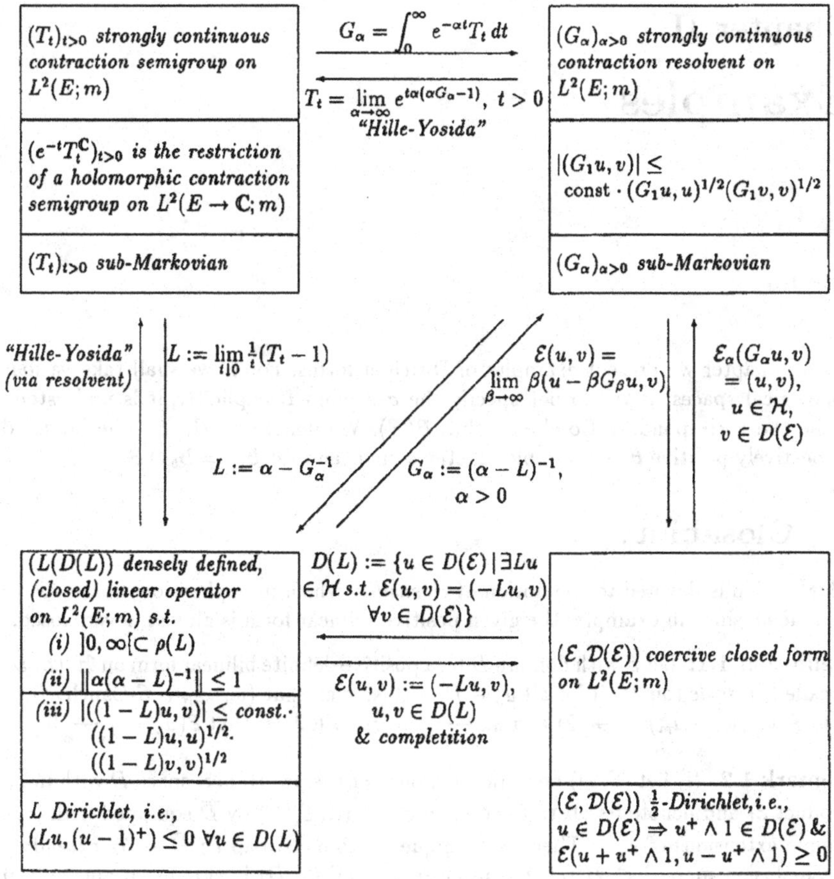
$(T_t)_{t>0}$ strongly continuous contraction semigroup on $L^2(E;m)$
$(e^{-t}T_t^{\mathbb{C}})_{t>0}$ is the restriction of a holomorphic contraction semigroup on $L^2(E \to \mathbb{C}; m)$
$(T_t)_{t>0}$ sub-Markovian
$G_\alpha = \int_0^\infty e^{-\alpha t} T_t \, dt$
$T_t = \lim_{\alpha\to\infty} e^{t\alpha(\alpha G_\alpha - 1)}$, $t > 0$
"Hille-Yosida"
$(G_\alpha)_{\alpha>0}$ strongly continuous contraction resolvent on $L^2(E;m)$
$|(G_1 u, v)| \leq$ const $\cdot (G_1 u, u)^{1/2} (G_1 v, v)^{1/2}$
$(G_\alpha)_{\alpha>0}$ sub-Markovian
"Hille-Yosida" (via resolvent)
$L := \lim_{t\downarrow 0} \frac{1}{t}(T_t - 1)$
$L := \alpha - G_\alpha^{-1}$
$G_\alpha := (\alpha - L)^{-1}$, $\alpha > 0$
$\mathcal{E}(u,v) = \lim_{\beta\to\infty} \beta(u - \beta G_\beta u, v)$
$\mathcal{E}_\alpha(G_\alpha u, v) = (u,v)$, $u \in \mathcal{H}$, $v \in D(\mathcal{E})$
$(L, D(L))$ densely defined, (closed) linear operator on $L^2(E;m)$ s.t.
(i) $]0,\infty[\subset \rho(L)$
(ii) $\|\alpha(\alpha - L)^{-1}\| \leq 1$
(iii) $|((1-L)u, v)| \leq$ const. $\cdot ((1-L)u,u)^{1/2} \cdot ((1-L)v,v)^{1/2}$
L Dirichlet, i.e., $(Lu, (u-1)^+) \leq 0 \; \forall u \in D(L)$
$D(L) := \{u \in D(\mathcal{E}) \,|\, \exists Lu \in \mathcal{H}$ s.t. $\mathcal{E}(u,v) = (-Lu, v) \; \forall v \in D(\mathcal{E})\}$
$\mathcal{E}(u,v) := (-Lu, v)$, $u, v \in D(L)$ & completition
$(\mathcal{E}, \mathcal{D}(\mathcal{E}))$ coercive closed form on $L^2(E;m)$
$(\mathcal{E}, \mathcal{D}(\mathcal{E}))$ $\frac{1}{2}$-Dirichlet, i.e., $u \in D(\mathcal{E}) \Rightarrow u^+ \wedge 1 \in D(\mathcal{E})$ & $\mathcal{E}(u + u^+ \wedge 1, u - u^+ \wedge 1) \geq 0$

Chapter II

Examples

In this chapter we present examples of Dirichlet forms. For E we shall take various topological spaces. If we do not specify the σ-algebra $\mathcal{B}$ explicitly, it is understood to be the corresponding Borel-σ-algebra $\mathcal{B}(E)$. We denote by $\mathcal{B}_b$, $\mathcal{B}^+$ the bounded respectively positive $\mathcal{B}$-measurable functions on E and set $\mathcal{B}_b^+ := \mathcal{B}_b \cap \mathcal{B}^+$.

1 Closability

This section is devoted to the notion of *closability* which in applications turns out to be crucial, since in examples the given positive bilinear form is almost never closed.

Definition 1.1. Let $\mathcal{E}$ with domain D be a positive definite bilinear form on $\mathcal{H}$. $(\mathcal{E}, D)$ is called *closable* (on $\mathcal{H}$) if for all $u_n \in D$, $n \in \mathbb{N}$, such that $(u_n)_{n\in\mathbb{N}}$ is $\mathcal{E}$-Cauchy (i.e., $\mathcal{E}(u_n - u_m, u_n - u_m) \underset{n,m\to\infty}{\longrightarrow} 0$) and $u_n \underset{n\to\infty}{\longrightarrow} 0$ in $\mathcal{H}$, it follows that $\mathcal{E}(u_n, u_n) \underset{n\to\infty}{\longrightarrow} 0$.

Remark 1.2. (i) Let $(\mathcal{E}, D)$ be as in 1.1. Consider the pre-Hilbert space D with inner product $\tilde{\mathcal{E}}_1$ and denote its abstract completion (w.r.t. $\tilde{\mathcal{E}}_1^{1/2}$) by $\overline{D}$ equipped with (the unique extension) of $\tilde{\mathcal{E}}_1^{1/2}$. There is a unique continuous map $i : \overline{D} \to \mathcal{H}$ extending the inclusion map $D \subset \mathcal{H}$ (cf. the proof of I.2.11). $(\mathcal{E}, D)$ is closable if and only if $i : \overline{D} \to \mathcal{H}$ is injective.
(ii) If $(\mathcal{E}, D)$ is closable, then its symmetric part $\tilde{\mathcal{E}}$ extends uniquely to $\overline{D}$ (cf. (i)). Suppose $(\mathcal{E}, D)$ satisfies the weak sector condition I.(2.3), then also $\mathcal{E}$ extends uniquely to $\overline{D}$. (Note that we already used this fact in the proof of I.2.11). We denote this extension again by $\mathcal{E}$. Clearly if D is dense in $\mathcal{H}$, $(\mathcal{E}, \overline{D})$ is the smallest coercive closed form (in the sense of I.2.4) extending $(\mathcal{E}, D)$. $(\mathcal{E}, \overline{D})$ is called the *closure* of $(\mathcal{E}, D)$.

Exercise 1.3. Let S be a negative definite linear operator on $\mathcal{H}$ such that $1 - S$ satisfies the sector condition I.(2.5). Define

$$\mathcal{E}(u, v) := (-Su, v)\ ;\quad u, v \in D(S)\ .$$

Then $(\mathcal{E}, D(S))$ is closable on $\mathcal{H}$. In particular, this holds for any negative definite linear operator on $\mathcal{H}$ which is symmetric (i.e., $(Su, v) = (u, Sv)$ for all $u, v \in D(S)$).

2 Starting point: operator

a) (a_{ij})-case

Let $E := U \subset \mathbb{R}^d$, $d \geq 1$, U open, and $m = dx :=$ *Lebesgue measure* on U. Let $a_{ij} : U \to \mathbb{R}$, $1 \leq i, j \leq d$, such that

(2.1) $\quad a_{ij} = a_{ji}$ for all $1 \leq i,j \leq d$ and $\sum_{i,j=1}^d a_{ij}(x)\xi_i\xi_j \geq 0$ for all $\xi_1, \ldots, \xi_d \in \mathbb{R}$, dx-a.e. $x \in U$.

(2.2) $\quad a_{ij} \in L^2_{loc}(U, dx), \frac{\partial}{\partial x_i} a_{ij} \in L^2_{loc}(U; dx), 1 \leq i, j \leq d,$

where the derivatives are taken in the sense of Schwartz distributions. Define the linear operator S on $L^2(U; dx)$ by

$$Su = \sum_{i,j=1}^d \frac{\partial}{\partial x_i}\left(a_{ij}\frac{\partial}{\partial x_j}\right) u\ , \quad u \in D(S) := C_0^\infty(U)\ . \tag{2.3}$$

Here $C_0^\infty(U)$ denotes the set of all infinitely differentiable functions on U with compact support. Note that (2.2) is necessary to have $Su \in L^2(U; dx)$ for every $u \in C_0^\infty(U)$. Define the positive definite bilinear form

$$\mathcal{E}(u, v) := (-Su, v) = \sum_{i,j=1}^d \int \frac{\partial u}{\partial x_i}\frac{\partial v}{\partial x_j} a_{ij} dx\ , \quad u, v \in C_0^\infty(U)\ .$$

Then by 1.3, $(\mathcal{E}, C_0^\infty(U))$ is closable on $L^2(U; dx)$. Since $C_0^\infty(U)$ is dense in $L^2(U; dx)$, its closure $(\mathcal{E}, D(\mathcal{E}))$ is a symmetric closed form on $L^2(U; dx)$, which is in fact a Dirichlet form since it is a special case of Subsection 3c) below.

b) Classical case with "minimal" domain

Consider a) with $a_{ij} := \frac{1}{2}\delta_{ij}$, i.e., $S = \frac{1}{2}\Delta$ with domain $C_0^\infty(U)$. We denote the corresponding symmetric closed form by $\mathbb{D}$, and its domain by $H_0^{1,2}(U)$ (since the completion of $C_0^\infty(U)$ w.r.t. $\mathbb{D}_1^{1/2}$ is by definition the $(1,2)$-*Sobolev space on U with Dirichlet boundary conditions*).

c) Powers of the Laplacian

Let $E = \mathbb{R}^d$, $m = dx$ and let " $\hat{}$ " resp. " $\check{}$ " denote *Fourier transform*, i.e.,

$$\hat{f}(x) = (2\pi)^{-d/2} \int \exp[i < x, y >_{\mathbb{R}^d}] f(y) dy\ ,$$

resp. its inverse. Define for $\alpha > 0$

$$(-\Delta)^\alpha u := \left(|x|^{2\alpha}\hat{u}\right)^\vee \ (\in L^2(\mathbb{R}^d; dx))\ ;\ u \in C_0^\infty(\mathbb{R}^d)\ .$$

Then $(-\Delta)^\alpha$ is a symmetric linear operator on $L^2(\mathbb{R}^d; dx)$ with dense domain $C_0^\infty(\mathbb{R}^d)$. Hence the form

$$\mathbb{D}^{(\alpha)}(u,v) := \frac{1}{2}\int \hat{u}\,\bar{\hat{v}}\,|x|^{2\alpha}\,dx \ , \ u,v \in C_0^\infty(\mathbb{R}^d) \ ,$$

is closable by 1.3, where "$-$" means complex conjugation. Its closure $(\mathbb{D}^{(\alpha)}, H^{\alpha,2}(\mathbb{R}^d))$ is hence a symmetric closed form on $L^2(\mathbb{R}^d; dx)$. If $0 < \alpha \leq 1$, it is a Dirichlet form by Subsection 3c) below and the following fact whose proof can be found in [Wl 82,p.97]. If $0 < \alpha < 1$, then $u \in H^{\alpha,2}(\mathbb{R}^d)$ if and only if

$$\iint \frac{|u(x)-u(y)|^2}{|x-y|^{2\alpha+d}}\,dxdy < \infty$$

and for $u,v \in H^{\alpha,2}(\mathbb{R}^d)$

$$\mathcal{E}(u,v) = c_{\alpha,d} \iint \frac{(u(x)-u(y))(v(x)-v(y))}{|x-y|^{2\alpha+d}}\,dxdy$$

for some constant $c_{\alpha,d} > 0$ (independent of u,v).

3 Starting point: bilinear form – finite dimensional case

a) Diagonal case

Let $E := U \subset \mathbb{R}^d$, U open, $m := \sigma \cdot dx$ for some $\sigma \in L^1_{loc}(U;dx)$, $\sigma \geq 0$ $dx-a.e.$ such that

(3.1) $$\int_V \sigma\ dx > 0 \ \text{ for all } \ V \subset U \ , V \text{ open} \ .$$

Let $\underline{\rho} := (\rho_1, \dots, \rho_d)$ with $\rho_i \in L^1_{loc}(U;dx)$, $\rho_i \geq 0$ $dx-a.e.$ and define for $u,v \in C_0^\infty(U)$

(3.2) $$\mathcal{E}_{\underline{\rho}}(u,v) := \sum_{i=1}^d \int \frac{\partial u}{\partial x_i}\frac{\partial v}{\partial x_i}\rho_i dx \ .$$

Then $\left(\mathcal{E}_{\underline{\rho}}, C_0^\infty(U)\right)$ is a densely defined symmetric positive definite bilinear form on $L^2(U;\sigma\cdot dx)$. We want to give conditions on ρ_i, σ so that $(\mathcal{E}_{\underline{\rho}}, C_0^\infty(U))$ is closable on $L^2(U;\sigma\cdot dx)$.
Define for $\rho \in \mathcal{B}^+(U)$

(3.3) $$R(\rho) := \left\{ x \in U \mid \int_{\{y\in U \mid |x-y|\leq\varepsilon\}} \rho^{-1}(y)dy < \infty \ \text{ for some } \ \varepsilon > 0 \right\} .$$

Here we use the convention that $\frac{a}{0} := (\text{sign} a)\cdot\infty$. Note that $R(\rho)$ is open and that $\rho > 0$ $dx - a.e.$ on $R(\rho)$. Obviously, $R(\rho)$ is the largest open set $V \subset U$ such that $\rho^{-1} \in L^1_{loc}(V;dx)$. Consider the following condition on ρ

(3.4) $$\rho = 0 \ dx - a.e \ . \text{ on } \ U \setminus R(\rho) \ .$$

Remark 3.1. Let $\rho : U \to [0,\infty[$ be Borel-measurable; then (3.4) is equivalent with

$$\text{(3.5)} \qquad \begin{array}{l} \text{for } dx\text{-a.e. } x \in \{\rho > 0\} \\ \qquad \int_{\{y \in U \mid |x-y| \le \varepsilon\}} \rho^{-1}(y)\, dy < \infty \text{ for some } \varepsilon > 0 \,. \end{array}$$

In particular, if ρ is lower semicontinuous or more generally if

$$\text{(3.6)} \qquad \begin{array}{l} \text{for } dx\text{-a.e. } x \in \{\rho > 0\} \\ \qquad \operatorname{ess\,inf}\{\rho(y) \mid |y-x| \le \varepsilon\} > 0 \text{ for some } \varepsilon > 0 \,, \end{array}$$

then ρ satisfies (3.5).

Lemma 3.2. *Let $\rho \in \mathcal{B}^+(U)$ satisfying (3.4). Then*

$$L^2(U; \rho \cdot dx) \subset L^1_{loc}(R(\rho); dx)$$

continuously.

Proof. Let $u \in L^2(U; \rho \cdot dx)$ and $K \subset R(\rho)$ compact. Then by the Cauchy-Schwarz inequality

$$\int_K |u| dx = \int_K |u| \rho\, \rho^{-1} dx \le \left(\int u^2 \rho\, dx\right)^{1/2} \left(\int_K \rho^{-1} dx\right)^{1/2} .$$

But

$$\int_K \rho^{-1}\, dx < \infty$$

since $K \subset R(\rho)$. □

Suppose that σ and $\rho_1, \ldots, \rho_d$ satisfy (3.4) and that $dx(R(\rho_i) \setminus R(\sigma)) = 0$. Then $(\mathcal{E}_{\underline{\rho}}, C_0^\infty(U))$ is closable on $L^2(U; \sigma \cdot dx)$ (cf. [RW 85]). To prove this, fix $1 \le i \le d$ and let $u_n \in C_0^\infty(U)$, $n \in \mathbb{N}$, such that $(u_n)_{n\in\mathbb{N}}$ is $\mathcal{E}_{\underline{\rho}}$-Cauchy and $u_n \xrightarrow[n\to\infty]{} 0$ in $L^2(U; \sigma \cdot dx)$, then $u_n \xrightarrow[n\to\infty]{} 0$ in $L^1_{loc}(R(\sigma); dx)$ by 3.2. Since $\frac{\partial u_n}{\partial x_i} \xrightarrow[n\to\infty]{} f_i$ in $L^2(U; \rho_i \cdot dx)$ for some $f_i \in L^2(U; \rho_i \cdot dx)$ we have that $\frac{\partial u_n}{\partial x_i} \xrightarrow[n\to\infty]{} f_i$ in $L^1_{loc}(R(\rho_i); dx)$ by 3.2. Hence, if $v \in C_0^\infty(R(\rho_i) \cap R(\sigma))$ then

$$\int f_i v\, dx = \lim_{n\to\infty} \int_{R(\rho_i)} \frac{\partial u_n}{\partial x_i} v\, dx = -\lim_{n\to\infty} \int_{R(\sigma)} u_n \frac{\partial v}{\partial x_i}\, dx = 0\,.$$

Hence $f_i = 0$ dx-a.e. on $R(\rho_i) \cap R(\sigma)$, hence $f_i = 0$ $(\rho_i \cdot dx)$ – a.e. Consequently,

$$\lim_{n\to\infty} \mathcal{E}_{\underline{\rho}}(u_n, u_n) = 0\,.$$

Remark 3.3. The above arguments can also be used to prove the closability of

$$\mathcal{E}(u,v) = \int \langle \nabla u, \nabla v \rangle_x \rho\, dx\ ;\ u, v \in C_0^\infty(U)$$

on $L^2(U; \sigma \cdot dx)$ if U is a Riemannian manifold with volume element dx and inner product $\langle\, ,\, \rangle_x$ on the tangent space at x (cf. [ABR 89, Theorem 4.2]).

In the above case the closure of $(\mathcal{E}_\rho, C_0^\infty(U))$ on $L^2(U; \sigma \cdot dx)$ is a symmetric closed form which is a Dirichlet form by Subsection 3c) below.

b) (a_{ij})-case

Let $E := U$ be as in Subsection a) and let $a_{ij} \in L^1_{loc}(U; dx)$, $a_{ij} = a_{ji}$, $1 \le i, j \le d$. Assume that there exist $\sigma, \rho_i \in L^1_{loc}(U; dx)$, $\rho_i, \sigma > 0$ dx-a.e. such that

(3.7) $$(\mathcal{E}_{\underline{\rho}}, C_0^\infty(U)) \text{ is closable on } L^2(U; \sigma \cdot dx)$$

(cf. Subsection 3a)) and such that

(3.8) $$\text{for } dx\text{-a.e. } x \in U \quad \sum_{i,j=1}^d a_{ij}(x)\xi_i\xi_j \ge \sum_{i=1}^d \rho_i(x)\xi_i^2 \text{ for all } \xi_1, \dots, \xi_d \in \mathbb{R}\,.$$

Then the form

$$\mathcal{E}(u,v) := \sum_{i,j=1}^d \int \frac{\partial u}{\partial x_i}\frac{\partial v}{\partial x_j} a_{ij}\, dx\ ;\ u, v \in C_0^\infty(U)\,,$$

is closable on $L^2(U; \sigma \cdot dx)$ (cf. [RW 85]). Indeed, if $u_n \in C_0^\infty(U)$, $n \in \mathbb{N}$, such that $(u_n)_{n\in\mathbb{N}}$ is $\mathcal{E}$-Cauchy and $u_n \underset{n\to\infty}{\longrightarrow} 0$ in $L^2(U; \sigma \cdot dx)$ then by (3.8), $(u_n)_{n\in\mathbb{N}}$ is $\mathcal{E}_{\underline{\rho}}$-Cauchy, hence $\lim_{n\to\infty} \mathcal{E}_{\underline{\rho}}(u_n, u_n) = 0$ by (3.7) and we can find a subsequence $(u_{n_k})_{k\in\mathbb{N}}$ such that $\frac{\partial}{\partial x_i} u_{n_k} \underset{k\to\infty}{\longrightarrow} 0$ dx-a.e. for all $1 \le i \le d$. Hence by Fatou's Lemma and (3.8)

$$\begin{aligned}
\mathcal{E}(u_n, u_n) &= \int \lim_{k\to\infty} \sum_{i,j=1}^d \left(\frac{\partial}{\partial x_i}(u_n - u_{n_k})\frac{\partial}{\partial x_j}(u_n - u_{n_k})a_{ij}\right) dx \\
&\le \liminf_{k\to\infty} \mathcal{E}(u_n - u_{n_k}, u_n - u_{n_k})
\end{aligned}$$

which is arbitrarily small if n is large. The closure of $(\mathcal{E}, C_0^\infty(U))$ is a Dirichlet form by Subsection 3c) below.

Exercise 3.4. Prove that $(\mathcal{E}, C_0^\infty(U))$ is in particular closable on $L^2(U; \sigma \cdot dx)$ if for every $K \subset U$, K compact, there exists a constant $c_K > 0$ such that

(3.9) $$\text{for } dx\text{-a.e. } x \in K \quad \sum_{i,j=1}^d a_{ij}(x)\xi_i\xi_j \ge c_K \|\xi\|^2_{\mathbb{R}^d} \text{ for all } \xi = (\xi_1, \dots, \xi_d) \in \mathbb{R}^d$$

(local (strict) ellipticity).

c) General regular symmetric case

Let $E := U$ be as in Subsection a) and m a positive Radon measure on U such that $\operatorname{supp}[m] = U$. Define for $u, v \in C_0^\infty(U)$

(3.10) $$\begin{aligned}
\mathcal{E}(u,v) &:= \sum_{i,j=1}^d \int \frac{\partial u}{\partial x_i}\frac{\partial v}{\partial x_j}\, d\nu_{ij} \\
&\quad + \int_{U\times U\setminus\Delta} (u(x) - u(y))(v(x) - v(y)) J(dx, dy) \\
&\quad + \int uv\, dk\,.
\end{aligned}$$

Here k is a positive Radon measure on U and J is a symmetric positive Radon measure on $U \times U \setminus \Delta$, where $\Delta := \{(x,x) | x \in U\}$, such that for all $u \in C_0^\infty(U)$

$$\int |u(x) - u(y)|^2 J(dxdy) < \infty . \tag{3.11}$$

For $1 \le i,j \le d$, ν_{ij} is a Radon measure on U such that

$$\begin{array}{c} \text{for every } K \subset U, K \text{ compact, } \nu_{ij}(K) = \nu_{ji}(K) \text{ and} \\ \sum_{i,j=1}^d \xi_i \xi_j \nu_{ij}(K) \ge 0 \text{ for all } \xi_1, \ldots, \xi_d \in \mathbf{R}^d . \end{array} \tag{3.12}$$

Then $(\mathcal{E}, C_0^\infty(U))$ is a densely defined symmetric positive definite bilinear form on $L^2(U;m)$.

Suppose that $(\mathcal{E}, C_0^\infty(U))$ is closable on $L^2(U;m)$ and let $(\mathcal{E}, D(\mathcal{E}))$ be its closure, then $(\mathcal{E}, D(\mathcal{E}))$ is a Dirichlet form. In order to prove this we need the following

Exercise 3.5. Show that for each $\varepsilon > 0$ there exists an infinitely differentiable function $\varphi_\varepsilon : \mathbf{R} \to [-\varepsilon, 1+\varepsilon]$ such that $\varphi_\varepsilon(t) = t$ for $t \in [0,1]$, $0 \le \varphi_\varepsilon(t) - \varphi_\varepsilon(s) \le t - s$ for all $t, s \in \mathbf{R}$, $t \ge s$, $\varphi_\varepsilon(t) = 1 + \varepsilon$ for $t \in [1+2\varepsilon, \infty[$ and $\varphi_\varepsilon(t) = -\varepsilon$ for $t \in]-\infty, -2\varepsilon]$.

Clearly, if $u \in C_0^\infty(U)$ and φ_ε is as in 3.5 then $\varphi_\varepsilon(u) \in C_0^\infty(U)$ and by the chain rule

$$\begin{aligned} \mathcal{E}(\varphi_\varepsilon(u), \varphi_\varepsilon(u)) &= \sum_{i,j=1}^d \int |\varphi_\varepsilon'(u)|^2 \frac{\partial u}{\partial x_i} \frac{\partial u}{\partial x_j} \, d\nu_{ij} \\ &\quad + \int_{U \times U \setminus \Delta} |\varphi_\varepsilon(u)(x) - \varphi_\varepsilon(u)(y)|^2 \, J(dxdy) + \int \varphi_\varepsilon(u)^2 \, dk \\ &\le \mathcal{E}(u,u) . \end{aligned}$$

By I.3.8 it follows that $(\mathcal{E}, D(\mathcal{E}))$ is a symmetric Dirichlet form. Observe that all examples considered above are of type (3.10), hence their closures are all Dirichlet forms. This is not a coincidence since conversely we have the following

Theorem 3.6. (Beurling-Deny formula) *Suppose $(\mathcal{E}, D(\mathcal{E}))$ is a symmetric Dirichlet form on $L^2(U;m)$ such that $C_0^\infty(U) \subset D(\mathcal{E})$. Then $(\mathcal{E}, C_0^\infty(U))$ can be expressed uniquely as in (3.10).*

For the proof we refer to[F 80, Section 2.2]. The three summands in (3.10) are called *diffusion part*, *jump part* and *killing part* respectively.

d) Non-symmetric cases

Let $d \ge 3$ and $E := U$, $U \subset \mathbf{R}^d$, open, and $m = dx$ on U. Let $a_{ij}, b_i, d_i, c \in L^1_{loc}(U; dx)$, $1 \le i,j \le d$, such that

$$\begin{array}{c} \text{for } dx\text{-a.e. } x \in U \\ \sum_{i,j=1}^d a_{ij}(x) \xi_i \xi_j \ge 0 \text{ for all } \xi = (\xi_1, \ldots, \xi_d) \in \mathbf{R}^d \end{array} \tag{3.13}$$

(i.e., *ellipticity*) and

(3.14) $c\,dx - \sum_{i=1}^{d} \frac{\partial d_i}{\partial x_i} \geq 0$ and $c\,dx - \sum_{i=1}^{d} \frac{\partial b_i}{\partial x_i} \geq 0$ (in the sense of Schwartz distributions, i.e., $\int \left(cu + \sum_{i=1}^{d} d_i \frac{\partial u}{\partial x_i}\right) dx, \int \left(cu + \sum_{i=1}^{d} b_i \frac{\partial u}{\partial x_i}\right) dx \geq 0$ for all $u \in C_0^\infty(U)$ with $u \geq 0$).

Define for $u, v \in C_0^\infty(U)$

$$\begin{aligned} \mathcal{E}(u,v) &= \sum_{i,j=1}^{d} \int \frac{\partial u}{\partial x_i}\frac{\partial v}{\partial x_j} a_{ij}\,dx + \sum_{i=1}^{d} \int u \frac{\partial v}{\partial x_i} d_i\,dx \\ &\quad + \sum_{i=1}^{d} \int \frac{\partial u}{\partial x_i} v b_i\,dx + \int uv\,c\,dx\,. \end{aligned} \tag{3.15}$$

Then $(\mathcal{E}, C_0^\infty(U))$ is a densely defined bilinear form on $L^2(U;dx)$ which is positive definite since for all $u \in C_0^\infty(U)$

$$\mathcal{E}(u,u) = \sum_{i,j=1}^{d} \int \frac{\partial u}{\partial x_i}\frac{\partial u}{\partial x_j} a_{ij}\,dx + \frac{1}{2}\int \left[\sum_{i=1}^{d} \frac{\partial u^2}{\partial x_i}(d_i + b_i) + u^2 3c\right] dx \geq 0$$

by (3.13), (3.14). Define $\tilde{a}_{ij} := \frac{1}{2}(a_{ij} + a_{ji})$ and $\check{a}_{ij} := \frac{1}{2}(a_{ij} - a_{ji})$, $1 \leq i,j \leq d$. Consider the following conditions:

(3.16) there exists $\nu \in]0,\infty[$ such that $\sum_{i,j=1}^{d} \tilde{a}_{ij}\xi_i\xi_j \geq \nu \|\xi\|^2_{\mathbb{R}^d}$ for all $\xi = (\xi_1, \ldots, \xi_d) \in \mathbb{R}^d$

and

(3.17) $|\check{a}_{ij}| \leq M \in]0,\infty[$ for all $1 \leq i,j \leq d$.

Proposition 3.7. *Let $(\mathcal{E}, C_0^\infty(U))$ be as in (3.15). Assume that (3.14), (3.16) and (3.17) hold and that*

(3.18) $c \in L^{d/2}_{loc}(U;dx)$, $b_i, d_i \in L^d_{loc}(U;dx)$ *and* $d_i - b_i \in L^d(U;dx) \cup L^\infty(U;dx)$, $1 \leq i \leq d$.

Then $(\mathcal{E}, C_0^\infty(U))$ is closable and its closure is a Dirichlet form on $L^2(U;dx)$.

Proof. [MR 92, Chap.II, Subsection 2d)]. □

Remark 3.8. Condition (3.16) can be relaxed. Under suitable assumptions on b_i, d_i and c, one has Dirichlet forms of the above type in sub-elliptic cases (cf.[RS 91b]).

Exercise 3.9. Let $(\mathcal{E}, C_0^\infty(U))$ be as in (3.15) such that (3.14), (3.16) and (3.17) are satisfied. Set $m := \left(\sum_{i=1}^{d} b_i^2 + d_i^2 + c + 1\right) dx$, where we assume that $c \in L^1_{loc}(U;dx)$ and b_i, $d_i \in L^2_{loc}(U;dx)$ for all $1 \leq i \leq d$. Prove that $(\mathcal{E}, C_0^\infty(U))$ is closable on $L^2(U;m)$ and that its closure $(\mathcal{E}, D(\mathcal{E}))$ is a Dirichlet form on $L^2(U;m)$.

4 Starting point: bilinear form – infinite dimensional case

a) Classical Dirichlet forms on infinite dimensional space

Let E be a separable real Banach space. In particular, $\mathcal{B}(E) = \sigma(E')$ by e.g. [Ba 70, Exposé n° 8, N° 7, Corollaire]. Let $m := \mu$ be a finite positive measure on $\mathcal{B}(E)$ such that $\operatorname{supp}\mu = E$ (i.e., $\mu(U) > 0$ for every $\emptyset \neq U \subset E$, U open). Let E' denote the dual of E and ${}_{E'}\langle\, ,\, \rangle_E : E' \times E \to \mathbb{R}$ the corresponding dualisation. Define a linear space of functions on E by

$$(4.1) \qquad \mathcal{F}C_b^\infty := \{f(l_1, \dots, l_m) \,|\, m \in \mathbb{N}\ ,\ f \in C_b^\infty(\mathbb{R}^m)\ ,\ l_1, \dots, l_m \in E'\}\ .$$

Here $C_b^\infty(\mathbb{R}^m)$ denotes the set of all infinitely differentiable (real-valued) functions on $\mathbb{R}^m$ with all partial derivatives bounded. By the Hahn-Banach theorem (cf. [Ch 69b]) $\mathcal{F}C_b^\infty$ separates the points of E; hence, since $\mathcal{B}(E) = \sigma(E')$,

$$(4.2) \qquad \mathcal{F}C_b^\infty \text{ is dense in } L^2(E;\mu)\ ,$$

by monotone class arguments (cf. [Sh 88, A.0.6]). Define for $u \in \mathcal{F}C_b^\infty$ and $k \in E$

$$\frac{\partial u}{\partial k}(z) := \frac{d}{ds} u(z + sk)|_{s=0}\ ,\ z \in E\ .$$

Observe that if $u = f(l_1, \dots, l_m)$, then

$$(4.3) \qquad \frac{\partial u}{\partial k} = \sum_{i=1}^{m} \frac{\partial f}{\partial x_i}(l_1, \dots, l_m)\, {}_{E'}\langle l_i, k\rangle_E \in \mathcal{F}C_b^\infty\ .$$

Define for $k \in E$

$$(4.4) \qquad \mathcal{E}_k(u,v) := \int \frac{\partial u}{\partial k}\frac{\partial v}{\partial k}\, d\mu\ ;\ u, v \in \mathcal{F}C_b^\infty\ ,$$

then $(\mathcal{E}_k, \mathcal{F}C_b^\infty)$ is a densely defined positive definite symmetric bilinear form on $L^2(E;\mu)$.

Definition 4.1. $k \in E$ is called *μ-admissible* if $(\mathcal{E}_k, \mathcal{F}C_b^\infty)$ is closable on $L^2(E;\mu)$.

For necessary and sufficient conditions for μ-admissibility we refer to [AR 90a, Sect. 3]. Now we shall see that $(\mathcal{E}_k, \mathcal{F}C_b^\infty)$ is closable if one has an integration by parts formula w.r.t. k.

Definition 4.2. $k \in E$ is called *well-μ-admissible* if there exists $\beta_k \in L^2(E;\mu)$ such that for all $u, v \in \mathcal{F}C_b^\infty$

$$(4.5) \qquad \int \frac{\partial u}{\partial k} v\, d\mu = -\int u \frac{\partial v}{\partial k}\, d\mu - \int uv\beta_k\, d\mu\ .$$

Exercise 4.3. Prove that the set of all well-μ-admissible elements form a linear subspace W of E and that $k \mapsto \beta_k$ is linear from W to $L^2(E;\mu)$.

Note that if $E = \mathbb{R}^d$ and $\mu = \rho \cdot dx$ for some nice function $\rho : \mathbb{R}^d \to [0,\infty[$ then $\beta_k = \frac{\partial}{\partial k}\ln\rho$. For a characterization of the well-μ-admissible elements in E we refer to [AKR 90].

Proposition 4.4. *If $k \in E$ is well-μ-admissible then it is μ-admissible.*

Proof. Define

$$S_k u := \frac{\partial}{\partial k}(\frac{\partial u}{\partial k}) + \beta_k \frac{\partial u}{\partial k} \ , \quad u \in \mathcal{F}C_b^\infty \ ,$$

then $(S_k, \mathcal{F}C_b^\infty)$ is a symmetric linear operator on $L^2(E;\mu)$ such that

$$\mathcal{E}_k(u,v) = (-S_k u, v) \ \text{ for all } u,v \in \mathcal{F}C_b^\infty \ .$$

Hence the assertion follows by 1.3. □

Fix a countable subset $K_0 \subset E$ such that

(4.6) $$\sum_{k \in K_0} {}_{E'}\langle l,k\rangle_E^2 < \infty \ \text{ for all } l \in E' \ .$$

Note that (4.6) holds automatically if K_0 is finite. Define

(4.7) $$\mathcal{E}_{K_0}(u,v) := \sum_{k \in K_0} \mathcal{E}_k(u,v) \ ; \ u,v \in \mathcal{F}C_b^\infty \ .$$

Observe that the sum in (4.7) converges since by (4.6) (and (4.3))

$$\sum_{k \in K_0} \mathcal{E}_k(u,u) \le \left(\sum_{i=1}^m \int (\frac{\partial f}{\partial x_i}(l_1,\ldots,l_m))^2 \, d\mu\right) \left(\sum_{i=1}^m \sum_{k \in K_0} {}_{E'}\langle l_i,k\rangle_E^2\right) < \infty$$

if $u = f(l_1,\ldots,l_m) \in \mathcal{F}C_b^\infty$.

Proposition 4.5. *Suppose K_0 is a (finite or) countable subset of E consisting of μ-admissible elements satisfying (4.6). Then $(\mathcal{E}_{K_0}, \mathcal{F}C_b^\infty)$ given by (4.7) is closable on $L^2(E;\mu)$ and its closure is a symmetric Dirichlet form.*

Proof. The closability follows because sums of closable forms are closable. Let φ_ε, $\varepsilon > 0$, be as in 3.5, then $\varphi_\varepsilon(u) \in \mathcal{F}C_b^\infty$ for every $u \in \mathcal{F}C_b^\infty$ and for each $k \in K_0$

$$\mathcal{E}_k(\varphi_\varepsilon(u), \varphi_\varepsilon(u)) = \int \varphi_\varepsilon'(u)^2 \left(\frac{\partial u}{\partial k}\right)^2 d\mu \le \mathcal{E}_k(u,u) \ .$$

Hence the last assertion follows by I.3.8. □

The Dirichlet forms arising in 4.5 are so-called *classical Dirichlet forms* introduced in[AR 90a] and studied intensively in [Ku 82], [AR 89 a,b, 90 a,b, 91], [R 90a,b], [S 90].

b) Gradient Dirichlet forms on infinite dimensional space

This subsection can be considered as a "coordinate free" version of the previous one. But we need a bit more structure. Assume that there exists a separable real Hilbert space $(H, \langle\ ,\ \rangle_H)$ densely and continuously embedded into E. Identifying H with its dual H' we have that

(4.8) $$E' \subset H \subset E \text{ densely and continuously}$$

and ${}_{E'}\langle\ ,\ \rangle_E$ restricted to $E' \times H$ coincides with $\langle\ ,\ \rangle_H$. Here E' is endowed with the strong topology (or equivalently with the operator norm $\|\ \|_{E'}$ if E is a Banach space). H should be thought of as a tangent space to E at each point. Observe that by (4.3) for $u \in \mathcal{F}C_b^\infty$ and $z \in E$ fixed, $h \mapsto \frac{\partial u}{\partial h}(z)$ is a continuous linear functional on H. Define $\nabla u(z) \in H$ by

$$(4.9) \qquad \langle \nabla u(z), h\rangle_H = \frac{\partial u}{\partial h}(z)\ ,\ h \in H\ .$$

Define

$$(4.10) \qquad \mathcal{E}(u,v) := \int \langle \nabla u, \nabla v\rangle_H \, d\mu\ ;\ \ u,v \in \mathcal{F}C_b^\infty\ ,$$

then $(\mathcal{E}, \mathcal{F}C_b^\infty)$ is a densely defined positive definite symmetric bilinear form on $L^2(E;\mu)$. The existence of the integral in (4.10) follows by

Remark 4.6. Clearly, $(\mathcal{E}, \mathcal{F}C_b^\infty)$ is a special case of (4.7). Since for every orthonormal basis K_0 of H, (4.6) holds and

$$\mathcal{E}(u,v) = \mathcal{E}_{K_0}(u,v) \ \text{ for all } u,v \in \mathcal{F}C_b^\infty\ .$$

4.6 and 4.5 imply:

Proposition 4.7. *Suppose there exists an orthonormal basis of H consisting of μ-admissible elements in E, then $(\mathcal{E}, \mathcal{F}C_b^\infty)$ given by (4.10) is closable on $L^2(E;\mu)$ and its closure is a symmetric Dirichlet form.*

c) Abstract Wiener spaces

In this section we look at the very special case where E, H, μ form an *abstract Wiener space* (E, H, μ) (cf. [Gr 65], [Kuo 75]) i.e., E is a separable real Banach space, H a separable real Hilbert space continuously and densely embedded into E, i.e.,

$$E' \subset H' \equiv H \subset E \text{ continuously and densely,}$$

and μ is a Gaussian measure on $\mathcal{B}(E)$ with covariance $\langle\ ,\ \rangle_H$, i.e., each $l \in E'$ is $N(0, \|l\|_H^2)$-distributed under μ and $\operatorname{supp}\mu = E$. Consider the linear map

$$(4.11) \qquad l \mapsto {}_{E'}\langle l, \cdot\rangle_E\ ,\ l \in E'$$

from E' to $L^2(E;\mu)$ and note that this is an isometry if E' is equipped with the norm inherited from H. Hence this map extends uniquely to an isometry

$$(4.12) \qquad h \mapsto X_h\ ,\ h \in H\ ,$$

from H to $L^2(E;\mu)$ (i.e., in particular, $X_l = {}_{E'}\langle l, \cdot\rangle_E$ if $l \in E' \subset H$).

Example 4.8. Let $E := \{f \in C([0,1], \mathbb{R}^d) \mid f(0) = 0\}$ equipped with the uniform norm, $H := \{f = (f_1, \ldots, f_d) \in E \mid f_i$ is absolutely continuous and $\int_0^1 f_i'(s)^2\, ds < \infty,\ 1 \le i \le d\}$ with inner product $\langle f, g\rangle_H = \sum_{i=1}^d \int_0^1 f_i'(s) g_i'(s)\, ds$ and $\mu :=$ Wiener measure on $\mathcal{B}(E)$. Then (E, H, μ) is an abstract Wiener space (cf. [Kuo 75]).

Theorem 4.9. *Each $h \in H$ is well-μ-admissible with $\beta_h = -X_h$.*

Proof. See e.g. [MR 92, II.3.11]. □

As a consequence we obtain by 4.4, 4.7

Corollary 4.10. *Define*

$$\mathcal{E}(u,v) := \int \langle \nabla u, \nabla v \rangle_H d\mu \quad ; u,v \in \mathcal{F}C_b^\infty \tag{4.13}$$

(cf. (4.9), (4.10)). Then $(\mathcal{E}, \mathcal{F}C_b^\infty)$ is closable on $L^2(E;\mu)$ and its closure $(\mathcal{E}, D(\mathcal{E}))$ is a symmetric Dirichlet form.

Remark 4.11. (i) $(\mathcal{E}, D(\mathcal{E}))$ in 4.10 is of fundamental importance in the Malliavin calculus (cf. [Ma 78]). If $(\overline{\nabla}, D(\overline{\nabla}))$ denotes the closure of $(\nabla, \mathcal{F}C_b^\infty)$ (which exists by 4.10,) then $D(\overline{\nabla}) = D(\mathcal{E})$ and

$$\mathcal{E}(u,v) = \int \langle \overline{\nabla} u, \overline{\nabla} v \rangle_H \, d\mu \; ; u,v \in D(\mathcal{E}) \; . \tag{4.14}$$

$\overline{\nabla}$ is just the Malliavin gradient (cf.[W 84]).
(ii) H in (E, H, μ) is called the *(generalized) Cameron-Martin space.* Examples where the Hilbert space appearing in (4.13) is replaced by some other Hilbert space can be found in [AR 90a], [RZ 90] and [R 90a].
(iii) For more examples, e.g. where μ is not (even absolutely continuous w.r.t.) a Gaussian measure, we refer to [AR 89a,b, 90a,b, 91].

d) Non-symmetric cases

Suppose E, H, μ are as in Subsection 4b) and let $(\mathcal{E}_\mu, \mathcal{F}C_b^\infty)$, defined by

$$\mathcal{E}_\mu(u,v) = \int \langle \nabla u, \nabla v \rangle_H \, d\mu \; ; \qquad u,v \in \mathcal{F}C_b^\infty$$

(cf. (4.9), (4.10)), be closable on $L^2(E;\mu)$. Let $\mathcal{L}_\infty(H)$ denote the set of all bounded linear operators on H with operator norm $\| \ \|$. Suppose $z \mapsto A(z)$, $z \in E$, is a map from E to $\mathcal{L}_\infty(H)$ such that $z \mapsto \langle A(z)h_1, h_2 \rangle_H$ is $\mathcal{B}(E)$-measurable for all $h_1, h_2 \in H$. Furthermore, assume that

$$\begin{aligned} &\text{there exists } \alpha \in]0,\infty[\text{ such that} \\ &\qquad \langle A(z)h, h \rangle_H \geq \alpha \|h\|_H^2 \text{ for all } h \in H \; . \end{aligned} \tag{4.15}$$

and that $\|\tilde{A}\|_\infty \in L^1(E;\mu)$ and $\|\check{A}\|_\infty \in L^\infty(E;\mu)$ where $\tilde{A} := \frac{1}{2}(A + \hat{A})$, $\check{A} := \frac{1}{2}(A - \hat{A})$ and $\hat{A}(z)$ denotes the adjoint of $A(z)$, $z \in E$. Let $c \in L^\infty(E;\mu)$ and $b, d \in L^\infty(E \to H;\mu)$ such that

$$\begin{aligned} &\int (\langle d, \nabla u \rangle_H + cu) \, d\mu \; , \quad \int (\langle b, \nabla u \rangle_H + cu) \, d\mu \geq 0 \\ &\text{for all } u \in \mathcal{F}C_b^\infty, \; u \geq 0 \; . \end{aligned} \tag{4.16}$$

(4.16) is e.g. fulfilled if d, b are in the domain of $\hat{\nabla}$, the adjoint of $\nabla : \mathcal{F}C_b^\infty \to L^2(E \to H; d\mu)$ and $\hat{\nabla} d + c \geq 0$, $\hat{\nabla} b + c \geq 0$. Define for $u, v \in \mathcal{F}C_b^\infty$

$$\begin{aligned} \mathcal{E}(u,v) &= \int \langle A(z)\nabla u(z), \nabla v(z)\rangle_H \, \mu(dz) + \int u \, \langle d, \nabla v\rangle_H \, d\mu \\ &\quad + \int \langle b, \nabla u\rangle_H \, v \, d\mu + \int u \, v \, c \, d\mu \,. \end{aligned} \tag{4.17}$$

Then $(\mathcal{E}, \mathcal{F}C_b^\infty)$ is a densely defined bilinear form on $L^2(E;\mu)$ which is positive definite since for all $u \in \mathcal{F}C_b^\infty$

$$\mathcal{E}(u,u) = \int \langle A\nabla u, u\rangle_H \, d\mu + \frac{1}{2}\int (\langle d + b, \nabla u^2\rangle_H + cu^2) d\mu \geq 0 \,. \tag{4.18}$$

Furthermore, $(\mathcal{E}, \mathcal{F}C_b^\infty)$ is closable on $L^2(E;\mu)$ and its closure $(\mathcal{E}, D(\mathcal{E}))$ is a Dirichlet form on $L^2(E;\mu)$ (cf. [MR 92, Chap.II, Subsection 3e)]).

5 Starting point: semigroup of kernels

Let $(E, \mathcal{B})$ be an arbitrary measurable space and let π be a *kernel on* $(E, \mathcal{B})$, i.e., $\pi : E \times \mathcal{B} \to [0, \infty[$ such that $\pi(z, \cdot)$ is a positive measure on $\mathcal{B}$ for each $z \in E$ and $z \mapsto \pi(z, A)$ is $\mathcal{B}$-measurable for every $A \in \mathcal{B}$. Define for $f : E \to \mathbb{R}$, $\mathcal{B}$-measurable, and $z \in E$

$$\pi f(z) := \pi(z, f) := \int f(y)\pi(z, dy) \tag{5.1}$$

if $\pi(z, f^+) \wedge \pi(z, f^-) < \infty$. π is called *sub-Markovian* if $\pi 1 \leq 1$ and *Markovian* if $\pi 1 = 1$. Let m be a positive measure on $(E, \mathcal{B})$. m is called *π-supermedian* if

$$\int \pi f \, dm \leq \int f \, dm \quad \text{for all } f \in \mathcal{B}^+ \,.$$

Two kernels $\pi, \hat{\pi}$ on $(E, \mathcal{B})$ are said to be *in duality w.r.t.* m if

$$\int \pi f \, g \, dm = \int f \, \hat{\pi} g \, dm \quad \text{for all } f, g \in \mathcal{B}^+ \,. \tag{5.2}$$

If $\pi = \hat{\pi}$ in (5.2) then π is called *m-symmetric*. Note that if there exists $\hat{\pi}$ satisfying (5.2) which is sub-Markovian then

$$\int \pi f \, dm = \int f \, \hat{\pi} 1 \, dm \leq \int f \, dm \quad \text{for all } f \in \mathcal{B}^+ \,,$$

i.e., m is π-supermedian. A (slightly weakened) converse to this statement is contained in the next proposition.

Proposition 5.1. *Let π be a sub-Markovian kernel on $(E, \mathcal{B})$ and let m be a positive measure on $(E, \mathcal{B})$ which is π-supermedian. Then there exists a bounded linear operator Π on $L^2(E; m)$ such that*

$$\pi f \text{ is an } m\text{-version of } \Pi f \quad \text{for all } f \in \mathcal{B}_b \cap L^2(E; m) \,. \tag{5.3}$$

Both Π and its adjoint $\hat{\Pi}$ are sub-Markovian contractions on $L^2(E; m)$.

Remark. Let $f \in \mathcal{B}_b \cap L^2(E;m)$. In 5.1 and henceforth (cf. also the following lemma) if we say "πf is an m-version of Πf" for some sub-Markovian kernel π resp. a bounded linear operator Π on $L^2(E;m)$, we mean $\pi\tilde{f}$ is an m-version of Πf for any $\mathcal{B}$-measurable representative $\tilde{f}$ of the class $f \in L^2(E;m)$. In particular, $\pi g = \pi\tilde{f}$ m-a.e. if $g = \tilde{f}$ m-a.e. .

Proof of 5.1. Let $f \in \mathcal{B}_b \cap L^2(E;m)$, then by Cauchy-Schwarz's inequality

$$\int (\pi f)^2 \, dm \leq \int \pi(f^2) \, dm \leq \int f^2 \, dm \; . \tag{5.4}$$

Hence $f = g$ m-a.e. implies $\pi f = \pi g$ m-a.e. for all $g \in \mathcal{B}_b$, and we can define Πf to be the m-class in $L^2(E;m)$ corresponding to πf. Since the bounded functions are dense in $L^2(E;m)$, Π extends by (5.4) to a unique bounded operator on $L^2(E;m)$ which is clearly a sub-Markovian contraction on $L^2(E;m)$. For its adjoint $\hat{\Pi}$ we have for any $f \in L^2(E;m)$, $0 \leq f \leq 1$, that

$$\int g \, \hat{\Pi} f \, dm = \int \Pi g \, f \, dm \quad \text{for all } g \in L^2(E;m) \; . \tag{5.5}$$

But if $g \in L^2(E;m)$, $g \geq 0$, then $\Pi g \geq 0$ by I.3.2, hence

$$0 \leq \int \Pi g \; f \; dm \leq \int \pi g \, dm \leq \int g \, dm \; .$$

Consequently, (5.5) implies that $0 \leq \hat{\Pi} f \leq 1$, i.e., $\hat{\Pi}$ is sub-Markovian. $\hat{\Pi}$ is a contraction since it is the adjoint of a contraction. □

Now let $(p_t)_{t>0}$ be a *semigroup of kernels on* $(E, \mathcal{B})$ (i.e., $p_t(p_s f)(z) = p_{t+s} f(z)$ for all $f \in \mathcal{B}^+$, $z \in E$ and $t, s > 0$) which is *sub-Markovian*, i.e., each p_t is sub-Markovian.

Proposition 5.2. *Let $(p_t)_{t>0}$ be as above and let m be a positive measure on $(E, \mathcal{B})$ such that m is p_t-supermedian for all $t > 0$. Let T_t, $t > 0$, be the corresponding sub-Markovian operators on $L^2(E;m)$. Then $(T_t)_{t>0}$ is a semigroup of contractions on $L^2(E;m)$. Suppose, in addition, that the following condition holds:*

(5.6) *There exists a dense subset $\mathcal{D} \subset L^2(E;m)$ such that* $p_t f \underset{t \to 0}{\longrightarrow} f$ *in m-measure for all $f \in \mathcal{D}$.*

Then $(T_t)_{t>0}$ is strongly continuous.

Proof. [MR 92, Chap.II, Proposition 4.3]. □

Consider the situation of 5.2. Let L be the generator of $(T_t)_{t>0}$ on $L^2(E;m)$. We know that if $1 - L$ satisfies the sector condition I.(2.5) (which is e.g. the case if p_t is m-symmetric, hence T_t is symmetric on $L^2(E;m)$ for each $t > 0$, see also I.2.17) then there exists a corresponding coercive closed form $(\mathcal{E}, D(\mathcal{E}))$ on $L^2(E;m)$. $(\mathcal{E}, D(\mathcal{E}))$ is a Dirichlet form by I.3.3, I.3.4 since both $(T_t)_{t>0}$ and $(\hat{T}_t)_{t>0}$ are sub-Markovian by 5.1.

Chapter III

Potential Theory of Dirichlet Forms

In this chapter we develop some analytic potential theory of Dirichlet forms. We try to keep the amount of material as small as possible but sufficient for understanding the probabilistic part of the theory treated in the next chapter. For the proofs not included here we again refer to [MR 92]. In this chapter we consider the following situation: let E be a Hausdorff topological space. Let $\mathcal{B}(E)$ be the σ-algebra consisting of its Borel subsets and m be a σ-finite positive measure on $(E, \mathcal{B}(E))$. Let $(\mathcal{E}, D(\mathcal{E}))$ be a fixed Dirichlet form on $L^2(E;m)$ with associated generator L, semigroups $(T_t)_{t>0}$, $(\hat{T}_t)_{t>0}$ and resolvents $(G_\alpha)_{\alpha>0}$, $(\hat{G}_\alpha)_{\alpha>0}$ on $L^2(E;m)$. Let $K \geq 1$ be a continuity constant of $(\mathcal{E}, D(\mathcal{E}))$ and, as before, let $\tilde{\mathcal{E}}$ denote its symmetric part. Recall that $\mathcal{E}_\alpha := \mathcal{E} + \alpha(\,,\,)$, $\alpha > 0$, where $(\,,\,)$ is the usual inner product in $L^2(E;m)$ and that $\|\;\| := (\,,\,)^{1/2}$.

1 Exceptional sets and capacities

Definition 1.1. (i) An increasing sequence $(F_k)_{k\in\mathbb{N}}$ of closed subsets of E is called an *$\mathcal{E}$-nest* if $\bigcup_{k\geq 1} \{u \in D(\mathcal{E}) \,|\, u = 0 \text{ } m\text{-a.e. on } E\backslash F_k\}$ is dense in $D(\mathcal{E})$ (w.r.t. $\tilde{\mathcal{E}}_1^{1/2}$).

(ii) A subset $N \subset E$ is called *$\mathcal{E}$-exceptional* if $N \subset \bigcap_{k\geq 1} F_k^c$ for some $\mathcal{E}$-nest $(F_k)_{k\in\mathbb{N}}$. We say that a property of points in E holds *$\mathcal{E}$-quasi-everywhere* (abbreviated $\mathcal{E}$-q.e.), if the property holds outside some $\mathcal{E}$-exceptional set.

Remark 1.2. Note that 1.1 only depends on the symmetric part of $(\mathcal{E}, D(\mathcal{E}))$.

Exercise 1.3. Show that every Borel $\mathcal{E}$-exceptional set has m-measure zero.

The description of "small sets" by 1.1 is essentially sufficient to formulate all subsequent results in this book. For the proofs and practical purposes, however, 1.1 is not very handy. In this section we therefore introduce and study capacities associated with the symmetric part $(\tilde{\mathcal{E}}, D(\mathcal{E}))$ of $(\mathcal{E}, D(\mathcal{E}))$ which can be used to characterize $\mathcal{E}$-nests

and whose zero sets are exactly the $\mathcal{E}$-exceptional sets (cf. Theorem 1.7 below). But these capacities are not unique.
Let $(\tilde{G}_\alpha)_{\alpha>0}$ be the resolvent corresponding to $(\tilde{\mathcal{E}}, D(\mathcal{E}))$ on $L^2(E;m)$.

Definition 1.4. Let $\varphi \in L^2(E;m)$, $0 < \varphi \leq 1$ and set $h := \tilde{G}_1\varphi$. Define for $U \subset E$, U open,

$$\tilde{\mathrm{Cap}}_h(U) := \inf\{\mathcal{E}_1(u,u) \mid u \in D(\mathcal{E}),\ u \geq h\ m\text{-a.e. on } U\}$$

and for any $A \subset E$

$$\tilde{\mathrm{Cap}}_h(A) := \inf\{\tilde{\mathrm{Cap}}_h(U) \mid A \subset U \subset E,\ U \text{ open}\}.$$

$\tilde{\mathrm{Cap}}_h$ is called the *h-scaled* (or *h-weighted*) *capacity* associated with $(\tilde{\mathcal{E}}, D(\mathcal{E}))$.

Theorem 1.5. *Let h be as in 1.4. $\tilde{\mathrm{Cap}}_h$ is a Choquet capacity on E, i.e., it has the following two properties:*

(i) If $(A_n)_{n\in\mathbb{N}}$ is an increasing sequence of subsets of E then

$$\tilde{\mathrm{Cap}}_h\Big(\bigcup_{n\geq1} A_n\Big) = \sup_{n\geq1} \tilde{\mathrm{Cap}}_h(A_n)\ .$$

(ii) If $(K_n)_{n\in\mathbb{N}}$ is a decreasing sequence of compact subsets of E then

$$\tilde{\mathrm{Cap}}_h\Big(\bigcap_{n\geq1} K_n\Big) = \inf_{n\geq1} \tilde{\mathrm{Cap}}_h(K_n)\ .$$

Furthermore, if $A_n \subset E$, $n \in \mathbb{N}$, then

$$\tilde{\mathrm{Cap}}_h\Big(\bigcup_{n\geq1} A_n\Big) \leq \sum_{n\geq1} \tilde{\mathrm{Cap}}_h(A_n)\ .$$

Remark 1.6. Suppose that E is a (topological) Souslin space (i.e., the continuous image of a complete separable metric space). Since $\tilde{\mathrm{Cap}}_h(E) < \infty$, 1.5 implies that for every $A \in \mathcal{B}(E)$

$$\tilde{\mathrm{Cap}}_h(A) = \sup\{\tilde{\mathrm{Cap}}_h(K) | K \subset A\ ,\ K \text{ compact}\}$$

(Choquet's Capacitability Theorem). This follows by [Bou 74, Chap.IX, Sect. 6, Def. 9, Théorème 6 and Proposition 10].

Theorem 1.7. *(i) An increasing sequence $(F_k)_{k\in\mathbb{N}}$ of closed subsets of E is an $\mathcal{E}$-nest if and only if $\lim_{k\to\infty} \tilde{\mathrm{Cap}}_h(F_k^c) = 0$.*

(ii) A subset $N \subset E$ is $\mathcal{E}$-exceptional if and only if $\tilde{\mathrm{Cap}}_h(N) = 0$.

Proof. [MR 92, III.2.11]. □

2 Quasi-continuity

In this section we fix $\varphi \in L^2(E;m)$, $0 < \varphi \leq 1$ m-a.e. and set

$$h := \tilde{G}_1\varphi \,. \tag{2.1}$$

Given an $\mathcal{E}$-nest $(F_k)_{k\in\mathbb{N}}$ we define

$$\begin{aligned} C(\{F_k\}) := \{f : A \to \mathbb{R} \mid & \bigcup_{k\geq 1} F_k \subset A \subset E\,, \\ & f_{|F_k} \text{ is continuous for every } k \in \mathbb{N}\}\,. \end{aligned} \tag{2.2}$$

Note that the notion introduced in the following definition only depends on the symmetric part of $(\mathcal{E}, D(\mathcal{E}))$ (cf. 1.2).

Definition 2.1. An $\mathcal{E}$-q.e. defined function f on E is called *$\mathcal{E}$-quasi-continuous* if there exists an $\mathcal{E}$-nest $(F_k)_{k\in\mathbb{N}}$ such that $f \in C(\{F_k\})$.

Exercise 2.2. Let S be a countable family of $\mathcal{E}$-quasi-continuous functions on E. Then there exists an $\mathcal{E}$-nest $(F_k)_{k\in\mathbb{N}}$ such that $S \subset C(\{F_k\})$.

Proposition 2.3. *Let $u \in D(\mathcal{E})$ such that it has an $\mathcal{E}$-quasi-continuous m-version $\tilde{u}$. Then for all $\lambda > 0$*

$$\tilde{\text{Cap}}_h\{|\tilde{u}| > \lambda\} \leq \frac{1}{\lambda^2}\mathcal{E}_1(u,u) \text{ (Chebychev-type inequality).}$$

Proof. [MR 92, III.3.4]. □

Let f, f_n, $n \in \mathbb{N}$, be ($\mathcal{E}$-q.e. defined) functions on E. We say that $(f_n)_{n\in\mathbb{N}}$ converges to f *$\mathcal{E}$-quasi-uniformly* if there exists an $\mathcal{E}$-nest $(F_k)_{k\in\mathbb{N}}$ such that $f_n \underset{n\to\infty}{\longrightarrow} f$ uniformly on each F_k.

Proposition 2.4. *Let $u_n \in D(\mathcal{E})$, which have $\mathcal{E}$-quasi-continuous m-versions $\tilde{u}_n$, $n \in \mathbb{N}$, such that $u_n \underset{n\to\infty}{\longrightarrow} u \in D(\mathcal{E})$ w.r.t. $\tilde{\mathcal{E}}_1^{1/2}$. Then there exists a subsequence $(u_{n_k})_{k\in\mathbb{N}}$ and an $\mathcal{E}$-quasi-continuous m-version $\tilde{u}$ of u such that $(\tilde{u}_{n_k})_{k\in\mathbb{N}}$ converges $\mathcal{E}$-quasi-uniformly to $\tilde{u}$.*

Proof. By 2.3 we may choose a subsequence $(u_{n_i})_{i\in\mathbb{N}}$ such that

$$\text{Cap}_{h,g}\left\{\left|\tilde{u}_{n_{i+1}} - \tilde{u}_{n_i}\right| > 2^{-i}\right\} \leq 2^{-i} \text{ for all } \quad i \in \mathbb{N}\,.$$

Let

$$A_k := \bigcup_{i\geq k}\{|\tilde{u}_{n_{i+1}} - \tilde{u}_{n_i}| > 2^{-i}\}\,, \quad k \in \mathbb{N}\,.$$

Let $(F'_k)_{k\in\mathbb{N}}$ be an $\mathcal{E}$-nest such that $\tilde{u}_n \in C(\{F'_k\})$ for each $n \in \mathbb{N}$, and define

$$F_k := F'_k \cap A_k^c\,, \quad k \in \mathbb{N}\,.$$

Then $(F_k)_{k\in\mathbb{N}}$ is an $\mathcal{E}$-nest and $(\tilde{u}_{n_i})_{i\in\mathbb{N}}$ converges uniformly on each F_k. Set

$$\tilde{u}(z) := \begin{cases} \lim_{i\to\infty}\tilde{u}_{n_i}(z) & \text{if} \quad z \in \bigcup_{k\geq 1} F_k \\ 0 & \text{else.} \end{cases}$$

Since $\tilde{\mathcal{E}}_1^{1/2}$-convergence implies the m-a.e. convergence of a subsequence and since $m\left(\bigcap_{k\geq 1} F_k^c\right) = 0$ by 1.3, $\tilde{u}$ is an m-version of u. □

Chapter IV

Markov Processes Associated with Dirichlet Forms

In this chapter we develop the probabilistic part of the theory of Dirichlet forms. We, however, only explain the results since the proofs are quite involved and are based on some techniques and methods which would require too much time to be presented here. We refer for details to [MR 92] (cf. also [AMR 92 a-d]) and rather concentrate on examples.
In this chapter for simplicity we assume E to be a (*metrizable*) *Lusin space* (i.e., a Borel subset of a complete separable metric space) though all results have their analogues for arbitrary Hausdorff topological spaces at least if $\mathcal{B}(E) = \sigma(C(E))$ (cf. [MR 92]).

1 Basics on Markov processes

Given a measurable space $(S, \mathcal{B})$, the *completion* of the σ-algebra $\mathcal{B}$ w.r.t. a probability measure P on $(S, \mathcal{B})$ is denoted by $\mathcal{B}^P$. An element B of $\mathcal{B}^* := \bigcap_{P \in \mathcal{P}(S)} \mathcal{B}^P$ is called a *universally measurable* set, where $\mathcal{P}(S)$ denotes the family of all probability measures on $(S, \mathcal{B})$. Given a sub-σ-algebra $\mathcal{C}$ of $\mathcal{B}$, we define $\mathcal{C}^P$, the *completion of* $\mathcal{C}$ *in* $\mathcal{B}^P$ w.r.t. $P \in \mathcal{P}(S)$, by

$$(1.1) \qquad \mathcal{C}^P = \{A \subset S \mid A \Delta B \subset N \text{ for some } B \in \mathcal{C}\,,\, N \in \mathcal{B}\,,\, P(N) = 0\}\,.$$

If $\mathcal{N}_P := \{N \in \mathcal{B}^P \mid P(N) = 0\}$ then $\mathcal{C}^P = \sigma\{\mathcal{C}, \mathcal{N}_P\}$ i.e., the smallest σ-algebra containing $\mathcal{C}, \mathcal{N}_P$.
Now fix a measurable space $(\Omega, \mathcal{M})$. Let $\mathcal{M}_t$, $t \in [0, \infty]$, be sub-σ-algebras of $\mathcal{M}$. $(\mathcal{M}_t) := (\mathcal{M}_t)_{t \in [0,\infty]}$ is called a *filtration* (on $(\Omega, \mathcal{M})$) if $\mathcal{M}_s \subset \mathcal{M}_t$ for $s \le t$ and

$$(1.2) \qquad \mathcal{M}_\infty = \bigvee_{t \in [0,\infty[} \mathcal{M}_t\,,$$

i.e., $\mathcal{M}_\infty$ is the smallest σ-algebra containing all $\mathcal{M}_t$, $t \in [0, \infty[$. Given a filtration $(\mathcal{M}_t)$ we set

$$\mathcal{M}_{t+} := \bigcap_{s>t} \mathcal{M}_s . \tag{1.3}$$

$(\mathcal{M}_t)$ is called *right continuous* if $\mathcal{M}_{t+} = \mathcal{M}_t$ for all $t \geq 0$.

Exercise 1.1. Show that $(\mathcal{M}_{t+})^P = (\mathcal{M}_t^P)_+$ for all $t \geq 0$ and any probability measure P on $(\Omega, \mathcal{M})$.

Remark 1.2. By convention a filtration $(\mathcal{M}_t)$ on a probability space $(\Omega, \mathcal{M}, P)$ is said to satisfy the *usual conditions* if $\mathcal{M}_t = \mathcal{M}_{t+}^P$ for all $t \geq 0$ (cf. [DM 78, IV.48]).

We adjoin an extra point Δ (the *cemetery*) to E as an isolated point to obtain a Hausdorff topological space $E_\Delta := E \cup \{\Delta\}$ with Borel σ-algebra $\mathcal{B}(E_\Delta)(= \mathcal{B}(E) \cup \{B \cup \{\Delta\} \mid B \in \mathcal{B}(E)\})$. Any function $f : E \to \mathbb{R}$ is considered as a function on E_Δ by putting $f(\Delta) = 0$. However, if E is locally compact then we shall sometimes also consider a different topology on E_Δ taking as neighbourhoods for Δ the complements of compact sets in E, i.e., E_Δ is then the *one-point-compactification* of E.

Exercise 1.3. Consider the topology on E_Δ where the neighbourhoods of Δ are the complements of compact sets in E. Prove that E_Δ is Hausdorff if and only if E is locally compact.

Definition 1.4. We say that $(X_t)_{t\geq 0}$ is a *stochastic process with state space E and life time ζ* (on $(\Omega, \mathcal{M})$) if

(i) $X_t : \Omega \to E_\Delta$ is an $\mathcal{M}/\mathcal{B}(E_\Delta)$-measurable map for all $t \in [0, \infty[$.

(ii) $\zeta : \Omega \to [0, \infty]$ is $\mathcal{M}$-measurable.

(iii) For all $\omega \in \Omega$, $X_t(\omega) \in E$ whenever $t < \zeta(\omega)$ and $X_t(\omega) = \Delta$ for all $t \geq \zeta(\omega)$.

A stochastic process $(X_t)_{t\geq 0}$ is said to be *measurable* if $(t, \omega) \mapsto X_t(\omega)$ is $\mathcal{B}([0,\infty[) \otimes \mathcal{M}/\mathcal{B}(E_\Delta)$-measurable. $(X_t)_{t\geq 0}$ is called $(\mathcal{M}_t)$-*adapted* for a filtration $(\mathcal{M}_t)$ on $(\Omega, \mathcal{M})$ if each X_t is $\mathcal{M}_t/\mathcal{B}(E_\Delta)$ measurable, $t \geq 0$. Below, for convenience we always set $X_\infty := \Delta$, unless otherwise specified.

Definition 1.5. A collection $\mathbf{M} := (\Omega, \mathcal{M}, (X_t)_{t\geq 0}, (P_z)_{z\in E_\Delta})$ is called a (time-homogeneous) *Markov process* (with state space E and life time ζ) if it has the following properties.

(M.1) There exists a filtration $(\mathcal{M}_t)$ on $(\Omega, \mathcal{M})$ such that $(X_t)_{t\geq 0}$ is an $(\mathcal{M}_t)$-adapted stochastic process with state space E and life time ζ.

(M.2) For each $t \geq 0$, there exists a *shift operator* $\theta_t : \Omega \to \Omega$ such that $X_s \circ \theta_t = X_{s+t}$ for all $s, t \geq 0$.

(M.3) P_z, $z \in E_\Delta$, are probability measures on $(\Omega, \mathcal{M})$ such that $z \mapsto P_z(\Gamma)$ is $\mathcal{B}(E_\Delta)^*$-measurable for each $\Gamma \in \mathcal{M}$ resp. $\mathcal{B}(E_\Delta)$-measurable if $\Gamma \in \sigma(X_s \mid s \in [0, \infty[)$, and $P_\Delta[X_0 = \Delta] = 1$.

(M.4) (*Markov property*) For all $A \in \mathcal{B}(E_\Delta)$ and any $t, s \geq 0$

$$P_z[X_{s+t} \in A \mid \mathcal{M}_s] = P_{X_s}[X_t \in A] \quad P_z\text{-}a.s.,\ z \in E_\Delta .$$

Given $\mathbf{M}$ as in 1.5 and a positive measure μ on $(E_\Delta, \mathcal{B}(E_\Delta))$ we define a positive measure P_μ on $(\Omega, \mathcal{M})$ by

$$(1.4) \qquad P_\mu(\Gamma) := \int_{E_\Delta} P_z(\Gamma)\mu(dz) \ , \ \Gamma \in \mathcal{M} \ .$$

Note that if μ is equivalent to a positive measure ν on $(E_\Delta, \mathcal{B}(E_\Delta))$ then P_μ is equivalent to P_ν (i.e., they have the same null sets). If $\mu \in \mathcal{P}(E_\Delta)$ we denote as usual the expectation and conditional expectation w.r.t. some sub-σ-algebra $\mathcal{G}$ of $\mathcal{M}$ w.r.t. P_μ by E_μ, $E_\mu[\,\cdot\,|\mathcal{G}]$ respectively. Let $(\mathcal{M}_t)$ be a filtration on $(\Omega, \mathcal{M})$. An $\mathcal{M}$-measurable function $\tau : \Omega \to [0, \infty]$ is called an $(\mathcal{M}_t)$-*stopping time* if $\{\tau \le t\} \in \mathcal{M}_t$ for all $t \ge 0$. For an $(\mathcal{M}_t)$-stopping time τ we define

$$(1.5) \qquad \mathcal{M}_\tau := \{\Gamma \in \mathcal{M} \,|\, \Gamma \cap \{\tau \le t\} \in \mathcal{M}_t \text{ for all } t \ge 0\} \ .$$

Clearly, $\mathcal{M}_\tau$ is a sub-σ-field of $\mathcal{M}$.

Definition 1.6. Let $\mathbf{M} = (\Omega, \mathcal{F}, (X_t)_{t \ge 0}, (P_z)_{z \in E_\Delta})$ be a Markov process with state space E, life time ζ and corresponding filtration $(\mathcal{M}_t)$. $\mathbf{M}$ is called a *right process* (w.r.t. $(\mathcal{M}_t)$) if it has the following additional properties.

(M.5) (*Normal property*) $P_z[X_0 = z] = 1$ for all $z \in E_\Delta$.

(M.6) (*Right continuity*) For each $\omega \in \Omega$, $t \mapsto X_t(\omega)$ is right continuous on $[0, \infty[$.

(M.7) (*Strong Markov property*) $(\mathcal{M}_t)$ is right continuous and for every $(\mathcal{M}_t)$-stopping time σ and every $\mu \in \mathcal{P}(E_\Delta)$

$$P_\mu \left[X_{\sigma+t} \in A | \mathcal{M}_\sigma\right] = P_{X_\sigma}[X_t \in A] \qquad P_\mu\text{-}a.s.$$

for all $A \in \mathcal{B}(E_\Delta)$, $t \ge 0$.

Let $\mathbf{M}$ be a Markov process w.r.t. a filtration $(\mathcal{M}_t)$. We set for $t \in [0, \infty]$

$$(1.6) \qquad \mathcal{F}_t^0 := \sigma\{X_s \,|\, 0 \le s \le t\}$$

$$(1.7) \qquad \mathcal{F}_t := \bigcap_{\mu \in \mathcal{P}(E_\Delta)} (\mathcal{F}_t^0)^{P_\mu} .$$

$(\mathcal{F}_t)$ is called the *natural filtration of* $\mathbf{M}$.

Remark 1.7. When dealing with a right process $\mathbf{M}$ w.r.t. some filtration $(\mathcal{M}_t)$ in the sequel, we shall always assume that $(\mathcal{M}_t)$ is the natural filtration $(\mathcal{F}_t)$ and set $\mathcal{F} := \mathcal{F}_\infty$, unless otherwise stated.

Let $\mathbf{M}$ be a right process with state space E and life time ζ. We use $E_z[\,\cdot\,]$, $E_z[\,\cdot\,|\mathcal{G}]$ for the expectation resp. conditional expectation given a sub-σ-algebra $\mathcal{G}$ of $\mathcal{F}$ w.r.t. P_z, $z \in E_\Delta$. Since $(X_t)_{t \ge 0}$ is measurable by (M.6),

$$(1.8) \qquad p_t f(z) := p_t(z, f) := E_z[f(X_t)] \ , \ \ z \in E \ , t \ge 0 \ , \ f \in \mathcal{B}(E)^+ \ ,$$

defines a sub-Markovian semigroup of kernels on $(E, \mathcal{B})$. We shall call $(p_t)_{t \ge 0}$ the *transition semigroup* of $\mathbf{M}$.

Definition 1.8. Let $\mathbf{M} = (\Omega, \mathcal{F}, (X_t)_{t\geq 0}, (P_z)_{z\in E_\Delta})$ be a right process with state space E and life time ζ.

(i) Let m be a σ-finite positive measure on $(E_\Delta, \mathcal{B}(E_\Delta))$. $\mathbf{M}$ is called an *m-special standard process* if for one (and hence all) measure(s) $\mu \in \mathcal{P}(E_\Delta)$ which are equivalent to m it has the following additional properties

(M.8) *(left limits up to ζ)* $X_{t-} := \lim_{\substack{s\uparrow t\\ s<t}} X_s$ exists in E for all $t \in]0, \zeta[$ P_μ-*a.s.* .

(M.9) *(quasi-left continuity up to ζ)* If τ, τ_n, $n \in \mathbb{N}$, are $(\mathcal{F}_t^{P_\mu})$-stopping times such that $\tau_n \uparrow \tau$ then $X_{\tau_n} \to X_\tau$ as $n \to \infty$ P_μ-*a.s.* on $\{\tau < \zeta\}$.

(M.10) *(special)* If τ, τ_n, $n \in \mathbb{N}$, are as in (M.9) then X_τ is $\bigvee_{n\in\mathbb{N}} \mathcal{F}_{\tau_n}^{P_\mu}$-measurable.

(ii) $\mathbf{M}$ is called a *special standard process* if it is a μ-special standard process for all $\mu \in \mathcal{P}(E_\Delta)$.

(iii) $\mathbf{M}$ is called a *Hunt process* if (M.8) and (M.9) holds with ζ replaced by ∞ and E by E_Δ for all $\mu \in \mathcal{P}(E_\Delta)$.

Definition 1.9. Let $(\mathcal{E}, D(\mathcal{E}))$ be a Dirichlet form on $L^2(E;m)$ and $(T_t)_{t>0}$, $(\hat{T}_t)_{t>0}$ the associated sub-Markovian strongly continuous contraction semigroups on $L^2(E;m)$.

(i) A right process $\mathbf{M}$ with state space E and transition semigroup $(p_t)_{t>0}$ is called *associated* (*coassociated* respectively) *with* $(\mathcal{E}, D(\mathcal{E}))$ if $p_t f$ is an m-version of $T_t f$ ($\hat{T}_t f$ respectively) for all $t > 0$, $f \in \mathcal{B}_b(E) \cap L^2(E;m)$. $\mathbf{M}$ is called *properly associated* (*properly coassociated* respectively) *with* $(\mathcal{E}, D(\mathcal{E}))$ if in addition, $p_t f$ is $\mathcal{E}$-quasi-continuous for all $t > 0$, $f \in \mathcal{B}_b(E) \cap L^2(E;m)$.

(ii) A pair $(\mathbf{M}, \hat{\mathbf{M}})$ is called *(properly) associated with* $(\mathcal{E}, \mathcal{D}(\mathcal{E}))$ if $\mathbf{M}$ is (properly) associated and $\hat{\mathbf{M}}$ is (properly) coassociated with $(\mathcal{E}, \mathcal{D}(\mathcal{E}))$.

Remark 1.10. (i) Note that by our notation convention (i.e., if $f \in \mathcal{B}_b(E)$ then f also denotes the associated m-class) p_t respects m-classes if $\mathbf{M}$ is associated with $(\mathcal{E}, D(\mathcal{E}))$ (cf. the Remark following II.5.1).
(ii) If $(\mathbf{M}, \hat{\mathbf{M}})$ is associated with $(\mathcal{E}, D(\mathcal{E}))$ and if $(p_t)_{t>0}$, $(\hat{p}_t)_{t>0}$ denote the corresponding transition semigroups, then clearly $(p_t)_{t>0}$ is in duality with $(\hat{p}_t)_{t>0}$ w.r.t. m. Hence m is p_t- and $\hat{p}_t$-supermedian for all $t > 0$ (cf. Chap.II, Sect. 5).

2 Quasi-regular Dirichlet forms

In this section we characterize the class of all Dirichlet forms (called *quasi-regular*) which are associated with right processes.

Definition 2.1. A Dirichlet form $(\mathcal{E}, D(\mathcal{E}))$ on $L^2(E;m)$ is called *quasi-regular* if:

(i) There exists an $\mathcal{E}$-nest $(E_k)_{k\in\mathbb{N}}$ consisting of compact sets.

(ii) There exists an $\tilde{\mathcal{E}}_1^{1/2}$-dense subset of $D(\mathcal{E})$ whose elements have $\mathcal{E}$-quasi-continuous m-versions.

(iii) There exist $u_n \in D(\mathcal{E})$, $n \in \mathbb{N}$, having $\mathcal{E}$-quasi-continuous m-versions $\tilde{u}_n$, $n \in \mathbb{N}$, and an $\mathcal{E}$-exceptional set $N \subset E$ such that $\{\tilde{u}_n | n \in \mathbb{N}\}$ separates the points of $E \setminus N$.

Remark 2.2. (i) We emphasize that the notion of quasi-regularity only depends on the symmetric part of $(\mathcal{E}, D(\mathcal{E}))$.
(ii) By III.1.7 property 2.1(i) is equivalent to the *tightness* of $\tilde{\mathrm{Cap}}_h$ (where h is as in III.(2.1)). $\tilde{\mathrm{Cap}}_h$ is called *tight* if there exist compact sets $K_n \subset E$, $n \in \mathbb{N}$, such that $\lim_{n\to\infty} \tilde{\mathrm{Cap}}_h(E \setminus K_n) = 0$. Hence 2.1(i) can be viewed as a substitute for local compactness. It is easy to see that any *regular* Dirichlet form on a locally compact separable metric state space E (in the sense of [F 80]) is quasi-regular. For details we refer to Subsection 3a) below.

Exercise 2.3. Let $(\mathcal{E}, D(\mathcal{E}))$ be a quasi-regular Dirichlet form on $L^2(E; m)$. Then each element $u \in D(\mathcal{E})$ has an $\mathcal{E}$-quasi-continuous m-version denoted by $\tilde{u}$ (which is in fact $\mathcal{E}$-q.e. unique).

Now we can formulate the existence theorem and its converse.

Theorem 2.4. *Let $(\mathcal{E}, D(\mathcal{E}))$ be a quasi-regular Dirichlet form on $L^2(E;m)$. Then there exists a pair $(\mathbf{M}, \hat{\mathbf{M}})$ of special standard processes which is properly associated with $(\mathcal{E}, D(\mathcal{E}))$.*

Remark 2.5. (i) 2.4 extends M. Fukushima's famous existence result (cf. [F 71, 73, 80]) in the "locally compact regular case" (cf. below).
(ii) Since the notion of quasi-regularity depends on the symmetric part of $(\mathcal{E}, D(\mathcal{E}))$ only, it follows from 2.4 that there also exists a special standard process $\tilde{\mathbf{M}}$ properly associated with $(\tilde{\mathcal{E}}, D(\tilde{\mathcal{E}}))$.

Theorem 2.6. *Suppose that there exists an m-special standard process $\mathbf{M} = (\Omega, \mathcal{F}, (X_t)_{t\geq 0}, (P_z)_{z\in E_\Delta})$ with state space E and life time ζ which is m-tight and associated with $(\mathcal{E}, D(\mathcal{E}))$. Then $(\mathcal{E}, D(\mathcal{E}))$ is quasi-regular, i.e., satisfies 2.1(i)-(iii). Moreover, $\mathbf{M}$ is properly associated with $(\mathcal{E}, D(\mathcal{E}))$.*

Corollary 2.7. *Suppose there exists an m-special standard process $\mathbf{M}$ with state space E associated with $(\mathcal{E}, D(\mathcal{E}))$. Then there exists a special standard process $\hat{\mathbf{M}}$ properly coassociated with $(\mathcal{E}, D(\mathcal{E}))$ which is in duality with $\mathbf{M}$ w.r.t. m (i.e., its transition function is in duality with that of $\mathbf{M}$ w.r.t. m).*

The cleanest formulation one has in terms of right processes.

Theorem 2.8. *A Dirichlet form $(\mathcal{E}, \mathcal{D}(\mathcal{E}))$ on $L^2(E;m)$ is quasi-regular if and only if there exists a pair $(\mathbf{M}, \hat{\mathbf{M}})$ of right processes associated with $(\mathcal{E}, \mathcal{D}(\mathcal{E}))$. In this case $(\mathbf{M}, \hat{\mathbf{M}})$ is always properly associated with $(\mathcal{E}, \mathcal{D}(\mathcal{E}))$.*

Applications of these results to the construction of certain processes will be given in Section 3 of this chapter and Chapters V, VI below. But one can also just study (right or m-)special standard processes in general using Dirichlet forms and vice versa. For example, one can characterize the sets which are not hit (or "touched") by the process P_m-a.s. .Let $\mathbf{M} = (\Omega, \mathcal{F}, (X_t)_{t\geq 0}, (P_z)_{z\in E_\Delta})$ be a special standard process with life time ζ associated with $(\mathcal{E}, D(\mathcal{E}))$. For $B \subset E$ we define the *first touching time* τ_B of B by

$$\tau_B = \inf\{0 \leq t < \zeta \,|\, X_t \in B \text{ or } X_{t-} \in B\} \wedge \zeta .$$

We again use the convention that $\inf \emptyset := \infty$.

Theorem 2.9. *A subset N of E is $\mathcal{E}$-exceptional if and only if $P_m[\tau_{\tilde{N}} < \zeta] = 0$ for some $\tilde{N} \in \mathcal{B}(E)$ with $N \subset \tilde{N}$.*

Concerning the probabilistic significance of $\mathcal{E}$-quasi-continuity one has

Theorem 2.10. *Let $u : E \to \mathbb{R}$ be $\mathcal{E}$-quasi-continuous, then for $\mathcal{E}$-q.e. $z \in E$ $u(X_{t-}) = u(X_t)_-$ for all $t \in]0, \zeta[$ P_z-a.s., and*

$$P_z[t \mapsto u(X_t) \text{ is right continuous on } [0,\infty[\text{ and } \\ t \mapsto u(X_{t-}) \text{ is left continuous on }]0,\zeta[\,] = 1 .$$

Also the path behaviour of the process can be described via the Dirichlet form. For a $\mathcal{B}(E)$-measurable function u on E we set

$$\text{(2.1)} \qquad \operatorname{supp}[u] := \operatorname{supp}[|u| \cdot m]$$

and call $\operatorname{supp}[u]$ the *support* of u. Note that $u = u'$ m-a.e. implies $\operatorname{supp}[u] := \operatorname{supp}[u']$. Hence $\operatorname{supp}[u]$ is well-defined by (2.1) for all $u \in L^2(E;m)$.

Definition 2.11. $(\mathcal{E}, \mathcal{D}(\mathcal{E}))$ is said to have the *local property* (or to be *local*) if

$$\mathcal{E}(u,v) = 0 \qquad \text{for all } u, v \in D(\mathcal{E}) \text{ with } \operatorname{supp}[u] \cap \operatorname{supp}[v] = \emptyset \\ \text{and } \operatorname{supp}[u], \operatorname{supp}[v] \text{ compact.}$$

Theorem 2.12. *$(\mathcal{E}, D(\mathcal{E}))$ has the local property if and only if*

$$P_z\,[t \mapsto X_t \text{ is continuous on } [0,\zeta[\,] = 1 \text{ for } \mathcal{E}\text{-q.e. } z \in E .$$

Remark 2.13. There is also a characterization of Hunt processes in terms of the associated Dirichlet form (cf. [MR 92, Chap.V, Sect. 2] and also [AMR 92d]).

Further illustrations of the fruitful interplay between Dirichlet forms and the corresponding Markov processes are contained in M. Fukushima's lectures. In particular, he will give an extension of the well-known Îto-decomposition of functions along the paths. This extension is meanwhile called *Fukushima decomposition* (cf. [F 80] and also V.2.2 below). We shall use this decomposition in an essential way in Chapter V below. Usually this decomposition as well as many other "classical" results on Dirichlet forms are proved only for *regular* (cf. the next section) Dirichlet forms on state spaces E which are *locally compact* separable metric spaces. To extend all these results to quasi-regular Dirichlet forms on Lusin (or more general) spaces we have developed a "regularization method" which we shall briefly describe in Section 4 of this chapter. But first we give some examples of quasi-regular Dirichlet forms.

3 Examples of quasi-regular Dirichlet forms

a) Regular Dirichlet forms

Assume E is a locally compact separable metric space and m a positive Radon measure on E. Let $(\mathcal{E}, D(\mathcal{E}))$ be a *regular* Dirichlet form on $L^2(E;m)$, i.e., $C_0(E) \cap D(\mathcal{E})$ is dense both in $D(\mathcal{E})$ w.r.t. $\tilde{\mathcal{E}}_1^{1/2}$ and in $C_0(E)$ w.r.t. the uniform norm $\| \|_\infty$. Then $(\mathcal{E}, D(\mathcal{E}))$ is quasi-regular. Indeed, 2.1(ii), (iii) are obvious. To show 2.1(i) let $(K_n)_{n\in\mathbb{N}}$ be a sequence of compact subsets of E such that K_{n+1} contains K_n in its interior for all $n \in \mathbb{N}$ and $E = \bigcup_{n\geq 1} K_n$. Clearly,

$$C_0(E) \cap D(\mathcal{E}) \subset \bigcup_{n\geq 1} D(\mathcal{E})_{K_n} ,$$

hence 2.1(i) holds by the regularity of $(\mathcal{E}, D(\mathcal{E}))$. This implies that 2.4 applies here to ensure the existence of a pair $(M, \hat{M})$ of (m-tight) special standard processes associated with $(\mathcal{E}, D(\mathcal{E}))$. We note that as in [Si 74], [F 80], [Ca-Me 75], [Le 82, 83], [O 88] it can be proved that they are indeed Hunt processes.
In particular, all this applies to those examples $(\mathcal{E}, D(\mathcal{E}))$ discussed in Chap.II, Sections 2,3, where E was an open subset U of $\mathbb{R}^d$ and $C_0^\infty(U)$ was dense in $D(\mathcal{E})$ w.r.t $\tilde{\mathcal{E}}_1^{1/2}$. In the case discussed in Chap.II, Subsection 2b) one just obtains *classical Brownian motion* on U with *absorbing boundary* .

b) Dirichlet forms on infinite dimensional state space

Let E be a separable real Banach space with dual E' and μ a positive finite measure on its Borel σ-algebra $\mathcal{B}(E)$ with $\operatorname{supp}\mu = E$. Let K_0 be a (finite or) countable set of μ-admissible elements (cf. II.4.1) such that

(3.1) there exists a constant $c > 0$ such that
$$\sum_{k\in K_0} {}_{E'}\langle l, k\rangle_E^2 \leq c^2 \|l\|_{E'}^2 \quad \text{for all } l \in E' .$$

(Note that (3.1) is, of course, always fulfilled if K_0 is finite.) In particular, condition II.(4.6) is satisfied, hence we know by II.4.5 that the closure of

$$\mathcal{E}(u,v) := \sum_{k\in K_0} \int \frac{\partial u}{\partial k}\frac{\partial v}{\partial k}\, d\mu \; ; \quad u,v \in \mathcal{F}C_b^\infty , \tag{3.2}$$

denoted by $(\mathcal{E}, D(\mathcal{E}))$ exists and is a Dirichlet form on $L^2(E;\mu)$. To show that $(\mathcal{E}, D(\mathcal{E}))$ is quasi-regular note that 2.1(ii) holds by definition. By the Hahn-Banach theorem (cf. e.g. [Ch 69b]) it follows that 2.1(iii) holds. In order to show 2.1(i) recall that by 2.2(ii) it is sufficient to prove that Cap := Cap_1 (i.e., $h = G_1 1 = 1$, which belongs to $\mathcal{F}C_b^\infty \subset D(\mathcal{E})$) is tight. This will be done below using a method due to B. Schmuland (cf. [RS 91a, Proposition 3.1]). We need the following lemma.

Lemma 3.1. *(i) Let $\varphi \in C^1(\mathbb{R})$ with bounded derivative φ'. Then $\varphi \circ u \in D(\mathcal{E})$ whenever $u \in D(\mathcal{E})$ and for every $k \in K_0$*

$$\frac{\partial}{\partial k}\varphi(u) = \varphi'(u)\frac{\partial u}{\partial k} ,$$

where $\frac{\partial}{\partial k}$ also denotes the closure of $\frac{\partial}{\partial k} : \mathcal{F}C_b^\infty \to L^2(E;\mu)$ on $L^2(E;\mu)$.

(ii) Let $u, v \in D(\mathcal{E})$ then for all $k \in K_0$

$$\frac{\partial}{\partial k}(u \vee v) = 1_{\{u>v\}}\frac{\partial u}{\partial k} + 1_{\{u<v\}}\frac{\partial v}{\partial k} + \frac{1}{2}1_{\{u=v\}}\left(\frac{\partial u}{\partial k} + \frac{\partial v}{\partial k}\right)$$

and

$$\frac{\partial}{\partial k}(u \wedge v) = 1_{\{u<v\}}\frac{\partial u}{\partial k} + 1_{\{u>v\}}\frac{\partial v}{\partial k} + \frac{1}{2}1_{\{u=v\}}\left(\frac{\partial u}{\partial k} + \frac{\partial v}{\partial k}\right) .$$

(iii) For all $u, v \in D(\mathcal{E})$

$$\sum_{k\in K_0}\left(\frac{\partial}{\partial k}(u \wedge v)\right)^2 \vee \sum_{k\in K_0}\left(\frac{\partial}{\partial k}(u \vee v)\right)^2 \leq \left(\sum_{k\in K_0}\left(\frac{\partial u}{\partial k}\right)^2\right) \vee \left(\sum_{k\in K_0}\left(\frac{\partial v}{\partial k}\right)^2\right) .$$

Proof. (i): The fact that $\varphi \circ u \in D(\mathcal{E})$ is a consequence of I.3.10(ii) since

$$|\varphi(u)(x) - \varphi(u)(y)| \leq \|\varphi'\|_\infty |u(x) - u(y)| \text{ for all } x, y \in E .$$

The rest is left as an exercise.
(ii): Let $(\delta_n)_{n\in\mathbb{N}}$ be a Dirac sequence (i.e., $\delta_n \in C_0^\infty(\mathbb{R})$, $\delta_n \geq 0$, $\int \delta_n ds = 1$, $\delta_n(s) = \delta_n(-s)$, $s \in \mathbb{R}$, and $\operatorname{supp}\delta_n \subset]-\frac{1}{n},\frac{1}{n}[$ for all $n \in \mathbb{N}$). Set $\varphi_n(t) := \int |t-s|\delta_n(s)\,ds$, $t \in \mathbb{R}$, $n \in \mathbb{N}$. Then $\varphi_n \in C^\infty(\mathbb{R})$, $\|\varphi_n'\|_\infty \leq 1$, $\varphi_n'(t) \underset{n\to\infty}{\longrightarrow} \operatorname{sign}(t)$ for all $t \in \mathbb{R}$ and $\varphi_n \underset{n\to\infty}{\longrightarrow} |\cdot|$ locally uniformly. Let $u \in D(\mathcal{E})$, u bounded, then $(\|\varphi_n(u)\|_\infty)_{n\in\mathbb{N}}$ is bounded, hence $\varphi_n(u) \underset{n\to\infty}{\longrightarrow} |u|$ in $L^2(E;\mu)$. The closedness of $(\mathcal{E}, D(\mathcal{E}))$ now implies that $\frac{\partial |u|}{\partial k} = \lim_{n\to\infty} \frac{\partial \varphi_n(u)}{\partial k}$ in $L^2(E;\mu)$ for all $k \in K_0$, since by (i) we have that

$$\frac{\partial \varphi_n(u)}{\partial k} = \varphi_n'(u)\frac{\partial u}{\partial k} \underset{n\to\infty}{\longrightarrow} \operatorname{sign}(u)\frac{\partial u}{\partial k} \text{ in } L^2(E;\mu) .$$

Consequently,

$$\frac{\partial |u|}{\partial k} = \operatorname{sign}(u)\frac{\partial u}{\partial k} \quad \text{for all } k \in K_0 .$$

An application of I.3.11 yields that this is true for all $u \in D(\mathcal{E})$. Using the identities $u \vee v = \frac{1}{2}(u+v) + \frac{1}{2}|u-v|$ and $u \wedge v = \frac{1}{2}(u+v) - \frac{1}{2}|u-v|$ we obtain (ii).
(iii) immediately follows from (ii). □

Proposition 3.2. Cap *is tight.*

Proof. Since E is separable we may choose a fixed countable dense set $\{y_m \mid m \in \mathbb{N}\}$ in E. By the Hahn-Banach theorem we can find $l_m \in E'$ so that $\|l_m\|_{E'} = 1$ and $l_m(y_m) = \|y_m\|_E$. Note that hence

$$\|z\|_E = \sup_{m\in\mathbb{N}} l_m(z) ;$$

(indeed, $l_m(z) \leq |l_m(z)| \leq \|z\|_E$ and for some subsequence $(m_n)_{n\in\mathbb{N}}$,

$$\|z\|_E = \lim_{n\to\infty} \|y_{m_n}\|_E = \lim_{n\to\infty} l_{m_n}(y_{m_n}) = \lim_{n\to\infty} l_{m_n}(z) \leq \sup_{m\in\mathbb{N}} l_m(z)) .$$

Let $\varphi \in C_b^\infty(\mathbb{R})$, φ increasing with $\varphi(t) = t$ for all $t \in [-1,1]$, $\|\varphi'\|_\infty \leq 1$, and for $m \in \mathbb{N}$ define $v_m : E \to \mathbb{R}$ by

$$v_m(z) = \varphi(\|z - y_m\|_E) \ , \quad z \in E \ .$$

Suppose we can show that

$$\text{(3.3)} \qquad w_n := \inf_{m \leq n} v_m, n \in \mathbb{N}, \text{ converges } \mathcal{E}\text{-quasi-uniformly to zero on } E \ .$$

Then (by III.1.7) for every $k \in \mathbb{N}$ there exists a closed set F_k such that $\mathrm{Cap}(F_k^c) < \frac{1}{k}$ and $w_n \underset{n\to\infty}{\longrightarrow} 0$ uniformly on F_k. Hence for all $0 < \epsilon < 1$ there exists $n \in \mathbb{N}$ such that $w_n \leq \epsilon$ on F_k, i.e., by the definition of w_n,

$$F_k \subset \bigcup_{m=1}^{n} B(y_m, \epsilon)$$

where $B(y,\epsilon) := \{z \in E \,|\, \|z-y\|_E \leq \epsilon\}$. Consequently, F_k is totally bounded, hence compact, and Cap is tight. To show (3.3) we first fix $m \in \mathbb{N}$ and note that if for $n \in \mathbb{N}$

$$u_n(z) := \sup_{j \leq n} \varphi(l_j(z - y_m)) \ , \ z \in E \ ,$$

then each u_n is a continuous version of an element in $D(\mathcal{E})$, $u_n \uparrow v_m$ on E as $n \to \infty$ hence in $L^2(E;\mu)$, and by 3.1(iii) (and induction)

$$\sum_{k \in K_0} \left(\frac{\partial u_n}{\partial k}\right)^2 \leq \sup_{j \leq n} \left(\sum_{k \in K_0} \varphi'(l_j(z - y_m))^2{}_{E'}\langle l_j, k\rangle_E^2 \right) \ .$$

Since $\|\varphi'\|_\infty \leq 1$ and $\|l_j\|_{E'} = 1$ for all $j \in \mathbb{N}$, it follows by (3.1) that this is bounded by c^2. Applying I.2.18 we conclude that $v_m \in D(\mathcal{E})$ and

$$\text{(3.4)} \qquad \sum_{k \in K_0} \left(\frac{\partial v_m}{\partial k}\right)^2 \leq c^2 \quad \text{for all } m \in \mathbb{N} \ .$$

By III.2.4 it follows that all v_m and hence w_n, $n \in \mathbb{N}$, are $\mathcal{E}$-quasi-continuous. Since $\{y_m \,|\, m \in \mathbb{N}\}$ is dense in E, $w_n \downarrow 0$ on E as $n \to \infty$ hence in $L^2(E;\mu)$. But by 3.1(iii) (and induction) and (3.4)

$$\sum_{k \in K_0} \left(\frac{\partial w_n}{\partial k}\right)^2 \leq \inf_{m \leq n} \sum_{k \in K_0} \left(\frac{\partial v_m}{\partial k}\right)^2 \leq c^2 \quad \text{for all } n \in \mathbb{N} \ .$$

Applying I.2.18 and III.2.4 we obtain that a subsequence of the Cesaro mean $\overline{w}_n = \frac{1}{n}\sum_{j=1}^n w_{n_j}$, $n \in \mathbb{N}$, of some subsequence $(w_{n_j})_{j \in \mathbb{N}}$ of $(w_n)_{n \in \mathbb{N}}$ converges to zero (in $D(\mathcal{E})$ and) $\mathcal{E}$-quasi-uniformly. But since $(w_n)_{n \in \mathbb{N}}$ is decreasing, $w_{n_n} \leq \frac{1}{n}\sum_{j=1}^{n} w_{n_j}$ and thus $w_n \underset{n\to\infty}{\longrightarrow} 0$ $\mathcal{E}$-quasi-uniformly, i.e., (3.3) is shown and the proof is complete. □

Remark. For an analogous result to 3.2 when E is replaced by the (non-metrizable) dual of a nuclear space we refer to [AR 89b, Sect.3.(c)].

By 2.4 we hence know that there exists a special standard process associated with $(\mathcal{E}, D(\mathcal{E}))$ which is in fact a diffusion with infinite life time. All this remains true for the more general (not necessarily symmetric) Dirichlet forms considered in Chap.II, Subsection 4d).

4 Regularization

Let $(E^\#, \mathcal{B}^\#)$ be a measurable space and let $i : E \to E^\#$ be a $\mathcal{B}(E)/\mathcal{B}^\#$-measurable map. Let $m^\# := m \circ i^{-1}$, i.e., the image measure of m under i. Define an isometry $i^* : L^2(E^\#; m^\#) \to L^2(E; m)$ by defining $i^*(u^\#)$ to be the m-class represented by $\tilde{u} \circ i$ for any $\mathcal{B}^\#$-measurable $m^\#$-version $\tilde{u}$ of $u^\# \in L^2(E^\#; m^\#)$. Note that the range of i^* is always closed, but, of course, in general strictly smaller than $L^2(E; m)$. Define

$$D(\mathcal{E}^\#) := \{u^\# \in L^2(E^\#; m^\#) | i^*(u^\#) \in D(\mathcal{E})\} \tag{4.1}$$

$$\mathcal{E}^\#(u^\#, v^\#) := \mathcal{E}(i^*(u^\#), i^*(v^\#))\ ;\ u^\#, v^\# \in D(\mathcal{E}^\#)\ . \tag{4.2}$$

Then clearly, $(\mathcal{E}^\#, D(\mathcal{E}^\#))$ is a positive definite bilinear form on $L^2(E^\#; m^\#)$ satisfying the weak sector condition, called the *image of* $(\mathcal{E}, D(\mathcal{E}))$ under i.

Exercise 4.1. (i) Show that the symmetric part of $(\mathcal{E}^\#, D(\mathcal{E}^\#))$ is closed. Show that if $D(\mathcal{E}^\#)$ is dense in $L^2(E^\#; m^\#)$ then $(\mathcal{E}^\#, D(\mathcal{E}^\#))$ is a Dirichlet form on $L^2(E^\#; m^\#)$, called the *image Dirichlet form* of $(\mathcal{E}, D(\mathcal{E}))$ under i.

(ii) Show that if i^* is onto then $D(\mathcal{E}^\#)$ is dense in $L^2(E^\#; m^\#)$.

One can prove the following (cf. [MR 92, VI.1.2, 1.4, 1.6])

Theorem 4.2. *There exists an $\mathcal{E}$-nest $(E_k)_{k\in\mathbb{N}}$ consisting of compact sets in E and a locally compact separable metric space $Y^\#$ such that:*

(i) $Y^\#$ is a local compactification of $Y := \cup E_k$ in the following sense: $Y^\#$ is a locally compact space containing Y as a dense subset and $\mathcal{B}(Y) = \{A \in \mathcal{B}(Y^\#) | A \subset Y\}$.

(ii) The trace topologies on E_k induced by E, $Y^\#$ respectively coincide for every $k \in \mathbb{N}$.

(iii) The image $(\mathcal{E}^\#, D(\mathcal{E}^\#))$ of $(\mathcal{E}, D(\mathcal{E}))$ under the inclusion map $i : Y \subset Y^\#$ is a regular Dirichlet form on $L^2(Y^\#; m^\#)$ where $m^\# := m \circ i^{-1}$ is a positive Radon measure on $Y^\#$.

Remark 4.3. Note that in 4.2 the terminology on image Dirichlet forms applies in the following sense. Since $E \setminus Y$ is $\mathcal{E}$-exceptional we may assume that $E = Y$ by just restricting every element in $L^2(E; m)$ to Y. Observe also that 4.2(i) implies that i^* is onto.

Corollary 4.4. *Let h be as in III.(2.1) and let $\tilde{\mathrm{Cap}}_h^\#$ be the corresponding capacity associated with $(\tilde{\mathcal{E}}^\#, D(\mathcal{E}^\#))$ on $E^\#$.*

(i) If $(F_k)_{k\in\mathbb{N}}$ is an $\mathcal{E}^\#$-nest, then $(F_k \cap E_k)_{k\in\mathbb{N}}$ is an $\mathcal{E}$-nest and vice versa.

(ii) $N^\# \subset E^\#$ is $\mathcal{E}^\#$-exceptional if and only if $N^\# \cap E$ is $\mathcal{E}$-exceptional. In particular, $\tilde{\mathrm{Cap}}_h^\#(E^\# \setminus E) = 0$.

(iii) A function $u^\# : E^\# \to \mathbb{R}$ is $\mathcal{E}^\#$-quasi-continuous if and only if $u^\#_{|E}$ is $\mathcal{E}$-quasi-continuous.

(iv) $\tilde{\mathrm{Cap}}_h^\#(A^\#) = \tilde{\mathrm{Cap}}_h(A^\# \cap E)$ for all $A^\# \subset E^\#$.

Definition 4.5. Let $\mathbf{M} = (\Omega, \mathcal{F}, (X_t)_{t\geq 0}, (P_z)_{z\in E_\Delta})$ be a right process with state space E and life time ζ. A Borel set $S \in \mathcal{B}(E)$ is said to be $\mathbf{M}$-*invariant* if there exists $\Omega_{E\setminus S} \in \mathcal{F}$ such that

$$\Omega_{E\setminus S} \supset \{\overline{X_0^t} \cap (E\setminus S) \neq \emptyset \text{ for some } 0 \leq t < \zeta\}$$

and $P_z[\Omega_{E\setminus S}] = 0$ for all $z \in S$. Here $\overline{X_0^t}(\omega)$ is the closure of $\{X_s(\omega) \mid s \in [0,t]\}$ in E.

Remark 4.6. Note that if S is $\mathbf{M}$-invariant, we can define a right process $\mathbf{M}|_S := (\Omega', \mathcal{F}', (X_t|_{\Omega'})_{t\geq 0}, (P_z|_{\mathcal{F}'})_{z\in S_\Delta})$ with state space S where $\Omega' := \Omega\setminus\Omega_{E\setminus S}$ and $\mathcal{F}' = \mathcal{F}\cap\Omega'$. $\mathbf{M}|_S$ is called the *restriction* of $\mathbf{M}$ to S.

Theorem 4.7. *Let $\mathbf{M} = (\Omega, \mathcal{F}, (X_t)_{t\geq 0}, (P_z)_{z\in E_\Delta})$ be a right process properly associated with the quasi-regular Dirichlet form $(\mathcal{E}, D(\mathcal{E}))$ on $L^2(E;m)$. Then there exists $N \subset E$, N $\mathcal{E}$-exceptional, such that $S := E\setminus N$ is $\mathbf{M}$-invariant and if $\overline{\mathbf{M}}$ is the trivial extension to $E^\#$ (cf. [MR 92, IV.3.23(i)]) of $\mathbf{M}$ restricted to S, then $\overline{\mathbf{M}}$ is a Hunt process properly associated with the regular Dirichlet form $(\mathcal{E}^\#, D(\mathcal{E}^\#))$ on $L^2(E^\#;m^\#)$, where $E^\#_\Delta$ is taken as the one point compactification of $E^\#$.*

By 4.2, 4.4, 4.7 one can transfer all results about classical Dirichlet forms on locally compact separable metric spaces and their associated Hunt processes to quasi-regular Dirichlet forms on Lusin (in fact arbitrary) state spaces and their associated right processes (cf. [MR 92 Chap.VI, Sect. 2] for details and [AMR 90] for the underlying ideas).

Chapter V

Application to SDE's with Infinite Dimensional State Space

Consider the situation of Chap II, Subsection 4 b), i.e., E is a separable real Banach space, μ a finite positive measure on $\mathcal{B}(E)$ with $\text{supp}\mu = E$, and $(H, <>_H)$ is a separable real Hilbert space such that

(C.1) $E' \subset H' \equiv H \subset E$ densely and continuously.

Assume

(C.2) there exists a dense linear space $K \subset E'(\subset H \subset E)$ consisting of well-μ-admissible elements in E.

Then by II. 4.7 the form

$$\mathcal{E}(u,v) = \int_E < \nabla u, \nabla v >_H d\mu \quad ; u, v \in \mathcal{F}C_b^\infty \tag{0.1}$$

is closable on $L^2(E;\mu)$ and its closure $(\mathcal{E}, D(\mathcal{E}))$ is a symmetric Dirichlet form.
In this chapter we want to prove that (under some mild assumptions) the diffusion process **M** associated with $(\mathcal{E}, D(\mathcal{E}))$ (cf. Chap. IV, Sect. 3) weakly solves a stochastic differential equation of type

$$\begin{aligned} dX_t &= dW_t + \beta(X_t)dt \\ X_0 &= z(\in E) \quad \text{under } P_z \end{aligned} \tag{0.2}$$

where $(W_t)_{t\leq 0}$ is an E-valued Brownian motion over H and $\beta : E \to E$.

1 Comparison of the finite and infinite dimensional case.

Let $E = \mathbb{R}^d$ and consider (0.2) in this case. Assume:

(1.1) There exists $\rho : \mathbb{R}^d \to \mathbb{R}_+$ weakly differentiable such that $\beta = \nabla \log \rho (= \nabla\rho/\rho) \in L^2(\mathbb{R}^d; \rho \cdot dz)$

where dz denotes d-dimensional Lebesgue measure. Assume for simplicity that $\int \rho\, dz < \infty$ (though $\rho \in L^1_{loc}(\mathbb{R}^d; dx)$ is, of course, enough). Then by II. 1.3 the form

$$\mathcal{E}(u,v) = \int_{\mathbb{R}^d} < \nabla u, \nabla v >_{\mathbb{R}^d} \rho\, dz; \quad u,v \in C_0^\infty(\mathbb{R}^d)$$

is closable on $L^2(\mathbb{R}^d; \rho \cdot dx)$ and by Chap. IV, Subsection 3a) its closure $(\mathcal{E}, D(\mathcal{E}))$ has an associated diffusion **M**. It was proved by Fukushima (cf. [F 81, 82, 84]) that **M** solves (0.2) in this case (i.e., with $E = H = \mathbb{R}^d$ and $\beta = \nabla \log \rho$) using the *Fukushima decomposition*. Since there is no Lebesgue measure on infinite dimensional space, there seems to be no direct analogue for the crucial assumption (1.1) at first sight. Below we shall see, however, that if (1.1) is reformulated in terms of "conditional densities" it carries over to the infinite dimensional case. We first need some preparations. Let E, μ be as at the beginning of this chapter.
For $k \in E \setminus \{0\}$ fixed, let E_k be a "complementing subspace", i.e., a closed subspace of E such that $E = E_k \oplus k\mathbb{R}$. For each $z \in E, z = x + sk$, where $x \in E_k, s \in \mathbb{R}$ are uniquely determined. Let $\pi_k : E \to E_k$ be the canonical projection. We can disintegrate μ w.r.t. $\pi_k : E \to E_k$, i.e. there exists a kernel $\rho_k : E_k \times \mathcal{B}(\mathbb{R}) \to [0,1]$ such that for all $u : E \to [0,\infty[$, $\mathcal{B}(E)$-measurable

$$(1.2) \qquad \int_E u(z)\mu(dz) = \int_{E_k}\int_{\mathbb{R}} u(x+sk)\rho_k(x,ds)\nu_k(dx)$$

where $\nu_k := \mu \circ \pi_k^{-1}$ (i.e., the image measure of μ under π_k). $\rho_k(\cdot, ds)$ is ν_k - a.e. uniquely determined. Correspondingly, $L^2(E;\mu)$ can be written as a direct integral of L^2-spaces over $\mathbb{R}$:

$$(1.3) \qquad L^2(E;\mu) = \int^{\oplus} L^2(\mathbb{R}; \rho_k(x,ds))\nu_k(dx)$$

in the sense that each $u \in L^2(E;\mu)$ corresponds to a "field of vectors" $(u_x)_{x\in E_k}$ where $u_x := u(x + \cdot k)$, $x \in E_k$ (cf. [AR 90a] for details and references). Now in case $E = \mathbb{R}^d$ and $\mu = \rho \cdot dz$ fix $k \in \mathbb{R}^d$ and disintegrate $\rho \cdot dz$ according to $\mathbb{R}^d = \mathbb{R}^{d-1} \oplus k\mathbb{R}$ as in (1.2), (1.3), i.e.,

$$(1.4) \qquad \rho(z) \cdot dz = \rho_k(x,ds)\nu_k(dx).$$

Then it is easy to see that $\nu_k(dx)$ is absolutely continuous w.r.t. $(d-1)$-dimensional Lebesgue measure with density $f(x) := \int \rho(x+sk)ds, x \in \mathbb{R}^{d-1}$, and for ν_k-a.e. $x \in \mathbb{R}^{d-1}, \rho_k(x,ds) = \rho_k(x,s)ds$ where

$$\rho_k(x,s) = 1_{\{f>0\}}(x)\frac{\rho(x+sk)}{f(x)}, s \in \mathbb{R}.$$

But then clearly

$$(1.5) \qquad \beta_k := < k, \beta >_{\mathbb{R}^d} = \frac{\partial}{\partial s}\log \rho_k.$$

Consequently, (1.1) reformulated in terms of the conditional densities $\rho_k(x,\cdot)$ becomes

$$(1.6) \qquad \beta_k := \frac{\partial}{\partial s}\log \rho_k \in L^2(\mathbb{R}^d; \rho \cdot dz) \text{ for all } k \in \mathbb{R}^d$$

and this carries over to infinite dimensions since we have the following

Theorem 1.1. *Let $k \in E \setminus \{0\}$. k is well-μ-admissible if and only if for ν_k-a.e. $x \in E_k, \rho_k(x,ds) = \rho_k(x,s)ds$ for some Borel measurable function $\rho_k(x,\cdot) : \mathbb{R} \to [0,\infty[$ such that $\frac{\partial}{\partial s}\rho_k(x,\cdot) \in L^1_{loc}(\mathbb{R}; ds)$ and $\hat{\beta}_k := (\frac{\partial}{\partial s}\rho_k(x,\cdot)/\rho_k(x,\cdot))_{x\in E_k} \in \int^{\oplus} L^2(\mathbb{R}; \rho_k(x,ds))\nu_k(dx) = L^2(E;\mu)$. Here E_k is some complementing subspace as above and $\rho_k(\cdot,ds), \nu_k$ are as in (1.2), (1.3). The derivative $\frac{\partial}{\partial s}$ is in the sense of Schwartz distributions on $\mathbb{R}$. In this case $\hat{\beta}_k = \beta_k$ (defined as in II. 4.2).*

Proof. The sufficiency part (which is what we need) is an easy excercise, and for the necessity part we refer to [AKR 90]. □

In 1.1 we use the conventions $\frac{a}{0} := (signa) \cdot (+\infty)$ for $a \in \mathbb{R}$ and $(\pm\infty) \cdot 0 = 0$. Because of 1.1 we made assumption (C.2) at the beginning of this chapter and did not merely assume that (0.1) is closable. In applications, however, β is given, so we have to construct μ such that the "components" β_k of β are just the $\hat{\beta}_k$ defined in 1.1 if that is possible (i.e., we need some kind of measure-theoretic "Poincaré Lemma"; cf. below for examples).

2 Fukushima decomposition and "componentwise" solution of the SDE

We consider the situation described at the beginning of this chapter. Subsequently we use the notions *additive functional, martingale additive functional* (abbreviated MAF) *of finite energy* and *continuous additive functional* (abbreviated CAF) *of zero energy* in the sense of [F 80, § 5.1, § 5.2]. We only recall that if $(M_t)_{t\geq 0}$ is a martingale additive functional of $\mathbf{M}$ of finite energy then for $\mathcal{E}$-q.e. $z \in E$ $(M_t, \mathcal{F}_t, P_z)_{t\geq 0}$ is a martingale (cf. also M. Fukushima's lectures).

Lemma 2.1. *Let $u \in L^2(E;\mu)$, then*

(i) $E_z[\int_0^t |u|(X_s)ds] < \infty, t \geq 0$, for $\mathcal{E}$-q. e. $z \in E$.

(ii) $N_t := \int_0^t u(X_s)ds, t \geq 0$, is a CAF of $\mathbf{M}$ of zero energy.

Proof. For $z \in E, t \geq 0$

$$E_z[\int_0^t |u|(X_s)ds] \leq e^t \int_0^\infty e^{-s} E_z[|u|(X_s)]ds.$$

The right hand side is an $\mathcal{E}$-quasi-continuous function (cf. [MR 92, IV. 2.8]) and (i) follows. In particular, $(N_t)_{t\geq 0}$ is well-defined (i.e., independent of the μ-version we chose for u). $(N_t)_{t\geq 0}$ is clearly a CAF of $\mathbf{M}$ which is by [F 80, (5.2.12)] of zero energy.
□

Theorem 2.2. (Fukushima-decomposition) *(i) Let $u \in D(\mathcal{E})$ and let $\tilde{u}$ be an $\mathcal{E}$-quasi-continuous μ-version of u. Then the additive functional $(\tilde{u}(X_t) - \tilde{u}(X_0))_{t\geq 0}$ of $\mathbf{M}$ can be uniquely represented as*

$$\tilde{u}(X_t) - \tilde{u}(X_0) = M_t^{[u]} + N_t^{[u]}, t \geq 0, P_z - \text{ a.e.}, \mathcal{E}\text{- q.e.} z \in E, \tag{2.1}$$

where $M_t^{[u]}, t \geq 0$ is a MAF of $\mathbf{M}$ of finite energy and $(N_t^{[u]})_{t\geq 0}$ is a CAF of $\mathbf{M}$ of zero energy.

(ii) If $u \in D(L)(\subset D(\sqrt{-L}) = D(\mathcal{E})$ where L is the generator of $(\mathcal{E}, D(\mathcal{E}))$ then

$$N_t^{[u]} = \int_0^t (Lu)(X_s)ds, t \geq 0.$$

(iii) If $u \in D(\mathcal{E})$ such that $\int < \nabla u, \nabla u >_H^2 d\mu < \infty$, then

$$(2.2) \qquad < M^{[u]} >_t = 2\int_0^t < \nabla u(X_s), \nabla u(X_s) >_H ds, t \geq 0.$$

Proof. (i) follows by regularization from [F 80, Theorem 5.2.2] (cf. [MR 92, VI.2.5]), (ii) is straightforward, and (iii) follows by regularization from [F 80, Theorems 5.1.3, 5.2.3] (cf. [AR 91, 4.5] for details). □

For simplicity we also assume from now on that

(C.3) $\int {}_{E'}< k, z >_E^2 \mu(dz) < \infty$ for all $k \in K$ (K as in (C.2)).

Define for $k \in K, u_k(z) := {}_{E'}< k, z >_E, z \in E$.

Exercise 2.3. Prove that $u_k \in D(\mathcal{E})$ for each $k \in K$.

Lemma 2.4. *Let $k \in K$. Then $u_k \in D(L)$ and $Lu_k = \beta_k$.*

Proof. We have for all $v \in D(\mathcal{E})$

$$\mathcal{E}(u_k, v) = \int < k, \nabla v >_H d\mu = \int \frac{\partial v}{\partial k} d\mu = -\int v\beta_k d\mu.$$

□

Proposition 2.5. *Let $k \in K$. Then the decomposition (2.1) for $u = u_k$ reads*

$$u_k(X_t) - u_k(X_0) = W_t^k + \int_0^t \beta_k(X_s)ds, t \geq 0, P_z - \text{ a.e.}, \mathcal{E}\text{-q.e. } z \in E$$

where $(W_t^k, \mathcal{F}_t, P_z)_{t\geq 0}$ is a one-dimensional $(\mathcal{F}_t)_{t\geq 0}$- Brownian motion starting at zero for $\mathcal{E}$-q.e. $z \in E$ if $\|k\|_H = \frac{1}{4}$.

Proof. By 2.2 (iii)

$$(2.3) < M^{[u_k]} >_t = 2\int_0^t < \nabla u_k(X_s), \nabla u_k(X_s) >_H ds = 2\int_0^t \|k\|_H^2 ds = 2t\|k\|_H^2.$$

Now the assertion is obvious by Levy's characterization of Brownian motion and 2.2 (ii). □

Corollary 2.6. *Let $k_1, ..., k_d \in K$ be an orthogonal system in H with norms equal to $\frac{1}{4}$. Then $\bar{W}_t := (W_t^{k_1}. \cdots, W_t^{k_d})$, is a d-dimensional $(\mathcal{F}_t)_{t\geq 0}$-Brownian motion starting at zero under P_z, $\mathcal{E}$-q.e. $z \in E$.*

Proof. (2.3) and polarization implies that for all $k, k' \in K$

$$< W^k, W^{k'} >_t = 2t < k, k' >_H, t \geq 0.$$

Hence again by Levy the assertion follows. □

Lemma 2.7. *There exist $K_0 \subset K$, K_0 separates the points of E and K_0 is an orthonormal basis of H.*

Proof. Since E' separates the points of E, so does K. Since E is Souslinean, we obtain by [Sch 73, Proposition 4, p. 105] that there exists a countable subset K_1 of K still separating the points of E, hence in particular having a linear span dense in H. Applying Gram-Schmidt orthogonalization to K_1 we obtain K_0. □
As an immediate consequence we now obtain a "componentwise" solution for an SDE of type (0.2) given by **M**.

Theorem 2.8. *Let $K_0 \subset K$ be as in 2.7. Then for $\mathcal{E}$-q.e. $z \in E$, $(\{_{E'}< k, X_t >_E |k \in K_0\}, \mathcal{F}_t, P_z)_{t\geq 0}$ weakly solves the following system of stochastic differential equations:*

$$(2.4) \qquad \begin{array}{rcl} dY_t^k & = & dW_t^k + \beta_k((Y_t^k)_{k\in K_0})dt, k \in K_0, \\ Y_0^k & = & _{E'}< k, z >_E \end{array}$$

where $\{(W_t^k)_{t\geq 0}|k \in K_0\}$ is a collection of independent one-dimensional $(\mathcal{F}_t)_{t\geq 0}$-Brownian motions starting at zero under P_z for $\mathcal{E}$-q.e. $z \in E$ (and where according to 2.7 we identify $z \in E$ with $(_{E'}< k, z >_E)_{k\in K_0}$).

3 (Weak) solution of the SDE

In the preceding section we have seen that **M** gives us a "componentwise" weak solution of the SDE (0.2). The proof followed by regularization more or less directly from the standard theory in [F 80]. It is much harder to "localize" the sample paths of the processes appearing in (2.4) which is the main objective of this section. In particular, the question whether $(W_t^k)_{t\geq 0}, k \in K_0$, in (2.4) are the components of a Brownian motion on E (over H), is answered by the following

Theorem 3.1. *Assume that for one (and hence all) $t > 0$ there exists a probability measure μ_t on $(E, \mathcal{B}(E))$ such that*

$$(3.1) \qquad \int \exp(i_{E'}< k, z >_E)\mu_t(dz) = \exp(-\frac{1}{2}t\|k\|_H^2) \text{ for all } k \in E'.$$

Then there exist maps $W : \Omega \to C([0, \infty[, E)$ and $N : \Omega \to C([0, \infty[, E)$ such that

(i) $\omega \mapsto W_t(\omega) := W(\omega)(t)$ and $\omega \mapsto N_t(\omega) := N(\omega)(t), \omega \in \Omega$, are both $\mathcal{F}_t/\mathcal{B}(E)$-measurable for all $t \geq 0$.

(ii) For $\mathcal{E}$-q.e. $z \in E$ under P_z, $(W_t)_{t\geq 0}$ is an $(\mathcal{F}_t)_{t\geq 0}$-Brownian motion starting at $0(\in E)$ with covariance $2 < , >_H$.

(iii) For each $k \in K$

$$_{E'}< k, N_t >_E = \int_0^t \beta_k(X_s)ds, t \geq 0, P_z - a.e. \text{ for } \mathcal{E}\text{-q.e. } z \in E.$$

(iv) $X_t = z + W_t + N_t, t \geq 0, P_z - a.e.$ for $\mathcal{E}$-q.e. $z \in E$.

The proof of 3.1 is technically rather difficult, so we omit here and refer instead to [AR 91, Sect. 6]. Below, we will only give a proof of 3.1 under the additional assumption (C.4) which simplifies the situation considerably.

Remark 3.2. (i) The assumption in 3.1 that the Gaussian measures satisfying (3.1) exist is of course, necessary. It just means that there exists a Brownian semigroup on E with covariance $2 <\ ,\ >_H$, i.e., there exists a Brownian motion on E over H. Hence 3.1 is the best result one could hope for.
(ii) For necessary and sufficient conditions for $\mu_t, t > 0$, as in (3.1) to exist we refer to [Gr 65] (see also [Kuo 75], [Y 89] and [AR 91]). Roughly speaking $\mu_t, t > 0$, exist on $(E, \mathcal{B}(E))$ if E is "big enough" by comparison with H.

Theorem 3.1 (iv) tells us that **M** weakly solves in a sense an SDE of type (0.2) except that we have not defined $\beta : E \to E$. And in fact, in our general situation, there is no guarantee that the map $k \mapsto \beta_k(z), k \in K \subset E'$, is represented by an element in E for μ-a.e. $z \in E$. Therefore, we introduce the following condition:

(C.4) There exists a $\mathcal{B}(E)/\mathcal{B}(E)$-measurable map $\beta : E \to E$ such that

(i) ${}_{E'}< k, \beta >_E = \beta_k$ μ-a.e. for each $k \in K (\subset E')$
(ii) $\int_E \|\beta\|_E^2 d\mu < \infty$.

(C.4) implies that the process $N = (N_t)_{t \geq 0}$ in 3.1 can be defined directly as a Bochner integral. Indeed, by 2.1 (i) we know that for $t \geq 0$

$$E_z[\int_0^t \|\beta\|_E(X_s)ds] < \infty \text{ for } \mathcal{E}\text{-q.e. } z \in E.$$

Hence, defining

$$\Omega_0 := \{\int_0^t \|\beta\|_E(X_s)ds < \infty \text{ for all } t > 0\} \tag{3.2}$$

we have that $P_z(\Omega_0) = 1$ for $\mathcal{E}$-q.e. $z \in E$. Consequently, if

$$N_t := 1_{\Omega_0} \int_0^t \beta(X_s)ds, t \geq 0, \tag{3.3}$$

(where the integral is in the sense of Bochner), we have that $N : \Omega \to C([0, \infty[, E), (N_t)_{t \geq 0}$ is $(\mathcal{F}_t)_{t \geq 0}$-adapted and for each $k \in K$

$${}_{E'}< k, N_t >_E = \int_0^t \beta_k(X_s)ds, t \geq 0, P_z - \text{a.e. for } \mathcal{E}\text{-q.e. } z \in E. \tag{3.4}$$

Proof. [3.1 assuming (C. 4)] Define $W_t := X_t - X_0 - N_t, t \geq 0$, where $(N_t)_{t \geq 0}$ is as in (3.3). Then $W : \Omega \to C([0, \infty[, E), (W_t)_{t \geq 0}$ is $(\mathcal{F}_t)_{t \geq 0}$-adapted and it follows by the uniqueness of the Fukushima decomposition (2.1) and by (3.4) that for each $k \in K$

$${}_{E'}< k, W_t >_E = W_t^k, t \geq 0, P_z - \text{a.e. for } \mathcal{E}\text{-q.e. } z \in E.$$

It is now fairly straightforward to prove that $(W_t)_{t \geq 0}$ also satisfies 3.1 (ii) (cf. [AR 91, Proof of 6.2]). □

Finally we give a sufficient condition for (C. 4) to hold which can be checked in many applications.

Proposition 3.3. *Assume that E is itself a real separable Hilbert space. Suppose that there exists another Hilbert space $(< H_0, < , >_{H_0})$ such that $E' \subset H_0$ densely by a Hilbert Schmidt map and such that there exists a constant $c \in]0, \infty[$ such that*

$$\int \beta_k^2 d\mu \leq c\|k\|_{H_0}^2 \text{ for all } k \in K. \tag{3.5}$$

Then (C. 4) holds.

Proof. [AR 91, Proposition 6.9]. □

In special cases one can also prove that our solution of (0.2) is unique and that it can be obtained by a Girsanov transformation. We refer for details to [RZ 90] and [ARZ 91], but concentrate now instead on examples.

4 Examples.

In these notes we only treat the linear case, i.e., β "comes from" a linear map on H. For the case of non-linear very singular β appearing in quantum field theory we refer to [AR 91, Section 7. II].

Given a separable real Hilbert space H and a self-adjoint operator A on H with $A \geq cId_H$ for some $c \in]0, \infty[$, we want to weakly solve an SDE which ("informally") can be written as:

$$\begin{aligned} dX_t &= dW_t - A(X_t)dt \\ X_0 &= z \end{aligned} \tag{4.1}$$

where $(W_t)_{t\geq 0}$ is a Brownian motion over H. If $dim\ H = +\infty, (W_t)_{t\geq 0}$ cannot be constructed to take values in H, but $W_t \in E, t \geq 0$, with E a separable real Banach space such that $H \subsetneq E$.

In order to apply our machinery we have to find μ. By Section 1 ("Poincaré type lemma") we know that "informally"

$$\mu(dz) = \exp[-\frac{1}{2} < Az, z >] \ "d^\infty z"$$

(where "$d^\infty z$" denotes "infinite dimensional Lebesgue measure"), i.e., μ should be a mean zero Gaussian measure with covariance operator A^{-1}. μ may only exist on a linear space E strictly larger than H. Then we apply our machinery to the closure of

$$\mathcal{E}(u, v) = \int_E < \nabla u, \nabla v >_H d\mu; \ u, v \in \mathcal{F}C_b^\infty.$$

But we also have to choose E large enough so that a Brownian motion on E over H exists. We can even choose E such that, in addition, condition (C.4) holds, as we shall see at the end of this section.

Choice of E: Let $\tilde{\mu}, \tilde{\mu}_t, t \geq 0$, be cylinder measures on H such that

$$\int_H \exp(i < h', h >_H)\tilde{\mu}(dh') = \exp(-\frac{1}{2}\|A^{-1/2}h\|_H^2) \text{ for all } h \in H, \tag{4.2}$$

and

$$\int_H \exp(i < h', h >_H)\tilde{\mu}_t(dh') = \exp(-\frac{1}{2}t\|h\|_H^2) \text{ for all } h \in H. \tag{4.3}$$

(These exist by the classical Bochner theorem on $\mathbb{R}^d$.) Choose any real Banach space E such that $H \subset E$, continuously and densely, and such that $\tilde{\mu}, \tilde{\mu}_t, t \geq 0$, "lift" to measures $\mu, \mu_t, \geq 0$, on $(E, \mathcal{B}(E))$. For sufficient conditions we refer e.g. to [Y 89, Theorem 3.2] (= Gross-Minlos-Sazonov theorem). For instance any real Hilbert space E such that $H \subset E$, continuously and densely by a Hilbert-Schmidt map, would do. Then

$$E' \subset H' \equiv H \subset E.$$

Now we shall check conditions (C. 1), (C. 2) and (C.3):
(C.1) is obvious. (C.2): Take $K := E' \cap D(A)$ (assuming that then K is dense in E', which can always be arranged by choosing E properly; cf. below). Then we have for each $k \in K (\subset H \subset E)$

$$(4.4) \qquad \begin{aligned} {}_{E'}< l, k >_E &= < l, k >_H = < A^{-1/2} l, A^{-1/2}(Ak) >_H \\ &= \int_E {}_{E'}< l, z >_E X_{Ak}(z)\mu(dz) \text{ for all } l \in E'(\subset H). \end{aligned}$$

Here we set for $h \in H, X_h := \lim_{n\to\infty} {}_{E'}< l_n, \cdot >_E$ in $L^2(E;\mu)$ where $l_n \xrightarrow[n\to\infty]{} h$ in H.

Exercise 4.1. Prove that (4.4) implies that μ is k-quasi-invariant and that the corresponding densities are given by

$$\frac{d\tau_{tk}(\mu)}{d\mu} = \exp[tX_{Ak} - \frac{1}{2}t^2\|Ak\|_H^2]$$

(cf. [AR 90a, Proposition 5.5 (i)]).

Using 4.1 it is easy to see that k is well-μ-admissible and that $\beta_k = -X_{Ak}$. Hence (C. 2) holds. (C.3): We have that $\int_E {}_{E'}< k, z >_E^2 \mu(dz) = \|A^{-1/2}k\|_H^2$ for all $k \in K$. Hence (C.3) also holds and we can apply 3.1 with the closure of

$$\mathcal{E}(u, v) = \int_E < \nabla u, \nabla v >_H d\mu; \; u, v \in \mathcal{F}C_b^\infty,$$

i.e., the associated diffusion process $\mathbf{M} = (\Omega, \mathcal{F}, (X_t)_{t\geq 0}, (P_z)_{z\in E})$ (which exists by Chap. IV, Sect. 3) satisfies the stochastic equation

$$(4.5) \qquad X_t = z + W_t + N_t, t \geq 0, P_z - \text{a.e. for } \mathcal{E}\text{-q.e.} z \in E$$

with $(W_t)_{t\geq 0}$ the corresponding Brownian motion on E over H and ${}_{E'}< k, N_t >_E = -\int_0^t X_{Ak}(X_s)ds$, $t \geq 0$, P_z-a.e. for $\mathcal{E}$-q.e. $z \in E$ for each $k \in K$.
In order to fulfil also (C.4) (and the assumption in (C.2) above) we have to construct E in a particular way. To this end, let $H_0 = D(A), < , >_{H_0} := < A\cdot, A\cdot >_H$, and let $\lambda_n \in]0, \infty[$ such that $\sum_{n=1}^\infty \lambda_n^2 < \infty$. Let $\{e_n | n \in \mathbb{N}\}$ be an orthonormal basis of H_0 and define $H_1 := \{h \in H_0 | \sum_{n=1}^\infty \lambda_n^{-2} < e_n, h >_{H_0}^2 < \infty\}$ with inner product

$$< \cdot, \cdot >_{H_1} := \sum_{n=1}^\infty \lambda_n^{-2} < e_n, \cdot >_{H_0} < e_n, \cdot >_{H_0}.$$

Then $(H_1, < , >_{H_1})$ is a separable real Hilbert space with orthonormal basis $\{\lambda_n e_n | n \in \mathbb{N}\}$. Hence $H_1 \subset H_0$ is dense and Hilbert-Schmidt. Define $E := H_1'$, then $E' = H_1 \subset$

$H_0 \subset H \equiv H' \subset H_0' \subset H_1' = E$, in particular $E' \subset D(A)$ (hence in (C.2) $K = E'$) and for all $k \in K$

$$\int \beta_k^2 \, d\mu = \int X_{Ak}^2 \, d\mu = \|A^{1/2}k\|_H^2 \leq c^{-1}\|k\|_{H_0}^2. \tag{4.6}$$

Hence we can apply 3.3 to conclude that (C. 4) holds. (Note that as before $\tilde{\mu}, \tilde{\mu}_t$ in (4.2), (4.3) still lift to measures on $(E, \mathcal{B}(E))$ since $H_0' \subset H_1'$ is Hilbert-Schmidt and hence, so is $H \subset E$.) Hence there exists $\beta : E \to E$, $\mathcal{B}(E)/\mathcal{B}(E)$-measurable such that ${}_{E'}\!< k, \beta >_E = \beta_k = -X_{Ak}$ μ-a.e. for all $k \in K$, $\int_E \|\beta\|_E^2 d\mu < \infty$, and in this case our diffusion process **M** in the weak sense satisfies the SDE

$$\begin{aligned} dX_t = dW_t \ &+ \ \beta(X_t)dt \\ X_0 \ &= \ z \end{aligned} \tag{4.7}$$

under P_z for $\mathcal{E}$-q.e. $z \in E$.

Remark 4.2. Note that (4.7) is the rigorous version of (4.1) which is informal since $(X_t)_{t \geq 0}$ in general does not take values in $D(A)$. β can be considered as a stochastic extension of A (cf. [I 84]) but which is not necessarily linear on E. However, if in addition $A(E') \subset E'$ and $A_{|E'} : E' \to E'$ is continuous w.r.t. $\| \ \|_{E'}$ then we have for the linear operator $A' : E \to E$ defined by ${}_{E'}\!< k, A'(z) >_E := {}_{E'}\!< Ak, z >_E$, $z \in E$, $k \in E'$, that $\beta(z) = A'(z)$ for μ-a.e. $z \in E$.

Chapter VI

Construction of Diffusions on Path and Loop Spaces of Compact Riemannian Manifolds

This chapter is a modification of [DR 92] which was motivated by the recent beautiful results in [AvB 91], [AiM 88]. It will be primarily written for the *path space case*. At the end we will explain the minor modifications needed to cover the *pinned loop space case*.

Let (M, g, ∇, o) be given, where M is a $d-$dimensional compact Riemannian manifold without boundary, g is a Riemannian metric on M, ∇ is a $g-$compatible covariant derivative, and $o \in M$ is a fixed base point. It will **always** be assumed that the covariant derivative ∇ is *torsion skew symmetric* - i.e., if T is the torsion tensor of ∇, then $g\langle T\langle X, Y\rangle, Z\rangle \equiv 0$ for all vector fields X, Y, and Z on M. We denote by $W(M)$ the set of continuous paths $\sigma : [0,1] \mapsto M$ such that $\sigma(0) = o$ and let $\mathcal{L}(M) = \{\sigma \in W(M) \,|\, \sigma(1) = o\}$ be the paths in $W(M)$ which also end at o. Both spaces we equip with the topology of uniform convergence. Wiener measure on $W(M)$ (the law of Brownian motion on M starting at $o \in M$) will be denoted by ν, and the pinned Wiener measure concentrated on $\mathcal{L}(M)$ will be denoted by ν_o. The corresponding real L^2-spaces are denoted by $L^2(\nu), L^2(\nu_0)$ respectively. Recall that the coordinate maps $\Sigma_s : W(M) \to M$ given by $\Sigma_s(\sigma) = \sigma(s)$ are M-valued semi-martingales relative to both of the measures ν and ν_o. Therefore it is possible to define a stochastic parallel translation operator $H_s(\sigma) : T_oM \to T_{\sigma(s)}M$ for ν or ν_o almost every path σ.

1 Path space case

In preparation for defining the *h-derivative* and the corresponding *gradient* we introduce the reproducing kernel Hilbert space H of functions $h : [0,1] \to T_oM$ such that $h(0) = 0$, h is absolutely continuous, and $(h,h)_H := \int_0^1 |h'(s)|^2 ds < \infty$,where $|v|^2 := g_o\langle v, v\rangle$ for $v \in T_oM$. A function $F : W(M) \mapsto \mathbb{R}$ is said to be a *smooth cylinder function* if F can be represented as $F(\sigma) = f(\sigma(s_1), \ldots, \sigma(s_n))$ where $f : M^n \mapsto \mathbb{R}$

is a smooth function and $0 \le s_1 \le s_2 \le \cdots \le s_n \le 1$. Let $\mathcal{F}C^\infty$ denote the set of all smooth cylinder functions. Note that $\mathcal{F}C^\infty$ is dense in $L^2(\nu)$ since it separates the points of $W(M)$. Given $h \in H$ and a smooth cylinder function $F(\sigma)$ as above the $h-$derivative of F is

$$\partial_h F(\sigma) := \sum_{i=1}^{n} g_{\sigma(s_i)}\langle \nabla_i f(\vec{\sigma}), H_{s_i}(\sigma) h(s_i)\rangle \tag{1.1}$$

where $\vec{\sigma} := (\sigma(s_1), \ldots, \sigma(s_n))$, $\nabla_i f(\vec{\sigma}) \in T_{\sigma(s_i)}M$ is the gradient of the function f relative to the i'th variable while holding the remaining variables fixed. We now restate Theorem 9.1. and Corollary 9.1. from [D 91].

Theorem 1.1. *The formula in (1.1) for $\partial_h F(\sigma)$ is well defined up to $\nu-$equivalence independent of the choices made in representing F as a smooth cylinder function and the chosen version for the process $H_s(\sigma)$. Furthermore, the adjoint ∂_h^* of ∂_h (relative to the $L^2(\nu)-$inner product) contains in its domain the class of smooth cylinder functions and hence is densely defined. The operator ∂_h^* acting on the smooth cylinder functions has the form $\partial_h^* = -\partial_h + z(h)$, where $z(h)$ is a function which is in $L^p(d\nu)$ for all p.*

Definition 1.2. Let F be a smooth cylinder function. The *gradient* of F is the ν a.e. defined function $DF : W(M) \to H$ which satisfies $\partial_h F(\sigma) = (DF(\sigma), h)_H$ for all $h \in H$.

Let $G : [0,1]^2 \to \mathbb{R}$ be the Green's function for the operator $-\frac{d^2}{ds^2}$ with Dirichlet boundary conditions at $s = 0$ and Neumann boundary conditions at $s = 1$. (Explicitly $G(s,t) := \min(s,t)$.) It is well known and easy to check that G is a reproducing kernel for H — that is for all $h \in H$ and $s \in [0,1]$, $h(s) = \int_0^1 \frac{\partial}{\partial u} G(s,u)\, h'(u) du$. Using this fact and (1.1) it is easy to verify that DF is given explicitly by

$$DF(\sigma)(s) = \sum_{i=1}^{n} G(s, s_i) H_{s_i}(\sigma)^{-1} \nabla_i f(\vec{\sigma}), \tag{1.2}$$

where $F(\sigma) = f(\vec{\sigma})$ as above.
Given two smooth cylinder functions F and K on $W(M)$ define a positive definite symmetric bilinear form on $L^2(\nu)$ by

$$\mathcal{E}(F, K) := \int_{W(M)} (DF(\sigma), DK(\sigma))_H \, \nu(d\sigma), \; F, K \in \mathcal{F}C^\infty. \tag{1.3}$$

Using the explicit expression (1.2) for DF and the fact G is a reproducing kernel it is easy to show that

$$\mathcal{E}(F, F) = \sum_{i,j=1}^{n} \int_{W(M)} G(s_i, s_j) g_o\langle H_{s_i}(\sigma)^{-1} \nabla_i f(\vec{\sigma}), H_{s_j}(\sigma)^{-1} \nabla_j f(\vec{\sigma})\rangle \, \nu(d\sigma), \tag{1.4}$$

where $F(\sigma) = f(\vec{\sigma})$ as above.

Lemma 1.3. *The densely defined quadratic form $(\mathcal{E}, \mathcal{F}C^\infty)$ is closable on $L^2(\nu)$ and its closure $(\mathcal{E}, D(\mathcal{E}))$ is a symmetric Dirichlet form on $L^2(\nu)$.*

Proof. Let $\{h_n\}_{n=1}^{\infty}$ be an orthonormal basis for H. Then $(DF(\sigma), DF(\sigma))_H = \sum_{n=1}^{\infty}(\partial_{h_n}F(\sigma))^2$ and so

$$\mathcal{E}(F,F) = \sum_{n=1}^{\infty} \|\partial_{h_n}F(\sigma)\|^2_{L^2(\nu)}. \tag{1.5}$$

The summands of (1.5) are closable on $L^2(\nu)$ by 1.1 and II.1.3, hence $(\mathcal{E}, \mathcal{F}C^\infty)$ is closable in $L^2(\nu)$. The fact that $(\mathcal{E}, D(\mathcal{E}))$ is a Dirichlet form is a direct consequence of the chain rule and I.3.8 (cf. II.4.3). □

We now come to the main theorem.

Theorem 1.4. *(i) $(\mathcal{E}, D(\mathcal{E}))$ is a local quasi-regular Dirichlet form on $L^2(\nu)$.*

(ii) There exists a diffusion process $\mathbf{M} = (\Omega, \mathcal{F}, (\mathcal{F}_t)_{t\geq 0}, (X_t)_{t\geq 0}, (P_z)_{z\in W(M)})$ associated with $(\mathcal{E}, D(\mathcal{E}))$, i.e. for all $u \in L^2(\nu)$, bounded, and $t > 0$,

$$\int u(X_t)dP_z = T_t u(z) \text{ for } \nu - \text{ a.e. } z \in W(M)$$

where $T_t := e^{tL}, t \geq 0$, and L is the generator of $(\mathcal{E}, D(\mathcal{E}))$.

Part (ii) follows from part (i) by IV.2.4 and IV.2.12. To show (i), since $\mathcal{F}C^\infty$ is dense in $D(\mathcal{E})$ w.r.t. $\mathcal{E}_1^{1/2}$ and $\mathcal{F}C^\infty$ separates the points of $W(M)$, according to the definition of quasi-regularity (cf. IV.2.1) and by IV.2.2(ii) we only have to prove

Proposition 1.5. *(i) Let $(G_\alpha)_{\alpha>0}$ be the resolvent of $(\mathcal{E}, D(\mathcal{E}))$ and let* Cap $:=$ Cap$_h$ *with $h := G_1 1 \equiv 1$. Then* Cap *is tight.*

(ii) $(\mathcal{E}, D(\mathcal{E}))$ is local.

Remark 1.6. It is easy to see that in Theorem 1.4, $T_t 1 = 1$ for all $t \geq 0$, hence $\mathbf{M}$ is conservative, and indeed $G_1 1 = 1$.

Now we turn to the proof of 1.5. We shall prove part (i) by adapting a method due to B. Schmuland (cf. [RS 91a] and also Chap.IV, Subsection 3b) above). We denote the closure of D also by D. As in IV.3.1 one proves using the chain rule for D that for all $F, K \in D(\mathcal{E})$

$$(D(F \vee K), D(F \vee K))_H^{1/2} \leq (DF, DF)_H^{1/2} \vee (DK, DK)_H^{1/2} \tag{1.6}$$

and

$$(D(F \wedge K), D(F \wedge K))_H^{1/2} \leq (DF, DF)_H^{1/2} \vee (DK, DK)_H^{1/2} \tag{1.7}$$

where $\vee, \wedge$ denotes sup, inf respectively. By the Whitney imbedding theorem we may assume without loss of generality that M is an imbedded submanifold of $\mathbb{R}^n$ for some integer n. Let x^i for $i = 1, 2, \ldots n$ denote the standard linear coordinates on $\mathbb{R}^n$, which may also be considered to be smooth functions on M. Let $\{s_k\}_{k=1}^{\infty}$ be a countable dense subset of [0,1]. For $\sigma, \tau \in W(M)$ define $\rho(\sigma, \tau) := \sup |x^i(\sigma(s_k)) - x^i(\tau(s_k))|$, where the sup is taken over all $i = 1, 2, \ldots, n$ and for all positive integers k. It is clear that ρ is compatible with the topology of uniform convergence on $W(M)$.

For the moment fix i, k, and τ and consider the smooth cylinder function $F(\sigma) := x^i(\sigma(s_k)) - x^i(\tau(s_k))$. By (1.2) we see that

$$(1.8)\quad (DF(\sigma), DF(\sigma))_H = G(s_k, s_k)\ g_o\langle H_{s_k}(\sigma)^{-1}\nabla x^i(\sigma(s_k)), H_{s_k}(\sigma)^{-1}\nabla x^i(\sigma(s_k))\rangle.$$

Using the facts that G is bounded by one and $H_{s_k}(\sigma)$ is an isometry, it follows from (1.8) that $(DF(\sigma), DF(\sigma))_H \leq C$, where $C := \max\{g\langle\nabla x^i(m), \nabla x^i(m)\rangle \mid m \in M \text{ and } i = 1, 2, \ldots, n\}$. Because M is compact it follows that $C < \infty$. Now by exactly the same arguments as in the proof of IV.3.2 it follows from (1.6) that

$$(1.9)\qquad \rho_\tau := \rho(\tau, \cdot) \in D(\mathcal{E}) \text{ for all } \tau \in W(M)$$

and

$$(1.10)\qquad (D\rho_\tau, D\rho_\tau)_H \leq C \text{ for all } \tau \in W(M)$$

Let $\{\tau_k \mid k \in \mathbb{N}\}$ be dense in $W(M)$. Set

$$w_n := \inf_{1 \leq k \leq n} \rho_{\tau_k}, \quad n \in \mathbb{N}.$$

Then $w_n \downarrow 0$ as $n \to \infty$ and by (1.7), (1.9) and (1.10) and the same arguments as in the proof of IV.3.2 it follows that

$$w_n \longrightarrow 0 \text{ quasi-uniformly on } W(M) \text{ as } n \to \infty,$$

i.e., there exist closed $F_m \subset W(M), m \in \mathbb{N}$, such that $w_n \longrightarrow 0$ uniformly on each F_m as $n \to \infty$ and $\lim_{m\to\infty} Cap(W(M) \setminus F_m) = 0$. By definition of $(w_n)_{n\in\mathbb{N}}$ it now follows that each F_m it totally bounded, hence compact, and 1.5(i) is proved.
To prove (ii) note that by a simple approximation argument

$$(1.11)\qquad D(u \cdot v) = u \cdot Dv + vDu \text{ for all bounded } u, v \in D(\mathcal{E})$$

Then 1.5(ii) is an immediate consequence of the following fact which is now a special case of [MR 92, Chap. VI, Proposition 1.7, see also Chap. VI, Example 1.13 (ii)] since we already know that $(\mathcal{E}, D(\mathcal{E}))$ is quasi-regular.

Proposition 1.7. *For every open $U \subset W(M)$ there exists $u \in D(\mathcal{E})$ such that $0 \leq u \leq 1_U$ (= indicator function of U) and $u > 0$ ν-a.e. on U.*

2 The pinned loop space case

As stated at the beginning, the above results are also true for the pinned loop space setting. The required modifications are as follows. Let $H_o = \{h \in H \mid h(1) = 0\}$, and $G_0(s,t) := \begin{cases} s(1-t) & \text{if } s \leq t \\ t(1-s) & \text{if } s \geq t \end{cases}$ which is the Green's function for the operator $-\frac{d^2}{ds^2}$ with Dirichlet boundary conditions at both $s = 0$ and $s = 1$. (Note: that G_o is a reproducing kernel for H_o.) It is shown in [D 92] that Theorem 1.1 still holds if Wiener measure ν is replaced by pinned Wiener measure ν_o and the function h is restricted to be in $H_o \cap C^1$. Definition 1.2 should be replaced by

Definition 2.1. The gradient of F is the ν_o-a.e. defined function $D_oF : \mathcal{L}(M) \to H_o$ such that $\partial_h F(\sigma) = (D_oF(\sigma), h)_H$ for all $h \in H_o$.

(1.2) still holds provided G is replaced by G_0. We now define $\mathcal{E}_0$ by the right member of (1.3) with $(D, H, W(M), \nu)$ replaced by $(D_o, H_o, \mathcal{L}(M), \nu_o)$. The proof of Lemma 1.3 is still valid provided we choose an orthonormal basis $\{h_n\}$ for H_o such that $\{h_n\} \subset H_o \cap C^1$. Thus we may define $(\mathcal{E}_0, D(\mathcal{E}_0))$ to be the form closure of $(\mathcal{E}_0, \mathcal{F}C^\infty{}_{|\mathcal{L}(M)})$. Note that $\mathcal{F}C^\infty{}_{|\mathcal{L}(M)}$ is dense in $L^2(\nu_0)$. The same proof given for Theorem 1.4 with $(D, H, W(M), \nu)$ replaced by $(D_o, H_o, \mathcal{L}(M), \nu_o)$ proves:

Theorem 2.2. *(i) $(\mathcal{E}_0, D(\mathcal{E}_0))$ is a local quasi-regular Dirichlet form on $L^2(\nu_0)$.*

(ii) There exists a conservative diffusion process **M** *with state space $\mathcal{L}(M)$ associated with $(\mathcal{E}_0, D(\mathcal{E}_0))$.*

For an analogue of Theorem 2.2 for the free loop space over $\mathbb{R}^d$ including rotational invariance we refer to [ALR 92].

References

[AvB 91] Airault , H., Van Biesen, J.: Le processus d'Ornstein-Uhlenbeck sur une sous-variété de l'espace de Wiener. Bull. Sc. math., 2e série, t. 115 (1991), 185-210.

[AiM 88] Airault , H., Malliavin, P.: Intégration géométrique sur l'espace de Wiener. Bull. Sc math., 2e série, t. 112 (1988), 3-52.

[ABR 89] Albeverio, S., Brasche, J., Röckner, M.: Dirichlet forms and generalized Schrödinger operators. In: Proc. Summer School Schrödinger operators, ed. H. Holden and A. Jensen. Lecture Notes in Physics **345**, 1-42. Berlin: Springer 1989.

[AKR 90] Albeverio, S., Kusuoka, S., Röckner, M.: On partial integration in infinite dimensional space and applications to Dirichlet forms. J. London Math. Soc. **42**, 122-136 (1990).

[ALR 92] Albeverio, S. Leandre, R., Röckner, M.: Construction of a rotational invariant diffusion on the free loop space. Preprint (1992). Publication in preparation.

[AMR 90] Albeverio, S., Ma, Z.M., Röckner, M.: A Beurling-Deny type structure theorem for Dirichlet forms on general state space. To appear in: Memorial Volume for R. Høegh-Krohn.

[AMR 92a] Albeverio, S., Ma, Z.M., Röckner, M.: Non-symmetric Dirichlet forms and Markov processes on general state space. C.R. Acad. Sci. Paris, t. **314**, Série I, 77-82 (1992).

[AMR 92b] Albeverio, S., Ma, Z.M., Röckner, M.:Regularization of Dirichlet spaces and applications. To appear in C.R. Acad. Sci. Paris, Série I, (1992).

[AMR 92c] Albeverio, S., Ma, Z.M., Röckner, M.: Local property of Dirichlet forms and diffusions on general state spaces. Preprint (1992). Publication in preparation.

[AMR 92d] Albeverio, S., Ma, Z.M., Röckner, M.: Characterization of (non-symmetric) Dirichlet forms associated with Hunt processes. Preprint (1992). Publication in preparation.

[AR 89a] Albeverio, S., Röckner, M.: Dirichlet forms, quantum fields and stochastic quantization. In: Stochastic analysis, path integration, and dynamics. Research Notes in Math. **200**, 1-21. Editors: K.D. Elworthy, J.C. Zambrini, Harlow: Longman 1989.

[AR 89b] Albeverio, S., Röckner, M.: Classical Dirichlet forms on topological vector spaces - construction of an associated diffusion process. Probab. Th. Rel. Fields **83**, 405-434 (1989).

[AR 90a] Albeverio, S., Röckner, M.: Classical Dirichlet forms on topological vector spaces - closability and a Cameron-Martin formula. J. Funct. Anal. **88**, 395-436 (1990).

[AR 90b] Albeverio, S., Röckner, M.: New developments in the theory and application of Dirichlet forms. In: Stochastic processes, physics and geometry, 27-76, Ascona/Locarno, Switzerland, 4-9 July 1988, Editors: S. Albeverio et al., Singapore: World Scientific 1990.

[AR 91] Albeverio, S., Röckner, M.: Stochastic differential equations in infinite dimensions: solutions via Dirichlet forms. Probab. Th. Rel. Fields **89**, 347-386 (1991).

[ARZ 91] Albeverio, S., Röckner, M., Zhang, T.S.: Girsanov transform for symmetric diffusions with infinite dimensional state space. Preprint (1991). Ann. Prob., in press.

[Ba 70] Badrikian, A.: Séminaire sur les fonctions aléatoires linéaires et les mesures cylindriques. Lecture Notes in Math. **139**. Berlin: Springer 1970.

[Be De 58] Beurling, A., Deny, J.: Espaces de Dirichlet. Acta Math. **99** , 203-224 (1958).

[Be De 59] Beurling, A., Deny, J.: Dirichlet spaces. Proc. Nat. Acad. Sci. U.S.A. **45**, 208-215 (1959).

[BH 86] Bouleau, N., Hirsch, F.: Formes de Dirichlet générales et densité des variables alétoires réelles sur l'espaces de Wiener. J. Funct. Anal. **69**, 229-259 (1986).

[Bou 74] Bourbaki, N.: Topologie générale. Chapitres 5 à 10. Paris: Hermann 1974.

[C 75] Cannon, J.T.: Convergence criteria for a sequence of semi groups. Appl. Anal. **5**, 23-31 (1975).

[Ca-Me 75] Carillo - Menendez, S.: Processus de Markov associé à une forme de Dirichlet non symétrique. Z. Wahrsch. verw. Geb. **33**, 139-154 (1975).

[Ch 69a] Choquet, G.: Lectures on Analysis I: Integration and topological vector spaces. London: Benjamin 1969.

[Ch 69b] Choquet, G.: Lectures on Analysis II: Representation Theory. London: Benjamin 1969.

[DM 78] Dellacherie, C., Meyer, P.A.: Probabilities and potential. Paris: Hermann 1978.

[De 70] Deny, J.: Méthodes Hilbertiennes et théorie potentiel. Potential Theory, Centro Internazionale Matematico Estivo, Edizioni Cremonese: Roma 1970.

[D 91] Driver, B. K.: A Cameron-Martin type quasi-invariance theorem for Brownian motion on a compact Riemannian manifold, (1991) to appear in J. Funct. Anal.

[D 92] Driver, B. K.: A Cameron-Martin type quasi-invariance theorem for pinned Brownian motion on a compact Riemannian manifold. UCSD-preprint (1992) to appear in Trans. of A.M.S.

[DR 92] Driver, B., Röckner, M.: Construction of diffusions on path and loop spaces of compact Riemannian manifolds. Preprint (1992). C. R. Acad. Sci., Paris, Série I, in press.

[F 71] Fukushima, M.: Dirichlet spaces and strong Markov processes. Trans. Amer. Math. Soc. **162**, 185-224 (1971).

[F 73] Fukushima, M.: On the generation of Markov processes by symmetric forms. Proceedings 2nd Japan - USSR symposium on probability theory. Lect. Notes in Math. **330**. Berlin: Springer 1973.

[F 80] Fukushima, M.: Dirichlet forms and Markov processes. Amsterdam-Oxford-New York: North Holland (1980).

[F 81] Fukushima, M.: On a stochastic calculus related to Dirichlet forms and distorted Brownian motion. Physical Reports **77**, 255-262 (1981).

[F 82] Fukushima, M.: On absolute continuity of multidimensional symmetrizable diffusions. In: Lecture Notes in Math. **923**, 146-176. Berlin - Heidelberg - New York: Springer 1982.

[F 84] Fukushima, M.: Energy forms and diffusion process. In: Mathematics and Physics, Lectures on recent results. Ed. Streit, L. Singapore: World Scientific Publishing Co. 1984.

[Go 85] Goldstein, J.A.: Semigroups of linear operators and applications. Oxford Math. Monographs. New York: O. U. Press 1985.

[Gr 65] Gross, L.: Abstract Wiener spaces. Proc. 5th Berkeley Symp. Math. Stat. Prob. **2**, 31-42 (1965).

[I 84] Ito, K.: Infinite dimensional Ornstein-Uhlenbeck processes. In: Stochastic Analysis. Ed.: K. Ito, 197-224. Amsterdam - Oxford - New York: North Holland 1984.

[Kuo 75] Kuo, H.: Gaussian measures in Banach spaces. Lect. Notes in Math. **463**, 1-224, Berlin-Heidelberg-New York: Springer 1975.

[Ku 82] Kusuoka, S.: Dirichlet forms and diffusion processes on Banach spaces. J. Fac. Sci. Univ. Tokyo, Sec. **IA 29**, 387-400 (1982).

[Le 82] LeJan, Y.: Dual Markovian semigroups and processes. In: "Functional Analysis in Markov processes", 47-75, ed. M. Fukushima. Lect. Notes Math. **923**. Berlin: Springer 1982.

[Le 83] LeJan, Y.: Quasi-continuous functions and Hunt processes. J. Math. Soc. Japan **35**, 37-42 (1983).

[MR 92] Ma, Z. M., Röckner, M.: An introduction to the theory of (non-symmetric) Dirichlet forms. Monograph, Springer, to appear September 1992.

[Ma 78] Malliavin, P.: Stochastic calculus of variation and hypoelliptic operators. Proc. of the International Symposium on Stochastic Differential Equations Kyoto 1976, Tokyo 1978.

[O 88] Oshima, Y.: Lectures on Dirichlet forms. Preprint Erlangen (1988).

[ReS 72] Reed, M., Simon, B.: Methods of modern mathematical physics I. Functional Analysis. New York - San Francisco - London: Academic Press 1972.

[ReS 75] Reed, M., Simon, B.: Methods of modern mathematical physics II. Fourier Analysis, self-adjointness. New York - San Francisco - London, Academic Press 1975.

[R 90a] Röckner, M.: Dirichlet forms on infinite dimensional state space and applications. Lectures held at Silivri Summer School. Preprint (1990). To appear.

[R 90b] Röckner, M.: Potential theory on non-locally compact spaces via Dirichlet forms. Preprint (1990). To appear as "main lecture" in: Proceedings "International conference on potential theory, Nagoya 1990".

[RS 91a] Röckner, M., Schmuland, B.: Tightness of general $C_{1,p}$-capacities on Banach space. Preprint (1991). To appear in J. Funct. Anal. .

[RS 91b] Röckner, M., Schmuland, B.: In preparation.

[RW 85] Röckner, M., Wielens, N.: Dirichlet forms – closability and change of speed measure In: "Infinite dimensional analysis and stochastic processes", Research Notes in Math. **124**, 119-144, Editor: S. Albeverio, Boston - London - Melbourne: Pitman 1985.

[RZ 90] Röckner, M., Zhang, T.S.: Uniqueness of generalized Schrödinger operators and applications. Preprint (1990). J. Funct. Anal. **105**, 187-231 (1992).

[S 90] Schmuland, B.: An alternative compactification for classical Dirichlet forms on topological vector spaces. Stochastics **33** (1990), 75-90.

[Sch 73] Schwartz, L.: Radon measures on arbitrary topological space and cylindrical measures. London: Oxford University Press 1973.

[Sh 88] Sharpe, M.T.: General theory of Markov processes. New York: Academic Press 1988.

[Si 74] Silverstein, M.L.: Symmetric Markov Processes. Lecture Notes in Math. **426**. Berlin - Heidelberg - New York: Springer 1974.

[St 64] Stampacchia, G.: Formes bilinéaires coercitives sur les ensembles convexes. C.R. Acad. Sc., Paris t. **258**, Série I, 4413-4416 (1964).

[W 84] Watanabe, S.: Lectures on stochastic differential equations and Malliavin calculus. Berlin - Heidelberg - New York - Tokyo: Springer 1984.

[Wl 82] Wloka, J.: Partielle Differentialgleichungen. Stuttgart: Teubner 1982.

[Y 89] Yan, J.A.: Generalizations of Gross' and Minlos' theorems. In Séminaire de Probabilités XXII, eds. J. Azema, P.A. Meyer, M. Yor. Lect. Notes in Math. **1372**, Springer 1989, 395-404.

Logarithmic Sobolev Inequalities for Gibbs States

Daniel W. Stroock

M.I.T.

October 14, 1992

Lecture I: Gibbs States in Finite Dimensions

Let M be a compact, connected RIEMANNian manifold with LÉVI–CIVITA connection ∇ and normalized RIEMANN measure λ. Given a $U \in C^\infty(M)$, the **Gibbs state with potential energy** U is the probability measure γ^U on M given by

$$\gamma^U(dx) = \frac{e^{-U(x)}}{Z}\lambda(dx), \tag{1.1}$$

where the constant Z is determined by the requirement that $\gamma^U(M) = 1$. The naïve physical principle underlying this terminology is that a dynamical system which when *free* (i.e. $U \equiv 0$) has λ as its equilibrium distribution will in the presence of the force field determined by the potential U have γ^U as its equilibrium distribution.

For example, let Δ denote the LAPLACE–BELTRAMI operator associated with ∇, and consider the *free heat flow semigroup* $\{P_t : t > 0\}$ on M. That is, $\{P_t : t > 0\}$ is the (unique) MARKOV semigroup of operators on $C(M)$ which is determined by the equation

$$[P_t\varphi](x) = \varphi(x) + \int_0^t [P_s \circ \Delta\varphi](x)\,ds, \quad t \in (0,\infty) \text{ and } \varphi \in C^\infty(M). \tag{1.2}$$

By elementary elliptic theory,

$$[P_t\varphi](x) = \int_M \varphi(y)P(t,x,dy)\,\lambda(dy) \quad \text{where } P(t,x,dy) = p(t,x,y)\,\lambda(dy)$$

and *the heat kernel* $p(t,x,y)$ is a strictly positive, smooth function on $(0,\infty)\times M^2$. Moreover, since, for all $T > 0$ and $\varphi,\ \psi \in C^\infty(M)$,

$$\frac{d}{dt}\mathbb{E}^\lambda[P_t\psi\, P_{T-t}\varphi] = \mathbb{E}^\lambda\Big[(\Delta P_t\psi)(P_{T-t}\varphi) - (P_t\psi)(\Delta P_{T-t}\varphi)\Big] = 0, \quad t \in (0,T),$$

it is clear that λ is $\{P_t : t > 0\}$**-reversible** in the sense that

$$\mathbb{E}^\lambda[\varphi P_T\psi] = \mathbb{E}^\lambda[\psi P_T\varphi], \quad T \in (0,\infty),$$

During the period of this research, the author was partially supported by NSF grant DMS 8913328.

first for smooth φ and ψ and then for all continuous ones. In particular, this means that $p(t,x,y) = p(t,y,x)$ for all $(t,x,y) \in (0,\infty) \times M \times M$. In addition, because $P_t 1 = 1$, we also see that

$$\mathbf{E}^{\lambda}[P_t\varphi] = \mathbf{E}^{\lambda}[\varphi], \quad t \in (0,\infty) \text{ and } \varphi \in C(\mathbf{M});$$

or equivalently, λ is $\{P_t : t > 0\}$**-invariant** in the sense that

$$\lambda = \lambda P_t \equiv \int P(t,x,\cdot\,)\,\lambda(dx), \quad t \in (0,\infty).$$

In fact, λ is the only $\{P_t : t > 0\}$-invariant μ in $\mathfrak{M}_1(M)$ (the space of BOREL probability measures on M). To see this, suppose that μ is a second one, and set $f(y) = \int p(1,x,y)\,\mu(dx)$. Clearly f is a strictly positive and smooth. In addition, because $\mu = \mu P_1$, $\mu(dy) = f(y)\,\lambda(dy)$; and so all that we have to do is check that $f = 1$, which, because $\mathbf{E}^{\lambda}[f] = 1$, comes down to showing that f is constant. To this end, note that

$$f(y) = \int f(x) p(1,x,y)\,\lambda(dx), \quad y \in \mathbf{M}.$$

Next, choose $y_0 \in M$ to be a point at which f achieves its minimum value. If f were non-constant, we would then have the contradiction

$$f(y_0) = \int f(y) p(1,x,y_0)\,\lambda(dx) > f(y_0).$$

We next *perturb* the free heat flow by the *conservative force field* $-\nabla U$ determined by the potential U. Thus, we consider the semigroup $\{P_t^U : t > 0\}$ determined by

$$P_t^U\varphi = \varphi + \int_0^t P_s^U \circ L^U\varphi\,ds, \quad t \in (0,\infty) \text{ and } \varphi \in C^\infty(M),$$

where L^U is the elliptic operator given by

$$L^U\varphi = \Delta\varphi - \nabla U\cdot\nabla\varphi$$

and we have used the (somewhat casual) notation $\mathbf{v}\cdot\mathbf{w}$ to denote the inner product determined by the RIEMANN metric on the tangent bundle $\mathbf{T}(M)$. Again elementary elliptic theory guarantees that

$$[P_t^U\varphi](x) = \int \varphi(y)\,P^U(t,x,dy) \quad \text{where } P^U(t,x,dy) = p^U(t,x,y)\,\gamma^U(dy)$$

and p^U is again a smooth, strictly positive function on $(0,\infty) \times M \times M$. Moreover, since an equivalent expression for L^U is

$$L^U\varphi = e^U\nabla\cdot(e^{-U}\nabla\varphi),$$

it is an easy matter to repeat the reasoning given above to see first that $p^U(t,x,y) = p^U(t,y,x)$, then that γ^U is $\{P_t^U : t > 0\}$-reversible, and finally that it is the only $\{P_t^U : t > 0\}$-invariant element μ of $\mathfrak{M}_1(M)$.

On the basis of the preceding uniqueness result, one can apply elementary ergodic theory to show that

$$\lim_{T\to\infty} \sup_{x\in M} \left\| \frac{1}{T}\int_0^T P^U(t,x,\cdot)\,dt - \gamma^U \right\|_{\text{var}} = 0,$$

where $\|\cdot\|_{\text{var}}$ denotes the variation norm for measures. In fact, by working just a little bit harder and capitalizing on the positivity of $p^U(t,x,y)$, one can use DOEBLIN's theory for the ergodic properties of MARKOV chains [10] to remove the averaging and assert that

$$\sup_{x\in M} \|P^U(T,x,\cdot) - \gamma^U\|_{\text{var}} \longrightarrow 0$$

exponentially fast as $T \to \infty$. However, because in the subsequent lectures we want to be working in an infinite dimensional setting, we will not use this line of reasoning but, instead, will adopt an approach which is better suited to analysis in infinite dimensions. In particular, our approach will take maximal advantage of the symmetry of $p^U(t,x,y)$ and minimal advantage of its regularity properties.

To understand how symmetry enters these considerations, observe that each of the operators P_t^U determines a unique self-adjoint extension $\overline{P_t^U}$ which is a contraction operator on $L^2(\gamma^U)$. Indeed, by JENSEN's inequality, for any $q \in [1,\infty)$:

$$|P_t^U\varphi|^q \le P_t^U|\varphi|^q,$$

and therefore, because $\gamma^U = \gamma^U P_t^U$,

$$\|P_t^U\varphi\|_{L^q(\gamma^U)} \le \|\varphi\|_{L^q(\gamma^U)}. \tag{1.3}$$

In particular, this proves that P_t^U admits a unique extension $\overline{P_t^U}$ as a contraction on $L^2(\gamma^U)$; and clearly the symmetry of P_t^U becomes the statement that $\overline{P_t^U}$ is self-adjoint. Moreover, $\{\overline{P_t^U} : t > 0\}$ inherits the semigroup property; and because $P_t^U\varphi \longrightarrow \varphi$ uniformly as $t \searrow 0$ first for all $\varphi \in C^\infty(M)$ and then for all $\varphi \in C(M)$, it is trivial to check that $\{\overline{P_t^U} : t > 0\}$ is strongly continuous in the sense that

$$\|\overline{P_t^U}\varphi - \varphi\|_{L^2(\gamma^U)} \longrightarrow 0 \quad \text{for each } \varphi \in L^2(\gamma^U).$$

Hence, by STONE's Theorem, there is a resolution of the identity in $L^2(\gamma^U)$ by orthogonal projections E_α, $\alpha \in [0,\infty)$, with the property that

$$\overline{P_t^U} = \int_{[0,\infty)} e^{-\alpha t}\,dE_\alpha. \tag{1.4}$$

Next, note that the argument which we used to prove that there is only one $\{P_t^U : t > 0\}$ invariant measure shows that $\mathbf{1}$ is (up to a change in sign) the unique eigenfunction for the P_t^U's with eigenvalue 1. Hence,

$$E_0\varphi = \mathbb{E}^{\gamma^U}[\varphi]\,\mathbf{1} \quad \text{for all } \varphi \in L^2(\gamma^U). \tag{1.5}$$

Clearly, (1.4) and (1.5) already imply that

$$\left\|\overline{P_t^U}\varphi - \mathbf{E}^{\gamma^U}[\varphi]\right\|_{L^2(\gamma^U)} \longrightarrow 0 \quad \text{for each } \varphi \in L^2(\gamma^U).$$

In fact, because we are working with a finite dimensional situation and therefore have a smooth density function $p^U(t,x.y)$, we can immediately say much more. Namely, observe that (by the semigroup property and symmetry)

$$\iint\limits_{M\times M} p^U(1,x,y)^2\,\gamma^U(dx)\gamma^U(dy) = \int_M p^U(2,x,x)\,\gamma^U(dx) < \infty,$$

which means that the operator $\overline{P_1^U}$ is HILBERT–SCHMIDT and therefore compact. Hence, the spectrum of $\overline{P_1^U}$ is completely discrete with an accumulation point only at 0. In particular, if we set

$$\alpha_1(U) = \inf\left\{\alpha > 0 : e^{-\alpha} \in \operatorname{spec}(\overline{P_1^U})\right\}, \tag{1.6}$$

then $\alpha_1(U) > 0$,

$$\overline{P_t^U}\varphi - \mathbf{E}^{\gamma^U}[\varphi] = \int_{[\alpha_1(U),\infty)} e^{-\alpha t}\, dE_\alpha\varphi,$$

and so

$$\left\|\overline{P_t^U}\varphi - \mathbf{E}^{\gamma^U}[\varphi]\right\|_{L^2(\gamma^U)} \le e^{-\alpha_1(U)t}\|\varphi\|_{L^2(\gamma^U)}, \quad t > 0 \text{ and } \varphi \in L^2(\gamma^U). \tag{1.7}$$

Finally, observe that, by the continuous form of MINKOWSKI's inequality,

$$\|P_1^U\varphi\|_{L^2(\gamma^U)} \le \int |\varphi(y)|\|p^U(1,\cdot,y)\|_{L^2(\gamma^U)}\,\gamma^U(dy) \le \max_{y\in M} p^U(2,y,y)^{\frac{1}{2}}\|\varphi\|_{L^1(\gamma^U)},$$

and so, by duality,

$$\|P_1^U\varphi\|_{C(M)} \le \max_{y\in M} p^U(2,y,y)^{\frac{1}{2}}\|\varphi\|_{L^2(\gamma^U)}.$$

Hence, by writing $P_{t+1}^U = P_t^U \circ P_1^U$, we see that

$$\left\|P_{t+1}^U\varphi - \mathbf{E}^{\gamma^U}[\varphi]\right\|_{C(M)} \le \max_{y\in M} p^U(2,y,y)^{\frac{1}{2}}\,e^{-\alpha_1(U)t}\|\varphi\|_{L^2(\gamma^U)}, \quad t \in (0,\infty),$$

which, because $\|\varphi\|_{L^2(\gamma^U)} \le \|\varphi\|_{C(M)}$ leads immediately to

$$\left\|P^U(t+1,x,\cdot) - \gamma^U\right\|_{\text{var}} \le \max_{y\in M} p^U(2,y,y)^{\frac{1}{2}}\,e^{-\alpha_1(U)t}, \quad (t,x) \in (0,\infty)\times M. \tag{1.8}$$

Similarly, by writing $P_{t+2}^U = P_1^U \circ P_t^U \circ P_1^U$, we arrive at

$$\sup_{(x,y)\in M^2}\left|p^U(t+2,x,y) - 1\right| \le \max_{x\in M} p^U(2,x,x)\,e^{-\alpha_1(U)t}, \quad t \in (0,\infty). \tag{1.9}$$

Unfortunately, only part of the preceding argument has a chance of surviving the transition to infinite dimensions. In particular, everything which relies on point-wise estimates for $p^U(t,x,y)$ must be replaced. To give a hint about what those replacements might be, first observe that the estimate in (1.7) can be obtained with much less information than we used. In fact, all that we needed to know is that

$$E_0\varphi = \mathbf{E}^{\gamma^U}[\varphi]\mathbf{1} \quad \text{and} \quad \alpha_1(U) \equiv \inf\{\alpha \in (0,\infty) : E_\alpha \neq 0\} > 0.$$

Equivalently, what we need is that

$$\|\varphi - \mathbf{E}^{\gamma^U}[\varphi]\|^2_{L^2(\gamma^U)} = \|\varphi - E_0\varphi\|^2_{L^2(\gamma^U)} \leq \frac{1}{\alpha_1(U)}\int_{[0,\infty)} \alpha\, d(E_\alpha\varphi,\varphi)_{L^2(\gamma^U)},$$

which comes down to the statement that

$$\alpha_1(U)\|\varphi - \mathbf{E}^{\gamma^U}[\varphi]\|^2_{L^2(\gamma^U)} \leq -\int_M \varphi\overline{L^U}\varphi\, d\gamma^U, \quad \varphi \in \mathrm{Dom}(\overline{L^U}),$$

where $\overline{L^U}$ is the generator of the semigroup $\{\overline{P_t^U} : t > 0\}$. But, because $\overline{L^U}$ is the closure in $L^2(\gamma^U)$ of $L^U \restriction C^\infty(M)$ and

$$\int_M \varphi L^U\varphi\, d\gamma^U = -\int_M |\nabla\varphi|^2\, d\gamma^U,$$

we now see that (1.7) is equivalent to the **Poincaré inequality**

$$\text{(1.10)} \qquad \alpha_1(U)\|\varphi - \mathbf{E}^{\gamma^U}[\varphi]\|_{L^2(\gamma^U)} \leq \mathcal{E}^U(\varphi,\varphi) \equiv \int_M |\nabla\varphi|^2\, d\gamma^U, \quad \varphi \in C^\infty(M).$$

In order to obtain alternative routes to estimates like the one in (1.8), we need to deal with quantities which are not so intimately tied to $L^2(\gamma^U)$ as those in (1.7). With this in mind, we introduce the **relative entropy** functional $\mathbf{H}(\mu|\nu)$ defined for $(\mu,\nu) \in \mathfrak{M}_1(M)^2$ by

$$\text{(1.11)} \qquad \mathbf{H}(\mu|\nu) = \begin{cases} \int_M f\log f\, d\nu & \text{if} \quad d\mu = f\, d\nu \\ \infty & \text{if} \quad \mu \not\ll \nu. \end{cases}$$

To see as soon as possible how entropy is related to estimates like the one in (1.8), first observe that

$$3(\xi-1)^2 \leq (4+2\xi)(\xi\log\xi - \xi + 1) \quad \text{for all } \xi \in [0,\infty).$$

Next suppose that $d\mu = f\,d\nu$ and note that

$$\begin{aligned} 3\|\mu-\nu\|^2_{\text{var}} &= 3\|f-1\|^2_{L^1(\nu)} \leq \|(4+2f)^{\frac{1}{2}}(f\log f - f + 1)^{\frac{1}{2}}\|^2_{L^1(\nu)} \\ &\leq \|4+2f\|_{L^1(\nu)}\|f\log f - f + 1\|_{L^1(\nu)} \leq 6\mathbf{H}(\mu|\nu). \end{aligned}$$

Hence,

(1.12) $$\|\mu-\nu\|_{\text{var}}^2 \le 2\mathbf{H}(\mu|\nu), \quad (\mu,\nu)\in \mathfrak{M}_1(M)^2.$$

Now let $\mu \in \mathfrak{M}_1(M)$ with $\mathbf{H}(\mu|\gamma^U) < \infty$ be given, set $f = \frac{d\mu}{d\gamma^U}$ and $f_t = P_t^U f$. Clearly f_t is both smooth and strictly positive. In addition, $f_t = \frac{d\mu P_t^U}{d\nu}$, and so

$$\frac{d}{dt}\mathbf{H}(\mu P_t^U|\gamma^U) = \int_M (L^U f_t)\log f_t \, d\gamma^U + \int_M L^U f_t \, d\gamma^U = -\int_M \frac{|\nabla f_t|^2}{f_t}\, d\gamma^U,$$

where we have used integration by parts to obtain the last equality. Equivalently (cf. (1.10)), we now know that

(1.13) $$\frac{d}{dt}\mathbf{H}(\mu P_t^U|\gamma^U) \le -4\mathcal{E}^U\left(f_t^{\frac{1}{2}}, f_t^{\frac{1}{2}}\right) \quad \text{where } f_t = \frac{d\mu P_t^U}{d\gamma^U}.$$

Thus, if we had an estimate of the form

(1.14) $$\kappa(U)\mathbf{H}(\mu|\gamma^U) \le 4\mathcal{E}^U\left(f^{\frac{1}{2}}, f^{\frac{1}{2}}\right) \quad \text{for } d\mu = f\, d\gamma^U,$$

we would then have that

$$\mathbf{H}(\mu P_t^U \mu|\gamma^U) \le e^{-4\kappa(U)t}\mathbf{H}(\mu|\gamma^U),$$

which, in conjunction with (1.12), would mean that

(1.15) $$\left\|\mu P_t^U - \gamma^U\right\|_{\text{var}} \le \sqrt{2\mathbf{H}(\mu|\gamma^U)}\, e^{-2\kappa(U)t}, \quad \text{for } t>0 \text{ and } \mu \in \mathfrak{M}_1(M).$$

In particular, because

$$B \equiv \sup_{y\in M}\int_M p^U(1,x,y)\log p^U(1,x,y)\,\gamma(dx) < \infty,$$

(1.15) leads to the conclusion that

$$\sup_{\mu\in\mathfrak{M}_1(M)} \left\|\mu P_t^U - \gamma^U\right\|_{\text{var}} \le \sqrt{2B}\, e^{-2\kappa(U)t},$$

which certainly guarantees that $\left\{P_t^U : t>0\right\}$ is uniformly ergodic.

Ever since the appearance of L. Gross's article [7], an inequality of the form in (1.13) has been known as **logarithmic Sobolev inequality**. Obviously, such an inequality contains a, albeit minimal, *coercivity* statement. Less immediately apparent is the ergodicity statement which the preceding line of reasoning has revealed.

Lecture II: Some Facts about Logarithmic Sobolev Inequalities

As I mentioned at the end of the preceding lecture, logarithmic Sobolev inequalities (cf. (1.13)) combine a minimal coercivity statement with a rather subtle ergodicity statement. To understand more precisely how these two are meshed, we will first affirm that the coercivity part of a logarithmic SOBOLEV inequality is strictly dominated by any classical SOBOLEV inequality.

Continuing with the notation introduced in Lecture I, recall that a classical SOBOLEV inequality is a statement of the form (cf. (1.10))

$$\|\varphi\|^2_{L^q(\gamma^U)} \le a_q(U)\mathcal{E}^U(\varphi,\varphi) + b_q(U)\|\varphi\|^2_{L^2(\gamma^U)}, \quad \varphi \in L^2(\gamma^U), \tag{2.1}$$

for some $q \in (2,\infty)$, $a_q(U) \in (0,\infty)$, and $b_q(U) \in [1,\infty)$. Next, let $\varphi \in C(M;(0,\infty))$ with $\|\varphi\|_{L^2(\gamma^U)} = 1$ be given, set $d\mu = \varphi^2\, d\gamma^U$, and use JENSEN's inequality to justify

$$\begin{aligned}\int \varphi^2 \log \varphi^2 \, d\gamma^U &= \frac{2}{q-2}\int \log \varphi^{q-2}\, d\mu \le \frac{2}{q-2}\log \int \varphi^{q-2}\, d\mu \\ &= \frac{q}{q-2}\log \|\varphi\|^2_{L^q(\gamma^U)} \le \frac{q}{q-2}\Big(\|\varphi\|^2_{L^q(\gamma^U)} - 1\Big).\end{aligned}$$

Hence, if

$$A_q(U) \equiv \frac{qa_q(U)}{q-2} \quad \text{and} \quad B_q(U) \equiv \frac{q\big(b_q(U)-1\big)}{q-2},$$

then (2.1) implies that

$$\int \varphi^2 \log\left(\frac{\varphi}{\|\varphi\|_{L^2(\gamma^U)}}\right)^2 d\gamma^U \le A_q(U)\mathcal{E}^U(\varphi,\varphi) + B_q(U)\|\varphi\|^2_{L^2(\gamma^U)}, \tag{2.3}$$

for all $\varphi \in C(M;(0,\infty))$.

By taking $\varphi = f^{\frac{1}{2}}$, one sees that (2.3) already captures the coercivity component of a logarithmic SOBOLEV inequality. However, because of the extra term on the right, it fails to capture the ergodic component. In fact, unless $b_1(U) = 1$, (2.1) by itself cannot lead to a logarithmic SOBOLEV inequality. However, as we are about to see, if one has, in addition to (2.3), a POINCARÉ inequality, then one can recover a logarithmic SOBOLEV inequality. As a preliminary step to seeing this, note that, because

$$\begin{aligned}\mathcal{E}^U(\varphi,\varphi) &= \lim_{t\searrow 0} \tfrac{1}{t}\Big(\varphi - P^U_t\varphi, \varphi\Big)_{L^2(\gamma^U)} \\ &= \lim_{t\searrow 0} \tfrac{1}{2t}\iint \big(\varphi(y)-\varphi(x)\big)^2 P^U(t,x,dy)\gamma^U(dx),\end{aligned} \tag{2.4}$$

one can easily show that (2.3) for $\varphi \in C(M;(0,\infty))$ implies itself first for all continuous φ's and then for all $\varphi \in L^2(\gamma^U)$. Second, observe that (2.3) together with (1.10) imply that

$$\int \varphi^2 \log\left(\frac{\varphi^2}{\|\varphi\|_{L^2(\gamma^U)}}\right) d\gamma^U \le \left(A_q(U) + \frac{B_q(U)}{\alpha_1(U)}\right)\mathcal{E}^U(\varphi,\varphi) \quad \text{if } \mathbb{E}^{\gamma^U}[\varphi] = 0.$$

Finally, given any $\nu \in \mathfrak{M}_1(M)$ and $\varphi \in L^2(\nu)$, set $\overline{\varphi} = \varphi - \mathbb{E}^{\nu}[\varphi]$. Then an elementary, but rather tedious, computation (cf. (6.1.26) in [5]) shows that:

$$\int \varphi^2 \log\left(\frac{\varphi}{\|\varphi\|_{L^2(\nu)}}\right)^2 d\nu \le \int \overline{\varphi}^2 \log\left(\frac{\overline{\varphi}^2}{\|\overline{\varphi}\|_{L^2(\nu)}}\right)^2 d\nu + 2\|\overline{\varphi}\|^2_{L^2(\nu)}.$$

Hence, we now conclude that (2.3) and (1.10) together imply that

$$\kappa(U) \int \varphi^2 \log\left(\frac{\varphi}{\|\varphi\|_{L^2(\gamma^U)}}\right)^2 \le \mathcal{E}^U(\varphi,\varphi), \quad \varphi \in L^2(\gamma^U), \tag{2.5}$$

where

$$\kappa(U)^{-1} \le A_q(U) + \frac{B_q(U)+2}{\alpha_1(U)};$$

and, by taking $\varphi^2 = \frac{d\mu}{d\gamma^U}$, we see that (2.5) certainly implies (1.14).

2.6 Remark: As the preceding development makes precise, the coercivity part of a logarithmic SOBOLEV inequality is at the bottom of the SOBOLEV ladder. On the other hand, the ergodic component is nearly on a par with a POINCARÉ inequality. To be more precise, because the estimate in (1.7) is (at the $L^2(\gamma^U)$ level) optimal, a comparison of (1.7) with (1.15) reveals that

$$2\kappa(U) \le \alpha_1(U). \tag{2.7}$$

Actually, it is not at all clear how often the inequality in (2.7) is an equality. In fact, so far as I know, there is no known non-degenerate (i.e. uniformly elliptic) example for which the equality fails and there are several examples for which it is known to hold. (See [9] for a degenerate example in which the inequality is strict and [1] and [6] for some non-degenerate examples when equality holds.) Although it seems unlikely that equality holds in *all* non-degenerate situations, it is possible that it does for a large class, and it would be very useful to know something about that class! Indeed, the constant $\alpha_1(U)$ comes from a quadratic variational problem which is inherently simpler than the variational problem of which $\kappa(U)$ is the solution.

At this point we know that a logarithmic SOBOLEV inequality can be obtained by the conjunction of a classical SOBOLEV inequality and a POINCARÉ inequality. However, for the program which we have in mind, it will be important to know how to obtain one logarithmic SOBOLEV inequality directly from another. To be precise, suppose that we have the estimate

$$\kappa \int \varphi^2 \log\left(\frac{\varphi}{\|\varphi\|_{L^2(\lambda)}}\right)^2 d\lambda \le \mathcal{E}(\varphi,\varphi) \equiv \int |\nabla\varphi|^2 \, d\lambda, \quad \varphi \in L^2(\lambda), \tag{2.8}$$

for some $\kappa \in (0,\infty)$ and ask what we can say on the basis of (2.8) about the $\kappa(U)$ appearing in (2.5). Because the integrand on the left hand side takes both signs it is not immediately clear how to proceed. On the other hand, one can get around this problem with the observation that, for any $\nu \in \mathfrak{M}_1(M)$, $\varphi \in L^2(\nu)$, and $t \in [0,\infty)$:

$$0 \le \int \varphi^2 \log\left(\frac{\varphi}{\|\varphi\|_{L^2(\nu)}}\right)^2 d\nu \le \int \left(\varphi^2 \log\varphi^2 - \varphi^2\log t^2 - \varphi^2 + t^2\right) d\nu$$

with the second inequality an equality precisely when $t = \|\varphi\|_{L^2(\nu)}$. In particular, this means that, for every $t \in [0,\infty)$, the integrand on the right is non-negative and that, by taking $t = \|\varphi\|_{L^2(\lambda)}$:

$$\begin{aligned}\int \varphi^2 \log\left(\frac{\varphi}{\|\varphi\|_{L^2(\gamma^U)}}\right)^2 d\gamma^U &\le \int \left(\varphi^2\log\varphi^2 - \varphi^2\log\|\varphi\|^2_{L^2(\lambda)} - \varphi^2 + \|\varphi\|^2_{L^2(\lambda)}\right) d\gamma^U \\ &\le \frac{e^{-\min U}}{Z}\int \varphi^2\log\left(\frac{\varphi}{\|\varphi\|_{L^2(\lambda)}}\right)^2 d\lambda \\ &\le \frac{e^{-\min U}}{\kappa Z}\int |\nabla\varphi|^2\, d\lambda \le \frac{e^{\operatorname{osc}(U)}}{\kappa}\int |\nabla\varphi|^2\, d\gamma^U,\end{aligned}$$

where $\operatorname{osc}(U) \equiv \max U - \min U$. In other words, we have now shown that (2.8) implies (2.3) with

$$\kappa(U) \ge \kappa e^{-\operatorname{osc}(U)}. \tag{2.9}$$

So far we have seen various ways to arrive at logarithmic SOBOLEV inequalities, and we have just seen a way to pass from one logarithmic SOBOLEV inequalities to another. However, as yet, we have seen very little to recommend logarithmic SOBOLEV inequalities. Indeed, the only reason that we have provided for considering them at all is that they give an elegant procedure for arriving at an estimate like the one in (1.15). On the other hand, before deriving (1.15) we proved stronger estimates, namely (1.8) and (1.9), on the basis of more familiar properties, all of which come quite easily from the existence of a classical SOBOLEV inequality. Thus, it is high time that we mention the one virtue possessed by logarithmic SOBOLEV inequalities and shared by no classical SOBOLEV inequality. Namely, although the q's for which (2.1) can hold are dictated by the relation

$$\frac{1}{q} \ge \frac{1}{2} - \frac{1}{\dim(M)},$$

the logarithmic SOBOLEV inequality is essentially *dimension free*. In fact, at a somewhat formal level, the logarithmic SOBOLEV can be thought of as the limit, as $q \searrow 2$, of classical SOBOLEV inequalities, which on the basis of the preceding relation, means that it is the SOBOLEV inequality which has a chance of holding even in infinite dimensions. Having made these heuristic remarks, we will close this lecture by giving them some substance.

Let M_1 and M_2 be a pair of RIEMANNian manifolds with associated connections ∇_1 and ∇_2, and suppose that, for $j \in \{1,2\}$, ν_j is an element of $\mathfrak{M}_1(M_j)$ for which one has the logarithmic SOBOLEV inequality

$$\kappa_j \int_{M_j} \varphi_j^2 \log\left(\frac{\varphi_j}{\|\varphi_j\|_{L^2(\nu_j)}}\right)^2 d\nu_j \leq \int_{M_j} |\nabla_j \varphi_j|^2 \, d\nu_j \tag{2.10}$$

for all $\varphi_j \in L^2(\nu_j)$. Next, set $M = M_1 \times M_2$, give M the product RIEMANNian structure, and take $\nu = \nu_1 \times \nu_2$. Given $\Phi \in C^\infty(M;(0,\infty))$, set

$$\psi(x_2) = \left(\int_{M_1} \Phi(x_1,x_2)^2 \, \nu_1(dx_1)\right)^{\frac{1}{2}},$$

and, as an application of (2.10) with $j = 1$ to $x_1 \in M_1 \longmapsto \Phi(x_1,x_2) \in (0,\infty)$, note that

$$\begin{aligned}\int_{M_1} &\Phi(x_1,x_2)^2 \log \Phi(x_1,x_2)^2 \, \nu_1(dx_1) \\ &\leq \kappa_1^{-1} \int_{M_1} \left|[\nabla_1 \Phi(\,\cdot\,,x_2)](x_1)\right|^2 \nu_1(dx_1) + \psi(x_2)^2 \log \psi(x_2)^2\end{aligned}$$

for each $x_2 \in M_2$. Hence, after integrating the preceding in $x_2 \in M_2$ with respect to ν_2 and then applying (2.10) with $j = 2$, we find that

$$\begin{aligned}\int_M \Phi^2 \log \Phi^2 \, d\nu &\leq \kappa_1^{-1} \int_M |\nabla_1 \Phi|^2 \, d\nu + \int_{M_2} \psi^2 \log \psi^2 \, d\nu_2 \\ &\leq \kappa_1^{-1} \int_M |\nabla_1 \Phi|^2 \, d\nu + \kappa_2^{-1} \int_{M_2} |\nabla_2 \psi|^2 \, d\nu_2 + \|\Phi\|_{L^2(\nu)}^2 \log \|\Phi\|_{L^2(\nu)}^2.\end{aligned}$$

Finally, by SCHWARZ's inequality,

$$|\nabla_2 \psi(x_2)| \leq \psi(x_2)^{-1} \int_{M_1} [\Phi\,|\nabla_2 \Phi|](x_1,x_2)\, \nu_1(dx_1) \leq \left(\int_{M_1} |\nabla_2 \Phi(\,\cdot\,,x_2)|^2 \, d\nu_1\right)^{\frac{1}{2}};$$

and so, on the basis of (2.10), we have arrive at

$$\kappa_1 \wedge \kappa_2 \int_M \Phi^2 \log\left(\frac{\Phi}{\|\Phi\|_{L^2(\nu)}}\right)^2 d\nu \leq \int_M |\nabla \Phi|^2 \, d\nu, \tag{2.11}$$

first for smooth positive Φ's and then, just as before, for all $\Phi \in L^2(\nu)$. In other words, we have now shown that *logarithmic* SOBOLEV *inequalities are preserved under tensor products, and that the logarithmic* SOBOLEV *constant for the product is the minimum of the constants for the factors.* In particular, this certainly opens the possibility of having a logarithmic SOBOLEV inequality in infinite dimensions. In fact, all that one has to do is take the infinite tensor product of any compact, connected RIEMANNian manifold with itself.

Lecture III: Gibbs States in Infinite Dimensions

As saw at the end of the preceding lecture, logarithmic SOBOLEV inequalities can hold in infinite dimensional settings. However, so far, the only settings which we know how to handle are, from a physical stand-point, trivial. That is, mathematically rigorous models of statistical mechanics involve infinite systems (gases) of *interacting particles*, whereas, tensor products model systems consisting of *non-interacting particles*. In order to get to interacting systems, we will deal with models which, for reasons which should become clear in a minute, are called **lattice gases**. However, before we can describe them, it will be necessary to be less casual than we have been heretofore and will have to introduce a disgusting amount of notation. In the belief that it is less painful to get it all written down as quickly as possible, we begin this lecture with a catalogue of the basic notation with which we will be working throughout.

The Lattice: The lattice underlying our models will be the d-dimensional square lattice $\mathbb{Z}^d$ for some fixed $d \in \mathbb{Z}^+$; and, for $\mathbf{k} = (k^1, \ldots, k^d) \in \mathbb{Z}^d$, we will use the norm $|\mathbf{k}| \equiv \max_{1 \le i \le d} |k^i|$. Given $\Lambda \subseteq \mathbb{Z}^d$, we will use $\Lambda\complement \equiv \mathbb{Z}^d \setminus \Lambda$ to denote the complement of Λ, $|\Lambda|$ to denote the cardinality of Λ, and $\mathbf{k}+\Lambda$ to denote the translate $\{\mathbf{k}+\mathbf{j} : \mathbf{j} \in \Lambda\}$ of Λ by $\mathbf{k} \in \mathbb{Z}^d$. Finally, we will occasionally use the notation $\Lambda \subset\subset \mathbb{Z}^d$ to mean that $|\Lambda| < \infty$, and $\mathfrak{F}$ will stand for the set of all non-empty $\Lambda \subset\subset \mathbb{Z}^d$.

The Spin and Configuration Spaces: The spin space M for our model will be either a finite set (with the discrete topology) or a compact, connected, finite dimensional RIEMANNian manifold; and our configuration space will be the product space $\mathbf{M} \equiv M^{\mathbb{Z}^d}$ endowed with the product topology. Given a non-empty $X \subseteq \mathbb{Z}^d$, we will use $\mathbf{x} \in \mathbf{M} \longmapsto \mathbf{x}_X \in M^X$ to denote the natural projection taking $\mathbf{M}$ onto M^X, $B_X(\mathbf{M})$ and $C_X(\mathbf{M})$ to denote the set of functions on $\mathbf{M}$ of the form $\mathbf{x} \in \mathbf{M} \longmapsto \varphi(\mathbf{x}_X) \in \mathbb{R}$ as φ runs over, respectively, the set $B(M^X)$ of bounded, BOREL-measurable and the set $C(M^X)$ of continuous functions on M^X; and $\mathcal{F}_X$ will denote the σ-algebra over $\mathbf{M}$ generated by the elements of $B_X(\mathbf{M})$. When $X = \{\mathbf{k}\}$, we will use $x_\mathbf{k}$ in place of $\mathbf{x}_{\{\mathbf{k}\}}$; and when $X = \mathbb{Z}^d$, it is clear that $\mathcal{F}_X$ is precisely the BOREL field $\mathcal{B}_\mathbf{M}$ over $\mathbf{M}$, and we will simply write $\mathcal{F}$ instead of $\mathcal{F}_{\mathbb{Z}^d}$. Also, we will say that a measurable or continuous $f : \mathbf{M} \longrightarrow \mathbb{R}$ is *local* and will write $f \in B_0(\mathbf{M})$ or $f \in C_0(\mathbf{M})$ if it is an element of $B_X(\mathbf{M})$ or $C_X(\mathbf{M})$ for some $X \in \mathfrak{F}$; and, for any bounded $f : \mathbf{M} \longrightarrow \mathbb{R}$, $\|f\|_u$ will be used to denote the *uniform norm* (i.e., "sup") norm of f. Finally, for each $\mathbf{k} \in \mathbb{Z}^d$, we define the *shift transformation* $\theta^\mathbf{k} : \mathbf{M} \longrightarrow \mathbf{M}$ so that $\left(\theta^\mathbf{k}\mathbf{x}\right)_\mathbf{j} = x_{\mathbf{k}+\mathbf{j}}$ for every $\mathbf{x} \in \mathbf{M}$ and every $\mathbf{j} \in \mathbb{Z}^d$.

For various constructions, it will be convenient to have introduced some additional notation. In the first place, given $\emptyset \neq X \subset \mathbb{Z}^d$, we define

$$\left(\mathbf{x}^X, \mathbf{y}^{X\complement}\right) \in M^X \times M^{X\complement} \longmapsto x^X \bullet y^{X\complement} \in \mathbf{M}$$

so that $\mathbf{x}^X \bullet \mathbf{y}^{X\complement}$ is the element $\mathbf{z} \in \mathbf{M}$ determined by

$$\mathbf{z}_X = \mathbf{x}^X \quad \text{and} \quad \mathbf{z}_{X\complement} = \mathbf{y}^{X\complement};$$

and, for $f: \mathbf{M} \longrightarrow \mathbb{R}$ and $\mathbf{y}^{X^\complement} \in M^{X^\complement}$, we define $f(\cdot\,|\mathbf{y}^{X^\complement})$ on M^X and $f_X(\cdot\,|\mathbf{y}^{X^\complement})$ on $\mathbf{M}$ by

$$\mathbf{x}^X \in M^X \longmapsto f(\mathbf{x}^X|\mathbf{y}^{X^\complement}) \equiv f(\mathbf{x}^X \bullet \mathbf{y}^{X^\complement})$$

and

$$\mathbf{x} \in \mathbf{M} \longmapsto f_X(\mathbf{x}|\mathbf{y}^{X^\complement}) \equiv f(\mathbf{x}_X \bullet \mathbf{y}^{X^\complement}).$$

Secondly, for $\mathbf{y} \in \mathbf{M}$, we write $f_X(\mathbf{x}|\mathbf{y})$ instead of $f_X(\mathbf{x}|\mathbf{y}_{X^\complement})$; and, when $X = \{\mathbf{k}\}$ we will use $f_{\mathbf{k}}(\cdot\,|\mathbf{y})$ in place of $f_{\{\mathbf{k}\}}(\cdot\,|\mathbf{y})$. Since both

$$(\mathbf{x}^X, \mathbf{y}^{X^\complement}) \in M^X \times M^{X^\complement} \longmapsto \mathbf{x}^X \bullet \mathbf{y}^{X^\complement} \in \mathbf{M} \quad \text{and} \quad (\mathbf{x}, \mathbf{y}) \in \mathbf{M}^2 \longmapsto \mathbf{x}_X \bullet \mathbf{y}_{X^\complement} \in \mathbf{M}$$

are continuous maps, all the preceding constructions preserve both continuity and measurability.

Measures and Partial Differences: For non-empty $X \subseteq \mathbb{Z}^d$, we use $\mathfrak{M}_1(M^X)$ to denote the space of Borel, probability measures μ on $(M^X, \mathcal{B}_M^X)$, and give $\mathfrak{M}_1(M^X)$ the topology of weak convergence, sic: $\{\mu_n\}_1^\infty \subseteq \mathfrak{M}_1(M^X)$ converges to μ, written $\mu_n \Longrightarrow \mu$, means that

$$\int_{M^X} \varphi\, d\mu_n \longrightarrow \int_{M^X} \varphi\, d\mu \quad \text{for every } \varphi \in C(M^X).$$

Next, given $\varphi \in B(M^X)$ and $\mu \in \mathfrak{M}_1(M^X)$, we will often write $\langle \varphi, \mu \rangle$ to denote $\int_{M^X} \varphi\, d\mu$; and, when $\emptyset \neq X \subset \mathbb{Z}^d$, $\mu \in \mathfrak{M}_1(M^X)$ and $\varphi \in B(\mathbf{M})$ ($\varphi \in C(\mathbf{M})$), we define the $\mathcal{F}_{X^\complement}$-measurable (continuous) mapping

$$\mathbf{x} \in \mathbf{M} \longmapsto \langle \varphi \rangle_{X,\mu}(\mathbf{x}) = \int_{M^X} f_X(\boldsymbol{\xi}^X|\mathbf{x})\, \mu(d\boldsymbol{\xi}^X).$$

Throughout, we will reserve λ to denote the ***standard reference probability measure*** on M. More specifically, depending on whether M is finite or a manifold, we will take λ to be either the normalized counting measure or the normalized RIEMANN measure on M. Further, will we write $\langle \varphi \rangle_X$ in place of $\langle \varphi \rangle_{X,\lambda^X}$ and $\langle \varphi \rangle_{\mathbf{k}}$ when $X = \{\mathbf{k}\}$. Finally, for each $Y \in \mathfrak{F}$, we define the *partial difference operator* $\partial_Y : B(\mathbf{M}) \longrightarrow B(\mathbf{M})$ by

$$\partial_Y \varphi = \langle \varphi \rangle_Y - \varphi, \tag{3.2}$$

use $\partial_{\mathbf{k}}$ when $Y = \{\mathbf{k}\}$, and set

$$|||\varphi||| = \sum_{\mathbf{k} \in \mathbb{Z}^d} \|\partial_{\mathbf{k}} \varphi\|_{\mathrm{u}} \quad \text{and} \quad \hat{C}(\mathbf{M}) = \left\{ \varphi \in C(\mathbf{M}) : |||\varphi||| < \infty \right\}.$$

Potentials and Gibbs States: A **potential** is a family $\mathcal{J} = \{J_X : X \in \mathfrak{F}\}$ where, for each $X \in \mathfrak{F}$, $J_X \in C_X(\mathbf{M})$. Throughout, we will be assuming that our potentials are *shift-invariant* in the sense that $J_{\mathbf{k}+X} = J_X \circ \theta^{\mathbf{k}}$, for all $X \in \mathfrak{F}$ and $\mathbf{k} \in \mathbb{Z}^d$. Further, we assume that $\mathcal{J}$ has *finite range* $R \in \mathbb{Z}^+$: that is, for each $X \in \mathfrak{F}$, $J_X \equiv 0$ if

$0 \in X \not\subseteq [-R,R]^d$. Given $\mathcal{J}$, we define the corresponding *local specification* $\mathfrak{E} = \mathfrak{E}(\mathcal{J})$ to be the collection of MARKOV operators

$$[\mathbb{E}^X\varphi](\xi) = \int_{\mathbf{M}} \varphi(\mathbf{y})\, E^X(d\mathbf{y}|\xi)$$
$$= \frac{1}{Z_X(\xi_{X^\complement})}\int_{\mathbf{M}} \varphi(\mathbf{y}_X \bullet \xi_{X^\complement}) \exp\left[-U^X(\mathbf{y}_X \bullet \xi_{X^\complement})\right] \boldsymbol{\lambda}(d\mathbf{y}),$$

where $\boldsymbol{\lambda} \equiv \lambda^{\mathbf{Z}^d}$,

$$U^X \equiv \sum_{\substack{A\in\mathfrak{F}\\ A\cap X\neq\emptyset}} J_A, \quad \text{and} \quad Z_X(\xi_{X^\complement}) \equiv \int_{\mathbf{M}} \exp\left[-U^X(\mathbf{y}_X \bullet \xi_{X^\complement})\right] \boldsymbol{\lambda}(d\mathbf{y}).$$

Finally, we say that $\mu \in \mathfrak{M}_1(\mathbf{M})$ is a **Gibbs state for** $\mathfrak{E}(\mathcal{J})$ and write $\mu \in \mathfrak{G}(\mathcal{J})$ if

$$\langle \mathbb{E}^X\varphi, \mu\rangle = \langle \varphi, \mu\rangle \quad \text{for all } X \in \mathfrak{F} \text{ and } \varphi \in C_0(\mathbf{M}).$$

3.3 Remark: It is important to understand the intuitive idea out of which the preceding definition of a GIBBS state grew. Namely, what one really would have liked to do is take

$$U = \sum_{A\in\mathfrak{F}} J_A \quad \text{and} \quad d\gamma^U = \frac{e^{-U}}{Z}\, d\boldsymbol{\lambda}.$$

However, this utter non-sense! On the other, if one proceeds at a formal level and asks what would be the conditional expectation value under γ^U of $f \in C_0(\mathbf{M})$ given $\mathcal{F}_{X^\complement}$, one realizes that it would be precisely $\mathbb{E}^X f$. Hence, all that we have done is reverse things and said that *a* GIBBS *state is a measure which gives the correct conditional expectation values.* Notice that this makes excellent sense physically since the measure $\mathbb{E}^X(\,\cdot\,|\mathbf{y})$ is *the* GIBBS *state inside* X *with potential* U^X *given that the configuration outside* X *is* $\mathbf{y}_{X^\complement}$.

Note that $\mathfrak{G}(\mathcal{J})$ is non-empty, compact, and convex. In fact, if $\{X_n\}_1^\infty \subseteq \mathfrak{F}$ is any non-decreasing sequence which exhausts $\mathbf{Z}^d$ and $\{\mathbf{y}_n\}_1^\infty \subseteq \mathbf{M}$, then any convergent subsequence of the measures $E^{X_n}(\,\cdot\,|\mathbf{y}_n)$ must be a GIBBS state. Furthermore, because $\mathbf{M}$ and therefore $\mathfrak{M}_1(\mathbf{M})$ are compact, a convergent subsequence will always exist. Thus $\mathfrak{G}(\mathcal{J}) \neq \emptyset$. At the same time, this reasoning also shows that $\mathfrak{G}(\mathcal{J})$ is compact. Finally, because the determining relations are affine, it is clear that $\mathfrak{G}(\mathcal{J})$ is convex.

Having introduced $\mathfrak{G}(\mathcal{J})$ and seen that it is a non-empty, convex, compact set in $\mathfrak{M}_1(\mathbf{M})$, one immediately realizes that there is a long list of questions which one ought to ask about the elements of $\mathfrak{G}(\mathcal{J})$; and, at the top of this list, is the question of *uniqueness*: when does $\mathfrak{G}(\mathcal{J})$ contain precisely one element? In terms of the physics which one is trying to model, $\mathfrak{G}(\mathcal{J})$ being a singleton means that $\mathcal{J}$ *determines a unique equilibrium distribution*, or, equivalently, that the system has only one *phase*; and on physical grounds what one expects is that there will be a single phase when $\mathcal{J}$ is sufficiently small (this corresponds to the *temperature* being sufficiently high). Mathematically, this expectation can be rationalized by first noting that $\mathfrak{G}(\mathcal{J}) = \{\boldsymbol{\lambda}\}$ when $\mathcal{J}$ is trivial (i.e., all the J_X's vanish identically) and then, even when $\mathcal{J}$ is not trivial, hoping to be saved by some sort of *perturbation* (i.e., analytic continuation) procedure. However, as one quickly realizes, the perturbation here is enormous and, as such, falls well outside the range of usual perturbation methods.

With these preliminary remarks and admonitions, we are, at last, ready to explain, by way of an example, the sort of analysis that we will be doing in the rest of these lectures. Namely, we will give over the remainder of this lecture to an explanation of a criterion, introduced originally by DOBRUSHIN and SHLOSMAN, which guarantees that $\mathfrak{G}(\mathcal{J})$ contains precisely one element. The *uniqueness condition of* DOBRUSHIN *and* SHLOSMAN (cf. [3]) says that there should exist a $Y \in \mathfrak{F}$ with $\mathbf{0} \in Y$ and a family of non-negative numbers $\alpha_{\mathbf{k}\mathbf{j}}$ for $\mathbf{k} \notin Y$ and $\mathbf{j} \in Y$ such that

$$\begin{aligned} &\left\|\partial_{\mathbf{k}}\mathbb{E}^Y\varphi - \mathbb{E}^Y\partial_{\mathbf{k}}\varphi\right\|_{\mathrm{u}} \leq \sum_{\mathbf{j}\in Y}\alpha_{\mathbf{k}\mathbf{j}}\left\|\partial_{\mathbf{j}}\varphi\right\|_{\mathrm{u}}, \quad \varphi \in C(\mathbf{M}), \\ &\text{and } \beta \equiv 1 - \frac{1}{|Y|}\sum_{\mathbf{k}\notin Y}\sum_{\mathbf{j}\in Y}\alpha_{\mathbf{k}\mathbf{j}} > 0; \end{aligned} \tag{3.4}$$

and their basic conclusion is that there is then only one element of $\mathfrak{G}(\mathcal{J})$.

3.5 Remark: One should begin by convincing oneself that (3.4) is a reasonable *perturbation condition.* To this end, take $Y = \{0\}$ and set

$$\rho(\mathbf{x}) = \frac{e^{-U^{\{0\}}(\mathbf{x})}}{Z_{\{0\}}(\mathbf{x}_{\{0\}^{\mathbf{C}}})}.$$

Then

$$\begin{aligned} \left[\partial_{\mathbf{k}}\mathbb{E}^{\{0\}}\varphi\right](\mathbf{x}) - \left[\mathbb{E}^{\{0\}}\partial_{\mathbf{k}}\varphi\right](\mathbf{x}) &= \int_M \left(\int_M ((\rho_{\mathbf{k}}(\xi|\mathbf{x}) - \rho(\mathbf{x}))\varphi_{\mathbf{k}}(\xi|\mathbf{x})\,\lambda(dx)\right)\lambda(dx_{\mathbf{0}}) \\ &= \int_M \left(\int_M (\rho_{\mathbf{k}}(\xi|\mathbf{x}) - \rho(\mathbf{x}))\varphi_{\mathbf{k}}(\xi|\mathbf{x})\,\lambda(dx_{\mathbf{0}})\right)\lambda(d\xi) \\ &= -\int_M \left(\int_M (\rho_{\mathbf{k}}(\xi|\mathbf{x}) - \rho(\mathbf{x}))\left(\partial_{\mathbf{0}}\varphi\right)_{\mathbf{k}}(\xi|\mathbf{x})\,\lambda(dx_{\mathbf{0}})\right)\lambda(d\xi), \end{aligned}$$

where, in the last equality, we used the fact that $\int_M \rho_{\mathbf{k}}(\xi|\mathbf{x})\,dx_{\mathbf{0}} = 1$ for all $\xi \in M$. In particular, this means that

$$\left|\left[\partial_{\mathbf{k}}\mathbb{E}^{\{0\}}\varphi\right](\mathbf{x}) - \left[\mathbb{E}^{\{0\}}\partial_{\mathbf{k}}\varphi\right](\mathbf{x})\right| \leq \max_{\xi\in M}\left(\frac{\rho_{\mathbf{k}}(\xi|\mathbf{x})}{\rho(\mathbf{x})} - 1\right)\left\|\partial_{\mathbf{0}}\varphi\right\|_{\mathrm{u}},$$

which leads to the condition in (3.4) as soon as the potential $\mathcal{J}$ is sufficiently small.

To understand what advantage might be gained by considering more general Y's, let $\mathfrak{F} \ni Y \ni \mathbf{0}$ be given, set

$$\rho(\mathbf{x}) = \frac{e^{-U^Y(\mathbf{x})}}{Z_Y(\mathbf{x}_{Y^{\mathbf{C}}})},$$

and, just as in the preceding, arrive at

$$\begin{aligned} &\left[\partial_{\mathbf{k}}\mathbb{E}^Y\varphi\right](\mathbf{x}) - \left[\mathbb{E}^Y\partial_{\mathbf{k}}\varphi\right](\mathbf{x}) \\ &\quad = \int_M \left(\int_{M^Y} (\rho(\mathbf{x}) - \rho_{\mathbf{k}}(\xi|\mathbf{x}))(\partial_Y\varphi)_{\mathbf{k}}(\xi|\mathbf{x})\,\lambda^Y(d\mathbf{x}^Y)\right)\lambda(d\xi). \end{aligned} \tag{3.6}$$

Next, let $\{\mathbf{j}_i\}_1^n$ be an enumeration of the elements in Y, set $Y_m = \{\mathbf{j}_1, \dots, \mathbf{j}_m\}$ for $1 \le m \le n$, and note that

$$\partial_Y \psi = \partial_{\mathbf{j}_1} \psi + \sum_{m=2}^{n} \langle \partial_{\mathbf{j}_m} \psi \rangle_{Y_{m-1}}.$$

Hence, the right hand side of (3.6) becomes the sum of terms having the form

$$\int_M \left(\int_{M^{\check{Y}_m}} \langle \rho - \rho_{\mathbf{k}}(\xi|\cdot) \rangle_{Y_{m-1}}(\mathbf{x}) \langle \partial_{\mathbf{j}_m} \varphi_{\mathbf{k}}(\xi|\cdot) \rangle_{Y_{m-1}}(\mathbf{x})\, \lambda^{\check{Y}_m}(d\mathbf{x}_{\check{Y}_m}) \right) \lambda(d\xi),$$

where $\check{Y}_m \equiv Y \setminus Y_{m-1}$; and each of these is dominated by

$$\|\partial_{\mathbf{j}_m} \varphi\|_{\mathrm{u}} \int_M \left(\int_{M^{\check{Y}_m}} \left| \langle \rho - \rho_{\mathbf{k}}(\xi|\cdot) \rangle_{Y_{m-1}}(\mathbf{x}) \right| \lambda^{\check{Y}_m}(d\mathbf{x}_{\check{Y}_m}) \right) \lambda(d\xi).$$

But clearly, if $\mathbf{k}$ is far away from Y_{m-1} or Y_{m-1} is most of Y, then one should expect that $\langle \rho - \rho_{\mathbf{k}}(\xi|\cdot) \rangle_{Y_{m-1}}$ will be nearly 0. Hence, by making a judicious choice of enumeration (the choice can depend on $\mathbf{k}$), one can hope to make the majority of $\alpha_{\mathbf{k}\,\mathbf{j}}$'s sufficiently small that the resulting average will be small.

Finally, it should be evident from the preceding considerations that, when dealing with finite range potentials, the existence of $\alpha_{\mathbf{k}\,\mathbf{j}}$'s for which the first line of (3.4) holds is never in doubt. It is the estimate in the second line which is the real condition in the DOBRUSHIN–SHLOSMAN criterion.

Our proof of the DOBRUSHIN–SHLOSMAN uniqueness criterion will be dynamical. To be more precise, we will introduce a MARKOV semigroup on $C(\mathbf{M})$ for which every element of $\mathfrak{G}(\mathbf{M})$ is invariant; and we will then show that, under the condition in (3.4), this semigroup is ergodic in the sense that it tends to equilibrium as $t \to \infty$. In particular, this will prove that the semigroup has only one invariant measure, which means, of course, that $\mathfrak{G}(\mathcal{J})$ can have only one element. To describe the semigroup which we have in mind, define the operator $\mathcal{L}^Y$ on $C_0(\mathbf{M})$ by

$$\mathcal{L}^Y \varphi = \sum_{\mathbf{k} \in \mathbf{Z}^d} \left(\mathbf{E}^{\mathbf{k}+Y} \varphi - \varphi \right). \tag{3.7}$$

Clearly,

$$\|\mathcal{L}^Y \varphi\|_{\mathrm{u}} \le 2|Y|\, |||\varphi|||,$$

and so $\mathcal{L}^Y$ determines a unique extension as an operator from $\hat{C}(\mathbf{M})$ into $C(\mathbf{M})$. Furthermore, from the definition of GIBBS states, it is clear that

$$\int \mathcal{L}^Y \varphi\, d\mu = 0 \quad \text{for all } \varphi \in \hat{C}(\mathbf{M}) \text{ and } \mu \in \mathfrak{G}(\mathcal{J}). \tag{3.8}$$

What is less obvious, but also not too difficult (cf. the Remark 3.12 below), is the proof that $\mathcal{L}$ determines a unique MARKOV semigroup $\{P_t^Y : t > 0\}$ on $C(\mathbf{M})$ with the properties that

$$(3.9) \quad P_t^Y\varphi - \varphi = \int_0^t P_s^Y \mathcal{L}^Y \varphi \, ds \quad \text{and} \quad |||P_t^Y\varphi||| \le e^{At}|||\varphi|||, \quad t > 0 \text{ and } \varphi \in \hat{C}(\mathbf{M}),$$

for some $A \in \mathbb{R}$. Notice that, by combining (3.8) and (3.9), we have that

$$\frac{d}{dt}\int_M P_t^Y\varphi \, d\mu = 0 \quad \text{for all } t > 0,\ \varphi \in \hat{C}(\mathbf{M}), \quad \text{and } \mu \in \mathfrak{G}(\mathcal{J});$$

and, therefore, every $\mu \in \mathfrak{G}(\mathcal{J})$ is $\{P_t^Y : t > 0\}$- invariant.

So far, nothing that we have said about $\{P_t^Y : t > 0\}$ depends the condition $\beta > 0$ in (3.4). This is because (3.4) enters only when we attempt to prove ergodicity. In fact, what we are going to do is show that the constant A in (3.9) can be dominated by $-\beta|Y|$; from which ergodicity is more or less immediate since we will then know that

$$(3.10) \qquad \left|P_t^Y\varphi(\mathbf{x}) - \int \varphi\, d\mu\right| \le \int \left|P_t^Y\varphi(\mathbf{x}) - P_t^Y\varphi(\mathbf{y})\right| \mu(d\mathbf{y}) \le e^{-\beta|Y|t}|||\varphi|||.$$

for all $(t, \mathbf{x}) \in (0,\infty) \times \mathbf{M}$, $\varphi \in \hat{C}(\mathbf{M})$, and $\mu \in \mathfrak{G}(\mathcal{J})$. In order to obtain the required estimate, let $\varphi \in \hat{C}(\mathbf{M})$ be given, and observe that

$$\partial_{\mathbf{k}}\left(E^{\ell+Y}\varphi - \varphi\right) = \begin{cases} -\partial_{\mathbf{k}}\varphi & \text{if } \mathbf{k} \in \ell + Y \\ \left(E^{\ell+Y}\varphi_{,\mathbf{k}} - \varphi_{,\mathbf{k}}\right) + \mathbf{R}_{\mathbf{k},\ell}\varphi & \text{otherwise,} \end{cases}$$

where we have used $\varphi_{,\mathbf{k}}$ to denote $\partial_{\mathbf{k}}\varphi$ and $\mathbf{R}_{\mathbf{k},\ell}$ to denote the commutator $\left[\partial_{\mathbf{k}}, \mathbb{E}^{\ell+Y}\right]$. Thus, if $\varphi_t = P_t^Y\varphi$ and $\varphi_{t,\mathbf{k}} = \partial_{\mathbf{k}}\varphi_t$, then

$$\tfrac{d}{dt}\varphi_{t,\mathbf{k}} = \partial_{\mathbf{k}}\mathcal{L}^Y\varphi_t = \sum_{\ell \notin \mathbf{k}-Y}\left(\mathbb{E}^{\ell+Y}\varphi_{t,\mathbf{k}} - \varphi_{t,\mathbf{k}}\right) - |Y|\varphi_{t,\mathbf{k}} + \mathbf{R}_{\mathbf{k}}\varphi_t\,,$$

where

$$\mathbf{R}_{\mathbf{k}} \equiv \sum_{\ell \notin \mathbf{k}-Y} \mathbf{R}_{\mathbf{k},\ell}.$$

Now observe that, just like $\mathcal{L}^Y$, the operator given by

$$\mathcal{L}^{(\mathbf{k})}\varphi \equiv \sum_{\ell \notin \mathbf{k}-Y}\left(\mathbb{E}^{\ell+Y}\varphi - \varphi\right)$$

determines a MARKOV semigroup $\{P_t^{(\mathbf{k})} : t > 0\}$ on $C(\mathbf{M})$; and so, by trivial considerations, we see first that

$$\frac{d}{ds}\left[e^{|Y|s}P_{t-s}^{(\mathbf{k})}\varphi_{s,\mathbf{k}}\right] = e^{|Y|s}P_{t-s}^{(\mathbf{k})} \circ \mathbf{R}_{\mathbf{k}}\varphi_s \quad \text{for } s \in (0,t)$$

and then that

$$e^{|Y|t}\|\varphi_{t,k}\|_u \le \|\varphi_{,k}\|_u + \int_0^t e^{|Y|s}\|\mathbf{R}_k\varphi_t\|_u\, ds.$$

Hence, after summing over $k \in \mathbb{Z}^d$ and noting that, by (3.4),

$$\begin{aligned}\sum_{k\in\mathbb{Z}^d} \|\mathbf{R}_k\varphi_t\|_u &\le \sum_{k\in\mathbb{Z}^d} \sum_{\ell\notin k-Y} \sum_{j\in k-Y} \alpha_{k-\ell,k-j}\|\varphi_{t,j}\|_u \\ &= \sum_j \|\varphi_{t,j}\|_u \sum_{\ell\in j-Y} \sum_{k\notin \ell+Y} \alpha_{k-\ell,j-\ell} \le (1-\beta)|Y|\, \|\!|\varphi_t|\!\|,\end{aligned}$$

we come to

$$e^{|Y|t}\|\!|\varphi_t|\!\| \le \|\!|\varphi|\!\| + (1-\beta)|Y| \int_0^t e^{|Y|s}\|\!|\varphi_s|\!\|\, ds;$$

from which

$$\|\!|P_t^Y\varphi|\!\| \le e^{-\beta|Y|t}\|\!|\varphi|\!\|, \quad t>0 \text{ and } \varphi\in\hat{C}(\mathbf{M}) \tag{3.11}$$

is an easy step.

3.12 Remark The preceding argument provides the basis for a proof that there is precisely one MARKOV semigroup with the property that the first equation in (3.9) holds for every $\varphi \in C_0(\mathbf{M})$ and that that unique semigroup satisfies the estimate in the second part of (3.9). Indeed, the essential step in the proof requires one to show that if $\Gamma(N) \equiv [-N,N]^d \cap \mathbb{Z}^d$ and one replaces the operator $\mathcal{L}^Y$ in (3.7) with the bounded operator

$$\mathcal{L}^{(N)} = \sum_{\{k:k+Y\subseteq\Gamma(N)\}} \left(\mathbb{E}^{k+Y} - I\right),$$

then the associated MARKOV semigroup $\{P_t^{(N)} : t>0\}$ satisfies

$$\|\!|P_t^{(N)}\varphi|\!\| \le e^{At}\|\!|\varphi|\!\|, \quad t\in(0,\infty) \quad \text{and } \varphi\in C_0(\mathbf{M}),$$

with an $A \in \mathbb{R}$ which is independent of $N \in \mathbb{Z}^+$. But, by essentially the same argument as we have just given (only this time being a little more careless) one finds that

$$\|\!|P_t^{(N)}\varphi|\!\| \le e^{(1-\beta)|Y|t}\|\!|\varphi|\!\|, \quad t\in(0,\infty),$$

even if β is negative.

Lecture IV: More about the Dobrushin–Shlosman Criterion

As we showed in the preceding lecture, the condition in (3.4) guarantees that there is precisely one element of $\mathfrak{G}(\mathcal{J})$, and first goal in this lecture will be that it guarantees more. To be precise, for $N \in \mathbf{Z}^+$, let $\Gamma(N)$ be the cube $[-N, N]^d \cap \mathbf{Z}^d$ in $\mathbf{Z}^d$. What we will show is that, under the condition in (3.4), there exist $C \in [2, \infty)$ and $\epsilon > 0$ such that the local specification $\mathfrak{E}(\mathcal{J})$ satisfies the *mixing condition*

$$\text{(4.1)} \quad \left|[\mathbb{E}^{\Gamma(N)}\varphi](\xi) - [\mathbb{E}^{\Gamma(N)}\varphi](\eta)\right| \leq C|||\varphi||| \exp\left[-\epsilon \operatorname{dist}(X, \Gamma(N)\complement)\right]$$
$$\text{for } N \in \mathbf{Z}^+,\ (\xi, \eta) \in \mathbf{M}^2,\ X \subset \Gamma(N), \text{ and } \varphi \in C_X(\mathbf{M}).$$

In fact, our proof will show that C and ϵ can be chosen to depend only on the dimension d of the lattice and the range R of the potential $\mathcal{J}$ in addition to the number $\beta \in (0, 1]$ and set Y appearing in (3.4).

To prove (4.1), let $K(N)$ be the set of $\boldsymbol{\ell} \in \mathbf{Z}^d$ for which $\boldsymbol{\ell} + Y \subseteq \Gamma(N)$, and define

$$\mathcal{L}^{(N)}\varphi = \sum_{\boldsymbol{\ell} \in K(N)} \left(\mathbb{E}^{\boldsymbol{\ell}+Y}\varphi - \varphi\right) \quad \text{for } \varphi \in C(\mathbf{M}).$$

As distinguished from the situation with the unbounded operator $\mathcal{L}$ introduced in the Lecture III, it is completely elementary to show that $\mathcal{L}^{(N)}$ determines a unique MARKOV semigroup $\{P_t^{(N)} : t > 0\}$ on $C(\mathbf{M})$ with the property that

$$\text{(4.2)} \qquad P_t^{(N)}\varphi - \varphi = \int_0^t P_s^{(N)} \circ \mathcal{L}^{(N)}\varphi\, ds, \quad t \in (0, \infty) \text{ and } \varphi \in C(\mathbf{M}).$$

(In fact, $P_t^{(N)}$ is given explicitly in terms of $\mathcal{L}^{(N)}$ by a exponential power series.) Moreover, because

$$\text{(4.3)} \qquad \mathbb{E}^Y = \mathbb{E}^Y \circ \mathbb{E}^X \quad \text{for all } X, Y \in \mathfrak{F} \text{ with } X \subseteq Y,$$

it is easy to check first that $\mathbb{E}^{\Gamma(N)} \circ \mathcal{L}^{(N)} = 0$ and then that $\mathbb{E}^{\Gamma(N)} \circ P_t^{(N)} = \mathbb{E}^{\Gamma(N)}$ for all $t > 0$. In particular, this means that

$$[\mathbb{E}^{\Gamma(N)}\varphi](\xi) - [\mathbb{E}^{\Gamma(N)}\varphi](\eta) = \iint_{\mathbf{M}^2} \left([P_t^{(N)}\varphi](\mathbf{x}) - [P_t^{(N)}\varphi](\mathbf{y})\right) E^{\Gamma(N)}(d\mathbf{x}|\xi) E^{\Gamma(N)}(d\mathbf{y}|\eta),$$

and so

$$\text{(4.4)} \qquad \left|[\mathbb{E}^{\Gamma(N)}\varphi](\xi) - [\mathbb{E}^{\Gamma(N)}\varphi](\eta)\right| \leq 2|||P_t^{(N)}\varphi||| \quad \text{for all } t \in (0, \infty).$$

In view of (4.4), all that we have to do is learn how to choose t so that $|||P_t^{(N)}\varphi|||$ is dominated by an expression of the form on the right hand side of (4.1). In part, this will involve the same idea that we used in the derivation of (3.11); however, here, we will be forced to take the boundary of $\Gamma(N)$ into account, and this will introduce a new twist. Be that as, the argument begins in the same way. Namely, given $\varphi \in C_X(\mathbf{M})$, set $\varphi_t = P_t^{(N)}\varphi$, and proceed as we did in Lecture III to see that

$$(4.5)\qquad e^{\sigma^{(N)}(\mathbf{k})t}\left\|\varphi_{t,\mathbf{k}}\right\|_{\mathrm{u}} \leq \|\varphi_{,\mathbf{k}}\|_{\mathrm{u}} + \int_0^t e^{\sigma^{(N)}(\mathbf{k})s}\left\|\mathbf{R}_{\mathbf{k}}^{(N)}\varphi_s\right\|_{\mathrm{u}} ds, \quad t > 0 \text{ and } \mathbf{k} \in \mathbb{Z}^d,$$

where

$$\sigma^{(N)}(\mathbf{k}) = \left|K(N) \cap (\mathbf{k} - Y)\right| \quad \text{and} \quad \mathbf{R}_{\mathbf{k}}^{(N)} = \sum_{\substack{\ell \in K(N) \\ \ell \notin \mathbf{k}-Y}} \left[\partial_{\mathbf{k}}, \mathbf{E}^{\ell+Y}\right].$$

Hence, if $\tilde{K}(N) = \{\mathbf{k} \in K(N) : \mathbf{k} - Y \subseteq K(N)\}$ and we set

$$u^{(N)}(t) = \sum_{\mathbf{k} \in \tilde{K}(N)} \left\|\varphi_{t,\mathbf{k}}\right\|_{\mathrm{u}} \quad \text{and} \quad w^{(N)}(t) = \sum_{\mathbf{k} \notin \tilde{K}(N)} \left\|\varphi_{t,\mathbf{k}}\right\|_{\mathrm{u}},$$

then

$$u^{(N)}(t) \leq e^{-\beta|Y|t}|||\varphi||| + |Y| \int_0^t e^{\beta|Y|(s-t)}\, w^{(N)}(s)\, ds.$$

In order to handle $w^{(N)}$ (which is the boundary effect alluded to earlier), set $S_0 = \tilde{K}(N)\complement$, use induction to define the sets S_n for $n \geq 1$ so that

$$S_n = \bigcup \left\{\ell + Y : \operatorname{dist}\left(\ell + Y, S_{n-1}\right) \leq R\right\},$$

and put

$$w_n^{(N)}(t) = \sum_{\mathbf{k} \in S_n} \left\|\varphi_{t,\mathbf{k}}\right\|_{\mathrm{u}}.$$

If

$$n^{(N)}(X) = \min\left\{n \geq 0 : \; S_n \cap X \neq \emptyset\right\},$$

then, by (4.5),

$$w_{n-1}^{(N)}(t) \leq |Y| \int_0^t w_n^{(N)}(s)\, ds \quad \text{for } n \leq n^{(N)}(X).$$

On the other hand, again by (4.5),

$$|||\varphi_t||| \leq |||\varphi||| + |Y| \int_0^t |||\varphi_s|||\, ds.$$

Hence, since $w_{n^{(N)}(X)}^{(N)}(t)$ is obviously dominated by $|||\varphi_t|||$, we conclude that

$$(4.6)\qquad w^{(N)}(t) = w_0^{(N)}(t) \leq |||\varphi|||\,\mathbf{e}_{n^{(N)}(X)}\left(|Y|t\right),$$

where

$$\mathbf{e}_n(s) \equiv e^s - \sum_0^{n-1} \frac{s^m}{m!} \leq e^s \left(\frac{se}{n}\right)^n.$$

In particular, after we combine these, we find that

$$|||\varphi_t||| \leq |||\varphi||| \left(e^{-\beta|Y|t} + 2\mathbf{e}_{n^{(N)}(X)}(|Y|t)\right) \quad \text{for all } t > 0,$$

which leads immediately to the estimate

$$\min_{t\in(0,\infty)} |||\varphi_t||| \leq 3|||\varphi||| \exp[-\delta n^{(N)}(X)]$$

for some $\delta > 0$ depending only on $\beta|Y|$. Finally, since there exists a $\rho \in (0,1)$, depending only on Y and R, such that

$$n^{(N)}(X) \geq \rho \operatorname{dist}(X, \Gamma(N)\complement) - \tfrac{1}{\rho},$$

(4.1) now follows from (4.4) and the above.

The mixing estimate in (4.1) is a much stronger assertion than the uniqueness statement obtained in Lecture III. Indeed, uniqueness is an entirely qualitative statement, whereas (4.1) actually provides an estimate of how well integrals with respect to the GIBBS state can be approximated by measures on $M^{\Gamma(N)}$. For example, as an immediate consequence of (4.1) one can say that

$$\sup_{\xi\in\mathbf{M}} \left| \int_{\mathbf{M}} \varphi(\mathbf{y})\, E^{\Gamma(N)}(d\mathbf{y}|\xi) - \int_{\mathbf{M}} \varphi\, d\gamma \right| \leq C|||\varphi||| \exp\left[-\epsilon \operatorname{dist}(X, \Gamma(N)\complement)\right] \tag{4.7}$$

$$\text{for } N \in \mathbf{Z}^+,\ X \subset \Gamma(N) \text{ and } \varphi \in C_X(\mathbf{M}).$$

On the other hand, if one is optimistically inclined, then (4.1) is less than one might have hoped for. Indeed, ones intuition inclines one toward the belief that (4.1) should be replaced by an estimate in which the set X on which φ is localized need not be buried deeply inside $\Gamma(N)$ so long as it stays far away from sites $\mathbf{k} \notin \Gamma(N)$ at which the configurations ξ and η differ. However, DOBRUSHIN and SHLOSMAN claim (cf. [4]) that such a conclusion will not, in general, follow from (3.4). For this reason, they introduced a strengthened version of (3.4). Namely, what we will call the **Dobrushin–Shlosman mixing condition** states that there should again exist a $Y \in \mathfrak{F}$ and a family $\{\alpha_{\mathbf{kj}} : \mathbf{k} \notin Y \text{ and } \mathbf{j} \in Y\} \subseteq [0,\infty)$ such that again

$$\beta \equiv 1 - \frac{1}{|Y|} \sum_{\substack{\mathbf{k}\notin Y\\ \mathbf{j}\in Y}} \alpha_{\mathbf{kj}} > 0, \tag{4.8(a)}$$

but this time

$$\left\| [\partial_{\mathbf{k}}, \mathbb{E}^X]\varphi \right\|_{\mathrm{u}} \leq \sum_{\mathbf{j}\in X} \alpha_{\mathbf{kj}} \left\|\partial_{\mathbf{j}}\varphi\right\|_{\mathrm{u}} \quad \text{for all } \mathbf{k} \notin Y \text{ and } \mathfrak{F} \ni X \subseteq Y. \tag{4.8(b)}$$

Obviously, (4.8) implies (3.4). Moreover, from the point of view which we have adopted, the strength of (4.8) over (3.4) is easily understood. Namely, it allows us to avoid the *boundary* considerations which entered our derivation of (4.1) from (3.4). To be more precise, let $\Gamma \in \mathfrak{F}$ be given, and consider the semigroup $\{P_t^{\Gamma,Y} : t > 0\}$ which is determined by the operator

$$\mathcal{L}^{\Gamma,Y} = \sum_{\ell \in \mathbf{Z}^d} \left(\mathbf{E}^{(\ell+Y)\cap\Gamma}\varphi - \varphi\right) \quad \text{where } \mathbf{E}^{\emptyset} \text{ is the identity.}$$

By repeating the argument given in Lecture III to derive (3.11) (not the one given above to derive (4.1)), we see that

$$\text{(4.9)} \qquad \left\|\!\left| P_t^{\Gamma,Y}\varphi \right|\!\right\|_{\Gamma} \le e^{-\beta|Y|t} \left\|\!\left|\varphi\right|\!\right\|_{\Gamma}, \quad t > 0 \text{ and } \varphi \in C(\mathbf{M}),$$

where

$$\left\|\!\left|\psi\right|\!\right\|_X \equiv \sum_{\mathbf{k} \in X} \left\|\partial_{\mathbf{k}}\psi\right\|_{\mathrm{u}} \quad \text{for } X \in \mathfrak{F}.$$

Next, observe that if X and S_0 are subsets of Γ, S_n is defined for $n \ge 1$ by

$$S_n = \bigcup \left\{(\ell + Y)\cap\Gamma : \exists \mathbf{k} \in S_{n-1} \ \mathrm{dist}(\mathbf{k}, (\ell+Y)\cap\Gamma) \le R\right\}$$

(we take $\mathrm{dist}(\mathbf{k}, \emptyset) = \infty$), and

$$n^{\Gamma}(X, S_0) = \min\{n \ge 0 : S_n \cap X \ne \emptyset\},$$

then the argument which led to (4.6) can be repeated to prove that

$$\text{(4.10)} \qquad \left\|\!\left| P_t^{\Gamma,Y}\varphi \right|\!\right\|_{S_0} \le \mathbf{e}_{n^{\Gamma}(X,S_0)}(|Y|t) \left\|\!\left|\varphi\right|\!\right\|_X \quad \text{for } \varphi \in C_{X\cup\Gamma\complement}(\mathbf{M}).$$

Finally, note that

$$\left[P_t^{\Gamma,Y}\varphi\right]_{\Gamma}(\mathbf{x}|\boldsymbol{\xi}) = \left[P_t^{\Gamma,Y}\varphi_{\Gamma}(\,\cdot\,|\boldsymbol{\xi})\right]_{\Gamma}(\mathbf{x}|\boldsymbol{\xi});$$

and conclude that, for each $\boldsymbol{\xi} \in \mathbf{M}$, the operators $P_t^{\Gamma,Y,\boldsymbol{\xi}}$, $t > 0$ given by

$$P_t^{\Gamma,Y,\boldsymbol{\xi}}\varphi = \left[P_t^{\Gamma,Y}\varphi\right]_{\Gamma}(\,\cdot\,|\boldsymbol{\xi}) \quad \text{for } t > 0 \text{ and } \varphi \in C(M^{\Gamma})$$

is the MARKOV semigroup on $C(M^{\Gamma})$ determined by

$$\mathcal{L}^{\Gamma,Y,\boldsymbol{\xi}}\varphi = \left[\mathcal{L}^{\Gamma,Y}\varphi\right]_{\Gamma}(\,\cdot\,|\boldsymbol{\xi}).$$

In particular, if $(\boldsymbol{\xi}, \boldsymbol{\eta}) \in \mathbf{M}^2$,

$$S = \left\{\ell : \mathrm{dist}(\mathbf{k}, (\ell + Y)\cap\Gamma) \le R \text{ for some } \mathbf{k} \notin \Gamma \text{ with } \xi_{\mathbf{k}} \ne \eta_{\mathbf{k}}\right\},$$

and $\varphi \in C_\Gamma(\mathbf{M})$, then, for $0 < s < t$:

$$\begin{aligned}\frac{d}{ds} P_{t-s}^{\Gamma,Y,\eta} \circ P_s^{\Gamma,Y,\xi}\varphi &= \sum_{\ell \in \mathbf{Z}^d} P_{t-s}^{\Gamma,Y,\eta} \circ \Big(\big[\mathbb{E}^{(\ell+Y)\cap\Gamma} \circ P_s^{\Gamma,Y,\xi}\varphi\big]_\Gamma(\cdot\,|\xi) \\ &\qquad\qquad - \big[\mathbb{E}^{(\ell+Y)\cap\Gamma} \circ P_s^{\Gamma,Y,\xi}\varphi\big]_\Gamma(\cdot\,|\eta)\Big) \\ &= \sum_{\ell \in S} P_{t-s}^{\Gamma,Y,\eta} \circ \Big(\big[\mathbb{E}^{(\ell+Y)\cap\Gamma} \circ P_s^{\Gamma,Y,\xi}\varphi\big]_\Gamma(\cdot\,|\xi) \\ &\qquad\qquad - \big[\mathbb{E}^{(\ell+Y)\cap\Gamma} \circ P_s^{\Gamma,Y,\xi}\varphi\big]_\Gamma(\cdot\,|\eta)\Big),\end{aligned}$$

and therefore

$$\begin{aligned}\big\|P_t^{\Gamma,Y,\xi}\varphi - P_t^{\Gamma,Y,\eta}\varphi\big\|_{\mathrm{u}} &\le V^\Gamma(\xi,\eta)\int_0^t \Big(\sum_{\ell\in S} \big\|\partial_{(\ell+Y)\cap\Gamma} P_s^{\Gamma,Y}\varphi\big\|_{\mathrm{u}}\Big)\, ds \\ &\le V^\Gamma(\xi,\eta)|Y| \int_0^t \big|\!\big|\!\big| P_s^{\Gamma,Y}\varphi \big|\!\big|\!\big|_{S_0}\, ds,\end{aligned}$$

where

$$V^\Gamma(\xi,\eta) = \max_{\ell\in\Gamma} \big\|E^{(\ell+Y)\cap\Gamma}(\cdot\,|\xi) - E^{(\ell+Y)\cap\Gamma}(\cdot\,|\eta)\big\|_{\mathrm{var}} \tag{4.11}$$

and

$$S_0 = \bigcup\Big\{(\ell+Y)\cap\Gamma : \operatorname{dist}\big(\mathbf{k},(\ell+Y)\cap\Gamma\big) \le R \text{ for some } \mathbf{k}\in S\Big\}.$$

Hence, after combining this with (4.10), we see that

$$\big\|P_t^{\Gamma,Y,\xi}\varphi - P_t^{\Gamma,Y,\eta}\varphi\big\|_{\mathrm{u}} \le |Y| V^\Gamma(\xi,\eta) |\!|\!|\varphi|\!|\!|_X\, \mathbf{e}_{n^\Gamma(X,S_0)+1}\big(|Y|t\big) \quad \text{for } \varphi\in C_{X\cup\Gamma\complement}(\mathbf{M}).$$

Finally, since

$$\big[\mathbb{E}^\Gamma \circ P_t^{\Gamma,Y,\xi}\varphi\big](\xi) = \big[\mathbb{E}^\Gamma\varphi\big](\xi) \quad \text{for all } t\in(0,\infty) \text{ and } \xi\in\mathbf{M},$$

we can put this together with (4.9) to produce a $C \in [2,\infty)$ and an $\epsilon > 0$, depending only on R as well as the β and Y in (4.8(a)), with the property that, for all $X \subseteq \Gamma \in \mathfrak{F}$ and $S \subseteq \Gamma\complement$:

$$\begin{aligned}&\Big|\big[\mathbb{E}^\Gamma\varphi\big](\xi) - \big[\mathbb{E}^\Gamma\varphi\big](\eta)\Big| \le C\, V^\Gamma(\xi,\eta) |\!|\!|\varphi|\!|\!|_X \exp\Big[-\epsilon \operatorname{dist}(X,S)\Big] \\ &\qquad \text{when } \varphi\in C_{X\cup(\Gamma\cup S)\complement}(\mathbf{M}) \text{ and } (\xi,\eta)\in\mathbf{M}^2 \text{ have } \xi_{(\Gamma\cup S)\complement} = \eta_{(\Gamma\cup S)\complement}.\end{aligned} \tag{4.12}$$

Lecture V: Back to Logarithmic Sobolev

In the remaining two lectures we will relate the considerations in Lectures III and IV to those in Lectures I and II; and, in order to simplify the presentation, we will, from now on, be assuming that M *is a compact, connected, finite dimensional* RIEMANN*ian manifold with* LEVI–CIVITA *connection* ∇ and normalized RIEMANN measure λ. In particular, this means that, for any non-empty $\Gamma \subseteq \mathbf{Z}^d$,

$$\alpha_1 \left\| \varphi - \langle \varphi, \lambda^\Gamma \rangle \right\|^2_{L^2(\lambda^\Gamma)} \leq \sum_{\mathbf{k}\in\Gamma} \int_{M^\Gamma} \left| \nabla_{\mathbf{k}} \varphi \right|^2 d\lambda^\Gamma, \quad \varphi \in C^1(M^\Gamma), \tag{5.1}$$

where (cf. (1.6) with $U = 0$) $\alpha_1 = \alpha_1(0) > 0$ is the gap in the spectrum of the LAPLACE–BELTRAMI operator for the RIEMANNian structure on M and we have introduced the notation $\nabla_{\mathbf{k}}$ to stand for the operation

$$\left[\nabla_{\mathbf{k}} \varphi\right](\mathbf{x}) = \left[\nabla \varphi_{\mathbf{k}}(\,\cdot\,|\mathbf{x})\right](x_{\mathbf{k}}).$$

(The argument that the spectral gap for the LAPLACian on M^Γ is independent of Γ is a simpler version of the argument which led from (2.10) to (2.11).)

Now let $\mathcal{J}$ be a potential of the sort introduced in Lecture II, only this time add the assumption that the functions J_X are smooth, and define the operator $\mathcal{L}$ on

$$C_0^\infty(\mathrm{M}) \equiv \left\{ \varphi \in C_0(\mathrm{M}) : \varphi \text{ is smooth} \right\}$$

so that

$$\mathfrak{L}\varphi = \sum_{\mathbf{k}\in\mathbf{Z}^d} e^{U^{\mathbf{k}}} \nabla_{\mathbf{k}} \cdot \left(e^{-U^{\mathbf{k}}} \nabla_{\mathbf{k}} \varphi\right), \tag{5.2}$$

where $U^{\mathbf{k}} \equiv U^{\{\mathbf{k}\}} = \sum_{X \ni \mathbf{k}} J_X$. One can then show that there exists a unique MARKOV semigroup $\{\mathcal{P}_t : t > 0\}$ on $C(\mathrm{M})$ with the property that

$$\mathcal{P}_t \varphi - \varphi = \int_0^t \mathcal{P}_s \circ \mathfrak{L}\varphi \, ds, \quad t > 0 \text{ and } \varphi \in C_0^\infty(\mathrm{M}). \tag{5.3}$$

Furthermore, if

$$|||\varphi|||_1 \equiv \sum_{\mathbf{k}\in\mathbf{Z}^d} \left\| \nabla_{\mathbf{k}} \varphi \right\|_{\mathrm{u}} \quad \text{for } \varphi \in C_0^\infty(\mathrm{M})$$

and $\hat{C}^1(\mathrm{M})$ denotes the completion of $C_0^\infty(\mathrm{M})$ with respect of $\|\cdot\|_{\mathrm{u}} + |||\cdot|||_1$, then one can show that $\hat{C}^1(\mathrm{M})$ is $\{\mathcal{P}_t : t > 0\}$-invariant and that

$$|||\mathcal{P}_t \varphi|||_1 \leq e^{At} |||\varphi|||_1, \quad \varphi \in \hat{C}^1(\mathrm{M}) \tag{5.4}$$

for some $A \in \mathbb{R}$.

Since, for each $\mathbf{k} \in \mathbf{Z}^d$ and $\mathbf{x} \in \mathbf{M}$,

$$\int_{\mathbf{M}} \psi(\xi) e^{U^{\mathbf{k}}} \nabla_{\mathbf{k}} \cdot (e^{-U^{\mathbf{k}}} \nabla_{\mathbf{k}} \varphi)(\xi)\, E^{\mathbf{k}}(d\xi|\mathbf{x}) = - \int_{\mathbf{M}} [\nabla_{\mathbf{k}} \psi \cdot \nabla_{\mathbf{k}} \varphi](\xi)|^2\, E^{\mathbf{k}}(d\xi|\mathbf{x}),$$

it is a relatively easy matter to check first that, for every $\mu \in \mathfrak{G}(\mathcal{J})$,

$$\int_{\mathbf{M}} \psi \mathfrak{L} \varphi\, d\mu = - \sum_{\mathbf{k} \in \mathbf{Z}^d} \int_{\mathbf{M}} \nabla_{\mathbf{k}} \psi \cdot \nabla_{\mathbf{k}} \varphi\, d\mu, \quad \varphi, \psi \in C_0^\infty(\mathbf{M}),$$

next that

$$(\psi, \mathcal{P}_t \varphi)_{L^2(\mu)} = (\varphi, \mathcal{P}_t \psi)_{L^2(\mu)}, \quad \varphi, \psi \in C(\mathbf{M}), \tag{5.5}$$

and then that there is a unique, strongly continuous semigroup $\{\overline{\mathcal{P}_t}^\mu : t > 0\}$ of self-adjoint contractions on $L^2(\mu)$ whose restriction to $C(\mathbf{M})$ is $\{\mathcal{P}_t : t > 0\}$. In addition, if $\{E_\alpha^\mu : \alpha \in [0, \infty)\}$ is the resolution of the identity in $L^2(\mu)$ such that

$$\overline{\mathcal{P}_t} = \int_{[0,\infty)} e^{-\alpha t}\, dE_\alpha^\mu, \tag{5.6}$$

then the associated DIRICHLET form

$$\mathcal{E}^\mu(\varphi, \varphi) \equiv \int_{(0,\infty)} \alpha\, d(E_\alpha^\mu \varphi, \varphi)_{L^2(\mu)}$$

is the closure in $L^2(\mu)$ of its restriction to $\hat{C}^1(\mathbf{M})$ and

$$\mathcal{E}^\mu(\varphi, \varphi) = \sum_{\mathbf{k} \in \mathbf{Z}^d} \int_{\mathbf{M}} \nabla_{\mathbf{k}} \psi \cdot \nabla_{\mathbf{k}} \varphi\, d\mu \quad \text{for } \varphi \in \hat{C}^1(\mathbf{M}). \tag{5.7}$$

As our first application of the results in Lectures III and IV, suppose that the DOBRUSHIN–SHLOSMAN uniqueness condition in (3.4) holds. We know then that the $\mathfrak{G}(\mathcal{J})$ consists of exactly one element γ. Further, if $\{P_t^Y : t > 0\}$ is the semigroup introduced in Lecture III, then (cf. (3.11))

$$\left\| P_t^Y \varphi - \langle \varphi, \gamma \rangle \right\|_{\mathrm{u}} \le e^{-\beta |Y| t} |||\varphi|||, \quad \text{for } \varphi \in \hat{C}(\mathbf{M}) \text{ and } t > 0. \tag{5.8}$$

Next (cf. (3.7)) observe that, for φ, $\psi \in \hat{C}(\mathbf{M})$:

$$(\psi, \mathcal{L}^Y \varphi)_{L^2(\gamma)} = \sum_{\mathbf{k} \in \mathbf{Z}^d} \int_{\mathbf{M}^2} \psi(\mathbf{x})(\varphi(\mathbf{y}) - \varphi(\mathbf{x}))\, \gamma^{\mathbf{k}}(d\mathbf{x} \times d\mathbf{y}),$$

where

$$\gamma^{\mathbf{k}}(d\mathbf{x} \times d\mathbf{y}) \equiv E^{\mathbf{k}+Y}(d\mathbf{y}|\mathbf{x})\, \gamma(d\mathbf{x}).$$

Hence, because

$$\left(\mathbb{E}^X\psi,\varphi\right)_{L^2(\gamma)} = \left(\mathbb{E}^X\varphi,\psi\right)_{L^2(\gamma)}$$

and therefore $\gamma^{\mathbf{k}}$ is symmetric, we see that

$$\left(\psi,\mathcal{L}^Y\varphi\right)_{L^2(\gamma)} = -\frac{1}{2}\sum_{\mathbf{k}\in\mathbf{Z}^d}\int_{\mathbf{M}^2}\left(\psi(\mathbf{y})-\psi(\mathbf{x})\right)\left(\varphi(\mathbf{y})-\varphi(\mathbf{x})\right)\gamma^{k}(d\mathbf{x}\times d\mathbf{y}).$$

In particular, proceeding as before, this means that each P_t^Y admits a unique extension $\overline{P_t^Y}$ as a self-adjoint contraction on $L^2(\gamma)$,

$$\overline{P_t^Y} = \int_{[0,\infty)} e^{-\alpha t}\,dE_\alpha^Y, \quad t>0,$$

where $\left\{E_\alpha^Y : \alpha\in[0,\infty)\right\}$ is a resolution of the identity, and the associated DIRICHLET form

$$\mathcal{E}^Y(\varphi,\varphi) \equiv \int_{(0,\infty)} \alpha\, d\left(E_\alpha^Y\varphi,\varphi\right)_{L^2(\gamma)}$$

is given by

$$\text{(5.9)}\qquad \mathcal{E}^Y(\varphi,\varphi) = \frac{1}{2}\sum_{\mathbf{k}\in\mathbf{Z}^d}\int_{\mathbf{M}^2}\left(\varphi(\mathbf{y})-\varphi(\mathbf{x})\right)^2\gamma^{\mathbf{k}}(\mathbf{x}\times d\mathbf{y}) \quad \text{for } \varphi\in\hat{C}(\mathbf{M}).$$

Now let $\varphi\in\hat{C}^1(\mathbf{M})$ be given and set $\bar{\varphi}=\varphi-\langle\varphi,\gamma\rangle$. By (5.8) and the spectral representation for $\overline{P_t^Y}$, we know that

$$\begin{aligned}\int_{[0,\infty)} e^{-\alpha t}\, d\left(E_\alpha^Y\bar{\varphi},\bar{\varphi}\right)_{L^2(\gamma)} &= \left(\bar{\varphi},P_t^Y\bar{\varphi}\right)_{L^2(\gamma)}\\ &\le \|\varphi\|_{L^2(\gamma)}\left\|P_t^Y\varphi-\langle\varphi,\gamma\rangle\right\|_{\mathrm{u}} \le e^{-\beta t}\|\varphi\|_{L^2(\gamma)}|||\varphi|||.\end{aligned}$$

But this is possible only if $E_\alpha^Y\bar{\varphi}=0$ for all $\alpha\in[0,\beta)$, and so, since $E_0^Y1=1$, we now know that

$$\text{(5.10)}\qquad \mathcal{E}^Y(\varphi,\varphi)\ge\beta|Y|\left\|\varphi-\langle\varphi,\gamma\rangle\right\|_{L^2(\gamma)} \quad \text{for all } \varphi\in\hat{C}^1(\mathbf{M}).$$

At the same time, by (5.1) and elementary manipulations,

$$\begin{aligned}\frac{\alpha_1}{2}\int_{\mathbf{M}^2}\left(\varphi(\mathbf{y})-\varphi(\mathbf{x})\right)^2\gamma^{\mathbf{k}}(d\mathbf{x}\times d\mathbf{y}) &\le \alpha_1 e^{-\min U^{\mathbf{k}+Y}}\int_{\mathbf{M}}\frac{1}{Z_{\mathbf{k}+Y}}\left\langle\left(\partial_{\mathbf{k}+Y}\varphi\right)^2\right\rangle_{\mathbf{k}+Y}d\gamma\\ \le e^{-\min U^{\mathbf{k}+Y}}\sum_{\mathbf{j}\in\mathbf{k}+Y}\int_{\mathbf{M}}\frac{1}{Z_{\mathbf{k}+Y}}\left\langle|\nabla_{\mathbf{j}}\varphi|^2\right\rangle_{\mathbf{k}+Y}d\gamma &\le e^{\operatorname{osc}U^{\mathbf{k}+Y}}\sum_{\mathbf{j}\in\mathbf{k}+Y}\int_{\mathbf{M}}|\nabla_{\mathbf{j}}\varphi|^2\,d\gamma,\end{aligned}$$

which, by the shift invariance of $\mathcal{J}$ plus (5.9) and (5.7), means that

$$\text{(5.11)}\qquad \alpha_1\mathcal{E}^Y(\varphi,\varphi)\le|Y|e^{\operatorname{osc}U^Y}\mathcal{E}^{\gamma}(\varphi,\varphi) \quad \text{for all } \varphi\in\hat{C}^1(\mathbf{M}).$$

Finally, after combining (5.11) and (5.10), we arrive at the conclusion that (3.4) implies the POINCARÉ inequality

$$\alpha_1(\mathcal{J})\|\varphi - \langle\varphi,\gamma\rangle\|_{L^2(\gamma)} \le \mathcal{E}^\gamma(\varphi,\varphi) \quad \text{with } \alpha_1(\mathcal{J}) \ge \alpha_1\beta e^{-\operatorname{osc}(U^Y)} > 0. \tag{5.12}$$

We have now seen that (3.4) is sufficient to guarantee a POINCARÉ inequality for the infinite, interacting diffusion determined by $\mathfrak{L}$. However, without any further information, this fact is not enough to conclude that the semigroup $\{\mathcal{P}_t : t > 0\}$ is actually ergodic (i.e., that $\mathcal{P}_t\varphi \longrightarrow \langle\varphi,\gamma\rangle$ point-wise). Indeed, the POINCARÉ inequality is only $L^2(\gamma)$-information and, in the absence of *a priori* regularity estimates, cannot be used to get point-wise information. Thus, before we can hope to prove ergodicity, we had better see if we can prove a coercivity estimate; and, as we discussed in Lecture I, the only one that we can reasonably expect is a logarithmic SOBOLEV inequality. Unfortunately, except in the relatively trivial case when $d = 1$, I do not know how to derive a logarithmic SOBOLEV inequality from the condition in (3.4). On the other hand, as we will see below, we can derive one from the DOBRUSHIN–SHLOSMAN mixing condition in (4.8).

With the preceding in mind, we now replace (3.4) with the condition in (4.8). We then know that (4.9) holds for every $\Gamma \subseteq \mathbf{Z}^d$, and therefore the argument just given to derive (5.12) can be repeated to show that, for all non-empty $\Gamma \subseteq \mathbf{Z}^d$ and $\boldsymbol{\xi} \in \mathbf{M}$:

$$\sigma(\mathcal{J}) \int_{\mathbf{M}} \left(\varphi(\mathbf{x}) - \mathbf{E}^\Gamma\varphi(\boldsymbol{\xi})\right)^2 E^\Gamma(d\mathbf{x}|\boldsymbol{\xi}) \le \sum_{\ell\in\Gamma} \int_{\mathbf{M}} \left|\nabla_{\mathbf{k}}\varphi(\mathbf{x})\right|^2 E^\Gamma(d\mathbf{x}|\boldsymbol{\xi}), \tag{5.13}$$

where

$$\sigma(\mathcal{J}) \ge \alpha_1\beta \min_{\mathfrak{J}\ni X\subseteq Y} e^{-\operatorname{osc} U^X} > 0.$$

In addition to (5.13), we also know that (4.12) holds. Finally, because

$$\kappa_0 \int_M \varphi^2 \log\left(\frac{\varphi}{\|\varphi\|_{L^2(\lambda)}}\right)^2 d\lambda \le \int_M |\nabla\varphi|^2\, d\lambda$$

for some $\kappa_0 > 0$ and therefore (cf. (2.11))

$$\kappa_0 \int_{M^\Gamma} \varphi^2 \log\left(\frac{\varphi}{\|\varphi\|_{L^2(\lambda^\Gamma)}}\right)^2 d\lambda^\Gamma \le \sum_{\mathbf{k}\in\Gamma} \int_{M^\Gamma} |\nabla_{\mathbf{k}}\varphi|^2\, d\lambda^\Gamma$$

for all $\emptyset \ne \Gamma \subseteq \mathbf{Z}^d$, we know (cf. (2.9)) that, for all $\Gamma \in \mathfrak{F}$ and $\boldsymbol{\xi} \in \mathbf{M}$,

$$\begin{aligned} &\int_{M^\Gamma} \varphi(\mathbf{x})^2 \log\varphi(\mathbf{x})^2\, E^\Gamma(d\mathbf{x}|\boldsymbol{\xi}) \\ &\quad \le \frac{1}{\kappa(\Gamma)} \sum_{\mathbf{k}\in\Gamma} \int_{M^\Gamma} |\nabla_{\mathbf{k}}\varphi(\mathbf{x})|^2\, E^\Gamma(d\mathbf{x}|\boldsymbol{\xi}) + [\mathbf{E}^\Gamma\varphi^2](\boldsymbol{\xi}) \log[\mathbf{E}^\Gamma\varphi^2](\boldsymbol{\xi}), \end{aligned} \tag{5.14}$$

where

$$\kappa(\Gamma) \geq \kappa_0 e^{-\operatorname{osc} U^\Gamma}.$$

Our goal is to derive a logarithmic SOBOLEV inequality for γ on the basis of (4.12), (5.13), and (5.14). It should come as no surprise that the idea underlying our derivation will be to utilize the *approximate independence of widely separated blocks* afforded by (4.12); and during the rest of this lecture we will be deriving from (4.12) the estimates on which our program turns.

5.15 Lemma. *Given $Y \in \mathfrak{F}$ and $\mathbf{k} \notin Y$, set*

$$U_{\mathbf{k}}^Y = \nabla_{\mathbf{k}} U^Y = \sum_{\substack{A\cap Y \neq \emptyset \\ A \ni \mathbf{k}}} \nabla_{\mathbf{k}} J_A,$$

$$\overline{U}_{\mathbf{k}}^Y = U_{\mathbf{k}}^\Gamma - \mathbb{E}^Y U_{\mathbf{k}}^Y,$$

and, for $\Gamma \subseteq Y$,

$$\mathfrak{R}(Y,\Gamma,\mathbf{k}) \equiv \max_{\xi \in \mathbf{M}} \left|\mathbb{E}^\Gamma \overline{U}_{\mathbf{k}}^Y\right](\xi)\big|. \tag{5.16}$$

Then there is a $\delta > 0$, depending only on the $\epsilon > 0$ in (4.12) and the range R of $\mathcal{J}$, such that

$$\mathfrak{R}(Y,\Gamma,\mathbf{k}) \leq C(\mathcal{J}) \exp\left[-\delta \operatorname{dist}(\mathbf{k}, Y \setminus \Gamma)\right], \tag{5.17}$$

where $C(\mathcal{J}) \in (0,\infty)$ depends only on the C in (4.12), the range R, and $\sum_{A \ni 0} \|\nabla_0 J_A\|_{\mathrm{u}}$. In addition,

$$\left|\nabla_{\mathbf{k}} (\mathbb{E}^Y \varphi^2)^{\frac{1}{2}}\right| \leq (\mathbb{E}^Y |\nabla_{\mathbf{k}} \varphi|^2)^{\frac{1}{2}} + \frac{\mathfrak{R}(Y,\Gamma,\mathbf{k})}{\sqrt{\sigma(\mathcal{J})}} \left(\sum_{\ell \in Y \setminus \Gamma} \mathbb{E}^Y |\nabla_\ell \varphi|^2\right)^{\frac{1}{2}} \tag{5.18}$$

for positive $\varphi \in C_0^\infty(\mathbf{M}) \cap C_{\Gamma\complement}(\mathbf{M})$.

PROOF: Because $\mathbb{E}^Y \overline{U}_{\mathbf{k}}^Y = 0$ and $\mathbb{E}^Y = \mathbb{E}^Y \circ \mathbb{E}^\Gamma$, we have that

$$\begin{aligned}
\left|[\mathbb{E}^\Gamma \overline{U}_{\mathbf{k}}^Y](\xi)\right| &= \left|\int_{\mathbf{M}} \left([\mathbb{E}^\Gamma \overline{U}_{\mathbf{k}}^Y](\xi) - [\mathbb{E}^\Gamma \overline{U}_{\mathbf{k}}^Y](\eta)\right] E^Y(d\eta|\xi)\right| \\
&\leq \int_{\mathbf{M}} \left|[\mathbb{E}^\Gamma \overline{U}_{\mathbf{k}}^Y](\xi) - [\mathbb{E}^\Gamma \overline{U}_{\mathbf{k}}^Y](\eta)\right| E^Y(d\eta|\xi) \\
&\leq \max\left\{\left|[\mathbb{E}^\Gamma \overline{U}_{\mathbf{k}}^Y](\xi) - [\mathbb{E}^\Gamma \overline{U}_{\mathbf{k}}^Y](\eta)\right| : \xi_{Y\complement} = \eta_{Y\complement}\right\}.
\end{aligned}$$

Hence, by taking $S = Y \cap \Gamma\complement$ and $X = \bigcup\{A \cap Y : A \ni \mathbf{k} \text{ and } J_A \not\equiv 0\}$ in (4.12), we arrive at the estimate in (5.17).

Now let φ be a positive element of $C_0^\infty(\mathbf{M}) \cap C_{\Gamma\mathfrak{c}}(\mathbf{M})$. To prove (5.18), first observe that

$$2(\mathbb{E}^Y\varphi^2)^{\frac{1}{2}}\nabla_{\mathbf{k}}(\mathbb{E}^Y\varphi^2)^{\frac{1}{2}} = 2\mathbb{E}^Y(\varphi\nabla_{\mathbf{k}}\varphi) + \mathbb{E}^Y(\overline{U}^Y_{\mathbf{k}}\varphi^2).$$

Next, by SCHWARZ's inequality,

$$\left|\mathbb{E}^Y(\varphi\nabla_{\mathbf{k}}\varphi)\right| \le (\mathbb{E}^Y\varphi^2)^{\frac{1}{2}}(\mathbb{E}^Y|\nabla_{\mathbf{k}}\varphi|^2)^{\frac{1}{2}};$$

and, because $\mathbb{E}^Y\overline{U}^Y_{\mathbf{k}} = 0$ and $\varphi \in C_{\Gamma\mathfrak{c}}(\mathbf{M})$, another application of SCHWARZ yields

$$\begin{aligned}\left|\mathbb{E}^Y(\overline{U}^Y_{\mathbf{k}}\varphi^2)\right| &= \left|\mathbb{E}^Y\left[(\mathbb{E}^\Gamma\overline{U}^Y_{\mathbf{k}})\left(\varphi^2 - (\mathbb{E}^Y\varphi)^2\right)\right]\right| \\ &\le \mathfrak{R}(Y,\Gamma,\mathbf{k})\mathbb{E}^Y\left(|\varphi - \mathbb{E}^Y\varphi||\varphi + \mathbb{E}^Y\varphi|\right) \\ &\le 2\mathfrak{R}(Y,\Gamma,\mathbf{k})(\mathbb{E}^Y\varphi^2)^{\frac{1}{2}}\left(\mathbb{E}^Y(\varphi - \mathbb{E}^Y\varphi)^2\right)^{\frac{1}{2}}.\end{aligned}$$

Finally, apply (5.13) to the last expression and thereby arrive at (5.18). □

Lecture VI: Logarithmic Sobolev Completed

We continue with the assumption that the potential $\mathcal{J}$ satisfies the DOBRUSHIN–SHLOSMAN mixing condition in (4.8) ; and our primary goal in this section will be to complete the proof that the associated unique GIBBS state γ satisfies a logarithmic SOBOLEV inequality.

Let $L \in \mathbf{Z}^+$ (to be prescribed later) be given, and set

$$Y_{\mathbf{k}} = \mathbf{k} + \left[0, 2(L+R)\right]^d \cap \mathbf{Z}^d \quad \text{for } \mathbf{k} \in \mathbf{Z}^d.$$

Next, let $\{\mathbf{v}_m\}_0^{2^d-1}$ be an enumeration of $\{0,1\}^d$, take

$$\mathfrak{K}_{m+r2^d} = (L+R)\mathbf{v}_m + 2(L+2R)\mathbf{Z}^d \quad \text{for } 0 \le m \le 2^d - 1 \text{ and } r \in \mathbf{N},$$

and set

$$\Lambda_m = \bigcup_{\mathbf{k}\in\mathfrak{K}_m} Y_{\mathbf{k}} \quad \text{for } m \in \mathbf{N}.$$

Here is a picture of the relative positions of Λ_m and Λ_{m+1} when $d = 1$:

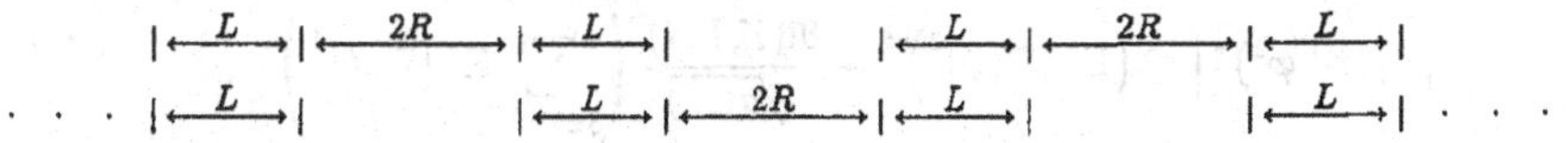

Note that, because the distance between $Y_{\mathbf{k}}$ and $Y_{\boldsymbol{\ell}}$ is greater than the range R of $\mathcal{J}$ when $\mathbf{k}$ and $\boldsymbol{\ell}$ are distinct elements of $\mathfrak{K}_m$,

$$\mathbb{E}^{Y_{\mathbf{k}}} \circ \mathbb{E}^{Y_{\boldsymbol{\ell}}} = \mathbb{E}^{Y_{\mathbf{k}} \cup Y_{\boldsymbol{\ell}}} = \mathbb{E}^{Y_{\boldsymbol{\ell}}} \circ \mathbb{E}^{Y_{\mathbf{k}}} \quad \text{for all } \mathbf{k}, \boldsymbol{\ell} \in \mathfrak{K}_m.$$

More generally, if $A \subset\subset \mathfrak{K}_m$, then

$$\mathbb{E}^{Y} = \prod_{\mathbf{k} \in A} \mathbb{E}^{Y_{\mathbf{k}}} \quad \text{where } Y = \bigcup_{\mathbf{k} \in A} Y_{\mathbf{k}},$$

and the product is independent of the order in which the factors are composed. In particular, this means that there is a unique MARKOV operator $\mathbb{E}^{\Lambda_m}$ from $B(\mathbf{M})$ into $B_{\Lambda_m{}^{\complement}}(\mathbf{M})$ with the property that, for each $X \in \mathfrak{F}$,

$$\mathbb{E}^{\Lambda_m} \varphi = \mathbb{E}^{Y} \varphi \quad \text{if } \varphi \in C_X(\mathbf{M}) \text{ and } Y = \bigcup_{\mathbf{k} \in A} Y_{\mathbf{k}},$$
$$\text{where } A \subset\subset \mathfrak{K}_m \text{ has the property that } Y_{\boldsymbol{\ell}} \cap X = \emptyset \text{ for all } \boldsymbol{\ell} \in \mathfrak{K}_m \setminus A.$$

In addition, $\mathbb{E}^{\Lambda_m}$ maps $C_0^{\infty}(\mathbf{M})$ into itself, and, because $\mathbb{E}^{\Lambda_m}$ comes from taking tensor products, we can use the same reasoning as led to (2.11) in order to see that (cf. (5.14)) with $\kappa(L) \equiv \kappa(Y_0)$

$$\mathbb{E}^{\Lambda_m}\left(\varphi^2 \log \varphi^2\right) \leq \frac{1}{\kappa(L)} \sum_{\mathbf{j} \in \Lambda_m} \mathbb{E}^{\Lambda_m} |\nabla_{\mathbf{j}} \varphi|^2 + \mathbb{E}^{\Lambda_m} \varphi^2 \log \mathbb{E}^{\Lambda_m} \varphi^2 \tag{6.1}$$

for all $m \in \mathbb{N}$ and $\varphi \in \hat{C}^1(\mathbf{M})$.

Now let φ be a positive element of $C^1(\mathbf{M}) \cap C_X(\mathbf{M})$ for some $X \in \mathfrak{F}$. Given $m \in \mathbb{N}$, choose $A \subset\subset \mathfrak{K}_m$ so that $X \cap Y_{\mathbf{k}} = \emptyset$ for all $\mathbf{k} \in \mathfrak{K}_m \setminus A$, and set $\Lambda = \bigcup_{\mathbf{k} \in A} Y_{\mathbf{k}}$. Clearly,

$$\operatorname{dist}(\boldsymbol{i}, \Lambda) > R \implies \nabla_{\boldsymbol{i}}\left(\mathbb{E}^{\Lambda_m} \varphi^2\right)^{\frac{1}{2}} = 0.$$

On the other hand, if $\boldsymbol{i} \notin \Lambda_m$ but $\operatorname{dist}(\boldsymbol{i}, \Lambda) \leq R$, then there is precisely one $\boldsymbol{\ell} \in A$ with the property $\operatorname{dist}(\boldsymbol{i}, Y_{\boldsymbol{\ell}}) \leq R$, in which case we set $Y = Y_{\boldsymbol{\ell}}$ and $Z = \Lambda \setminus Y$ and note that $\mathbb{E}^{\Lambda_m} \varphi^2 = \mathbb{E}^{Z} \circ \mathbb{E}^{Y} \varphi^2$. Hence, because $\mathfrak{R}(Z, \emptyset, \boldsymbol{i}) = 0$, (5.18) applied to $\left(\mathbb{E}^{Y} \varphi^2\right)^{\frac{1}{2}}$, yields first

$$\left|\nabla_{\boldsymbol{i}}\left(\mathbb{E}^{\Lambda_m} \varphi^2\right)^{\frac{1}{2}}\right|^2 \leq \mathbb{E}^{Z}\left|\nabla_{\boldsymbol{i}}\left(\mathbb{E}^{Y} \varphi^2\right)^{\frac{1}{2}}\right|^2,$$

and then

$$\left\langle \left|\nabla_{\boldsymbol{i}}\left(\mathbb{E}^{\Lambda_m} \varphi^2\right)^{\frac{1}{2}}\right|^2, \gamma \right\rangle \leq \left\langle \left|\nabla_{\boldsymbol{i}}\left(\mathbb{E}^{Y} \varphi^2\right)^{\frac{1}{2}}\right|^2, \gamma \right\rangle.$$

Next, given $r \in [0, \infty)$, take $\Gamma = \{\mathbf{j} \in Y : |\mathbf{j} - \boldsymbol{i}| \leq r\}$, set $\psi = \left(\mathbb{E}^{\Gamma} \varphi^2\right)^{\frac{1}{2}}$, and apply (5.18) to see that

$$\left|\nabla_{\boldsymbol{i}}\left(\mathbb{E}^{Y} \varphi^2\right)^{\frac{1}{2}}\right| \leq \left(\mathbb{E}^{Y} |\nabla_{\boldsymbol{i}} \psi|^2\right)^{\frac{1}{2}} + \frac{\mathfrak{R}(Y, \Gamma, \boldsymbol{i})}{\sqrt{\sigma(\mathcal{J})}} \left(\sum_{\mathbf{j} \in Y \setminus \Gamma} \mathbb{E}^{Y} |\nabla_{\mathbf{j}} \psi|^2\right)^{\frac{1}{2}},$$

which means that

$$\left\langle \left|\nabla_i (\mathbb{E}^{\Lambda_m}\varphi^2)^{\frac{1}{2}}\right|^2, \gamma\right\rangle \le 2\left\langle \left|\nabla_i \psi\right|^2, \gamma\right\rangle + \frac{2\mathfrak{R}(Y,\Gamma,\mathbf{i})^2}{\sigma(\mathcal{J})} \sum_{\mathbf{j}\in Y\setminus\Gamma} \left\langle \left|\nabla_{\mathbf{j}}\psi\right|^2, \gamma\right\rangle.$$

Finally, again by (5.18),

$$\left\langle \left|\nabla_i \psi\right|^2, \gamma\right\rangle \le 2\left\langle \left|\nabla_i \varphi\right|^2, \gamma\right\rangle + \frac{2\mathfrak{R}(\Gamma,\emptyset,\mathbf{i})^2}{\sigma(\mathcal{J})} \sum_{\mathbf{j}\in\Gamma} \left\langle \left|\nabla_{\mathbf{j}}\varphi\right|^2, \gamma\right\rangle$$

and, for each $\mathbf{j} \in Y \setminus \Gamma$,

$$\left\langle \left|\nabla_{\mathbf{j}} \psi\right|^2, \gamma\right\rangle \le 2\left\langle \left|\nabla_{\mathbf{j}} \varphi\right|^2, \gamma\right\rangle + \frac{2\mathfrak{R}(\Gamma,\emptyset,\mathbf{i})^2 \mathbf{1}_{\Gamma(R)}(\mathbf{j})}{\sigma(\mathcal{J})} \sum_{\boldsymbol{\ell}\in\Gamma} \left\langle \left|\nabla_{\boldsymbol{\ell}}\varphi\right|^2, \gamma\right\rangle,$$

where $\Gamma(R) = \{\mathbf{j} : \operatorname{dist}(\mathbf{j},\Gamma) \le R\}$. Hence, after combining these, we find that

$$\begin{aligned} &\left\langle \left|\nabla_i (\mathbb{E}^{\Lambda_m}\varphi^2)^{\frac{1}{2}}\right|^2, \gamma\right\rangle \\ &\le 4\left\langle \left|\nabla_i \varphi\right|^2, \gamma\right\rangle + \frac{4C(\mathcal{J})^2}{\sigma(\mathcal{J})}\left(1 + \frac{\mathfrak{R}(Y,\Gamma,\mathbf{i})^2 |\Gamma(R)|}{\sigma(\mathcal{J})}\right) \sum_{\mathbf{j}\in\Gamma} \left\langle \left|\nabla_{\mathbf{j}}\varphi\right|^2, \gamma\right\rangle \\ &\qquad + \frac{4\mathfrak{R}(Y,\Gamma,\mathbf{i})^2}{\sigma(\mathcal{J})} \sum_{\mathbf{j}\in Y\setminus\Gamma} \left\langle \left|\nabla_{\mathbf{j}}\varphi\right|^2, \gamma\right\rangle. \end{aligned}$$

In particular, because $|\Gamma(R)| \le (2(r+R))^d$, the preceding in conjunction with (5.17) gives us the estimate

$$\begin{aligned} &\left\langle \left|\nabla_i (\mathbb{E}^{\Lambda_m}\varphi^2)^{\frac{1}{2}}\right|^2, \gamma\right\rangle \\ &\le K(\mathcal{J})\left(\sum_{|\mathbf{j}-\mathbf{i}|\le r} \left\langle \left|\nabla_{\mathbf{j}}\varphi\right|^2, \gamma\right\rangle + e^{-\frac{\delta r}{2}} \sum_{r<|\mathbf{j}-\mathbf{i}|\le 2L+3R} \left\langle \left|\nabla_{\mathbf{j}}\varphi\right|^2, \gamma\right\rangle\right), \end{aligned} \tag{6.2}$$

where $K(\mathcal{J}) \in (0,\infty)$ depends only on R, the $\sigma(\mathcal{J})$ in (5.13), and the numbers $C(\mathcal{J})$ and δ in (5.17).

Given (6.2), we proceed as follows. Let Π_0 denote the identity map on $B(\mathbf{M})$, and define $\Pi_n = \mathbb{E}^{\Lambda_{n-1}} \circ \Pi_{n-1}$ for $n \in \mathbb{Z}^+$. Next, let φ be a positive element of $C_0^\infty(\mathbf{M})$, and set $\varphi_n = (\Pi_n\varphi^2)^{\frac{1}{2}}$ for $n \in \mathbb{N}$. From the preceding, we know that each φ_n is a positive element of $C_0^\infty(\mathbf{M})$. In addition, because $\varphi_n^2 = \mathbb{E}^{\Lambda_{n-1}}\varphi_{n-1}^2$, an application of (6.1) shows that

$$\langle \varphi_{n-1}^2 \log \varphi_{n-1}^2 \rangle \le \frac{1}{\kappa(L)} \sum_{\mathbf{i}\notin\Lambda_n} \left\langle \left|\nabla_i \varphi_{n-1}\right|^2, \gamma\right\rangle + \left\langle \varphi_n^2 \log \varphi_n^2, \gamma\right\rangle. \tag{6.3}$$

Hence, if we set

$$\Phi_n \equiv \sum_{i\in\mathbf{Z}^d} \left\langle \left|\nabla_i \varphi_n\right|^2, \gamma\right\rangle$$

and show that, for an appropriate choice of L, there exists a $B \in (0,\infty)$ such that

$$\sum_{m=0}^{\infty} \Phi_m \leq B\Phi_0, \tag{6.4}$$

then we will have shown that

$$\langle \varphi^2 \log \varphi^2, \gamma\rangle \leq \frac{B}{\kappa(L)}\left\langle \left|\nabla_i \varphi\right|^2, \gamma\right\rangle + \langle \varphi^2, \gamma\rangle \log\langle \varphi^2, \gamma\rangle, \tag{6.5}$$

which is equivalent to saying that γ satisfies a logarithmic SOBOLEV inequality with logarithmic SOBOLEV constant at least as large as $\frac{\kappa(L)}{B}$. Indeed, to see how (6.3) and (6.4) together lead to (6.5), it is sufficient to observe that, because of (5.12), (6.4) implies that

$$\left\|\varphi_n^2 - \langle \varphi_n^2, \gamma\rangle\right\|_{L^1(\gamma)} \leq 2\|\varphi_n\|_{L^2(\gamma)}\left\|\varphi_n - \langle \varphi_n, \gamma\rangle\right\|_{L^2(\gamma)} \leq 2\|\varphi\|_u \left(\frac{\Phi_n}{\alpha_1(\mathcal{J})}\right)^{\frac{1}{2}},$$

which, because $\langle \varphi_n^2, \gamma\rangle = \langle \varphi^2, \gamma\rangle$, means that last term in (6.3) tends to the last term in (6.5).

In view of the preceding, all that remains is the proof that (6.4) holds for an appropriate choice of L. To this end, we first use (6.2) with $r = 0$ to see that

$$\Phi_n \leq K(\mathcal{J})(4L + 6R)^d \Phi_{n-1}, \quad n \in \mathbf{Z}^+. \tag{6.6}$$

Next, given $\ell \in \mathbf{Z}^+$ and $i \notin \Lambda_{\ell 2^d}$, note that there is an $0 \leq n < 2^d - 1$ with the property that

$$\left\{\mathrm{j} : |\mathrm{j} - i| \leq \frac{L}{2}\right\} \subseteq \Lambda_{\ell 2^d - n - 1}. \tag{6.7}$$

After repeated application of (6.2), we find that

$$\begin{aligned}\left\langle \left|\nabla_i \varphi_{\ell 2^d}\right|^2, \gamma\right\rangle \leq K(\mathcal{J})^n (2r)^{n-1} &\sum_{|\mathrm{j}-i|\leq nr} \left\langle \left|\nabla_{\mathrm{j}} \varphi_{\ell 2^d - n}\right|^2, \gamma\right\rangle \\ &+ e^{-\frac{\delta r}{2}} K(\mathcal{J}) \sum_{m=1}^{n} \left(2rK(\mathcal{J})\right)^{m-1} \sum_{|\mathrm{j}-i|\leq m(2L+3R)} \left\langle \left|\nabla_{\mathrm{j}} \varphi_{\ell 2^d - m}\right|^2, \gamma\right\rangle\end{aligned}$$

for every $r \in [0,\infty)$. In particular, if $r = 2^{-(d+1)}L$, then (cf. (6.7))

$$|\mathrm{j} - i| \leq nr \implies \mathrm{j} \in \Lambda_{\ell 2^d - n - 1} \implies \nabla_{\mathrm{j}} \varphi_{\ell 2^d - n} = 0,$$

and therefore

$$\left\langle\left|\nabla_i\varphi_{\ell 2^d}\right|^2,\gamma\right\rangle$$

$$\leq \exp\left[-\frac{\delta L}{2^{d+2}}\right] K(\mathcal{J}) \sum_{m=1}^{n} \left(2rK(\mathcal{J})\right)^{m-1} \sum_{|\mathbf{j}-\mathbf{i}|\leq m(2L+3R)} \left\langle\left|\nabla_{\mathbf{j}}\varphi_{\ell 2^d - m}\right|^2,\gamma\right\rangle.$$

Hence, after putting this together with (6.6), we arrive at

$$\Phi_{\ell 2^d} \leq K'(\mathcal{J})(4L+6R)^{d2^d+1} \exp\left[-\frac{\delta L}{2^{d+2}}\right] \Phi_{(\ell-1)2^d}, \tag{6.8}$$

where $K'(\mathcal{J}) \in (0,\infty)$ depends only on $K(\mathcal{J})$ and d. Finally, by choosing L to be the smallest non-negative integer satisfying

$$K'(\mathcal{J})(4L+6R)^{d2^d+1} \exp\left[-\frac{\delta L}{2^{d+2}}\right] \leq \frac{1}{2}, \tag{6.9}$$

we see that (6.4) holds with a $C \in (0,\infty)$ depending only on $K(\mathcal{J})$ and this choice of L.

We have now proved the DOBRUSHIN–SHLOSMAN mixing condition in (4.8) implies that the associated unique $\gamma \in \mathfrak{G}(\mathcal{J})$ satisfies the logarithmic SOBOLEV inequality

$$\kappa(\mathcal{J}) \int \varphi^2 \log\left(\frac{\varphi^2}{\|\varphi\|^2_{L^2(\gamma)}}\right) d\gamma \leq \mathcal{E}^\gamma(\varphi,\varphi) \equiv \sum_{\mathbf{k}\in\mathbf{Z}^d} \left\langle\left|\nabla_{\mathbf{k}}\varphi\right|^2,\gamma\right\rangle, \tag{6.10}$$

where $\kappa(\mathcal{J})$ dominates a strictly positive number κ which depends only on the range R of $\mathcal{J}$, the dimension d of the lattice, the manifold M, the set Y and number $\beta \in [0,1)$ in (4.8), and the quantity

$$\|\mathcal{J}\| \equiv \sum_{A\ni\mathbf{0}} \left\|\nabla_{\mathbf{0}} J_A\right\|_{\mathrm{u}}. \tag{6.11}$$

In particular, this (cf. Remark 2.6) confirms the conclusion already drawn in (5.12) that γ satisfies the POINCARÉ inequality

$$\alpha_1(\mathcal{J})\left\|\varphi - \langle\varphi,\gamma\rangle\right\|^2_{L^2(\gamma)} \leq \mathcal{E}^\gamma(\varphi,\varphi), \tag{6.12}$$

or, equivalently,

$$\left\|\mathcal{P}_t\varphi - \langle\varphi,\gamma\rangle\right\|_{L^2(\gamma)} \leq e^{-\alpha_1(\mathcal{J})t}\|\varphi\|_{L^2(\gamma)}, \quad t\in(0,\infty) \text{ and } \varphi\in C(\mathbf{M}), \tag{6.13}$$

only now with the estimate $\alpha_1(\mathcal{J}) \geq 2\kappa(\mathcal{J})$. In fact, (cf. (1.15)), (6.10) implies that

$$\left\|\mu\mathcal{P}_t - \gamma\right\|_{\mathrm{var}} \leq e^{-2\kappa(\mathcal{J})t}\mathbf{H}(\mu|\gamma), \quad t\in(0,\infty) \text{ and } \mu\in\mathfrak{M}_1(\mathbf{M}). \tag{6.14}$$

Unfortunately, unlike their finite dimensional counterparts, neither (6.13) nor (6.14) can be used to get very much information about the ergodic properties of the semigroup $\{\mathcal{P}_t : t>0\}$. In particular, because the relative entropy $\mathbf{H}(\mu\mathcal{P}_1|\gamma)$ will, in general, be infinite unless $\mathbf{H}(\mu|\gamma)<\infty$, (6.14) cannot be used to check that $\{\mathcal{P}_t : t>0\}$ admits no invariant measures other than γ.

Actually, exactly the same line of reasoning which led from Lemma 5.15 to (6.10) can be used to show that, for all $\Gamma \in \mathfrak{F}$ and $\xi \in M$:

$$\kappa \int \varphi(\eta)^2 \log\left(\frac{\varphi(\eta)^2}{\|\varphi\|^2_{2,\gamma,\xi}}\right) E^\Gamma(d\eta|\xi) \leq \sum_{k\in\mathbf{Z}^d} [\mathbb{E}^\Gamma|\nabla_k\varphi|^2](\xi) \tag{6.15}$$

where we have introduced the notation $\|\psi\|_{q,\Gamma,\xi}$ to denote the L^q norm of ψ with respect to the measure $E^\Gamma(\cdot\,|\xi)$. As it turns out, (6.15) is the key with which we can vastly improve the statement in (6.13). Namely, we will spend the rest of this lecture showing that, for each $X \in \mathfrak{F}$ and $\theta \in (0,1)$, there is a constant $K(\theta, X) \in (0,\infty)$ with the property that

$$\|\mathcal{P}_t\varphi - \langle\varphi,\gamma\rangle\|_u \leq K(\theta,X)e^{-\theta\alpha_1(\mathcal{J})t} \sum_{k\in X} \|\nabla_k\varphi\|_u, \quad \varphi \in C^\infty_X(\mathbf{M}), \tag{6.16}$$

which is obviously much more than is needed to prove that γ is the only $\{\mathcal{P}_t : t > 0\}$-invariant $\mu \in \mathfrak{M}_1(\mathbf{M})$.

To see how to get from (6.15) to (6.16), set

$$\mathfrak{L}^\gamma\varphi = \sum_{k\in\Gamma} e^{U^k}\nabla_k(e^{-U^k}\nabla_k\varphi), \quad \varphi \in C^\infty_0(\mathbf{M}),$$

and let $\{\mathcal{P}^\Gamma_t : t > 0\}$ denote the corresponding MARKOV semigroup on $C(\mathbf{M})$. Clearly,

$$\frac{d}{dt}\mathbb{E}^\Gamma(\varphi\mathcal{P}^\Gamma_t\varphi) = -\sum_{k\in\Gamma}\mathbb{E}^\Gamma|\nabla_k\varphi|^2, \quad \varphi \in C^\infty_0(\mathbf{M}).$$

Hence, by L. GROSS's integration lemma, (6.15) implies that

$$\begin{gathered}\|\mathcal{P}^\Gamma_t\varphi\|_{q(t),\Gamma,\xi} \leq \|\varphi\|_{2,\Gamma,\xi}, \quad t \in (0,\infty) \text{ and } \varphi \in C(\mathbf{M}),\\ \text{where } q(t) \equiv 1 + e^{2\kappa t}.\end{gathered} \tag{6.17}$$

Secondly, (cf. Lemma 2.3 in [12]) elliptic regularity estimates allow one to produce an $A \in (0,\infty)$ with the property that

$$\|\mathcal{P}^\Gamma_1\varphi\|_u \leq e^{A|\Gamma|}\|\varphi\|_{1,\Gamma,\xi}, \quad \varphi \in C_\Gamma(\mathbf{M}) \text{ and } \xi \in \mathbf{M},$$

which, by an easy interpolation argument, means that

$$\|\mathcal{P}^\Gamma_1\varphi\|_u \leq e^{\frac{A|\Gamma|}{q}}\|\varphi\|_{q,\Gamma,\xi}, \quad q \in [1,\infty),\ \varphi \in C_\Gamma(\mathbf{M}), \text{ and } \xi \in \mathbf{M}. \tag{6.18}$$

Hence, by combining (6.17) with (6.18), we have that

$$\|\mathcal{P}^\Gamma_{s+1}\varphi\|_u \leq e^{\frac{A|\Gamma|}{q(s)}}\|\varphi\|_{2,\Gamma,\xi}, \quad s \in (0,\infty),\ \varphi \in C_\Gamma(\mathbf{M}), \text{ and } \xi \in \mathbf{M}. \tag{6.19}$$

Finally, by very much the same sort of argument as we used in the derivation of (4.6) (cf. Lemma 1.8 in [12] for a general statement about such estimates) one can show that there is an $B \in (0,\infty)$ with the property that, for all $X \in \mathfrak{F}$ and $\varphi \in C^\infty_X(\mathbf{M})$:

$$\begin{gathered}\{j : \operatorname{dist}(j, X) \leq nR\} \subseteq \Gamma \implies\\ \sup_{\xi\in\mathbf{M}} \|\mathcal{P}_t\varphi - [\mathcal{P}^\Gamma_t\varphi]_\Gamma(\cdot\,|\xi)\|_u \leq \mathbf{e}_n(Bt)\sum_{k\in X}\|\nabla_k\varphi\|_u,\end{gathered} \tag{6.20}$$

where $\mathbf{e}_n(s)$ is defined as it was in (4.6).

Now let $X \in \mathfrak{F}$ and $\theta \in (0,1)$ be given, and set $\Gamma(t) = \{\mathbf{k} : \mathrm{dist}(\mathbf{k}, X) \leq t^2 R\}$. Given $\varphi \in C_0^\infty(\mathbf{M})$ with $\langle \varphi, \gamma \rangle = 0$, we have, from (6.20) and (6.19), that

$$\left\| \mathcal{P}_{t+1}\varphi \right\|_{\mathrm{u}} \leq \mathrm{e}_{[t^2]}(B(t+1)) \sum_{\mathbf{k}\in X} \left\| \nabla_{\mathbf{k}}\varphi \right\|_{\mathrm{u}} + \exp\left[\frac{A|\Gamma(t)|}{q((1-\theta)t)} \right] \left\| \mathcal{P}_{\theta t}^{\Gamma(t)}\varphi \right\|_{2,\Gamma(t),\boldsymbol{\xi}}$$

for all $t \in (0,\infty)$ and $\boldsymbol{\xi} \in \mathbf{M}$. At the same time, by another application of (6.20),

$$\left\| \mathcal{P}_{\theta t}^{\Gamma(t)}\varphi \right\|_{2,\Gamma(t),\boldsymbol{\xi}} \leq \mathrm{e}_{[t^2]}(Bt) \sum_{\mathbf{k}\in X} \left\| \nabla_{\mathbf{k}}\varphi \right\|_{\mathrm{u}} + \left\| \mathcal{P}_{\theta t}\varphi \right\|_{2,\Gamma(t),\boldsymbol{\xi}}.$$

Hence, after combining these two and then integrating in $\boldsymbol{\xi}$ with respect to γ and using (6.13), we arrive at

$$\left\| \mathcal{P}_{t+1}\varphi \right\|_{\mathrm{u}} \leq 2\mathrm{e}_{[t^2]}(B(t+1)) \sum_{\mathbf{k}\in X} \left\| \nabla_{\mathbf{k}}\varphi \right\|_{\mathrm{u}} + \exp\left[\frac{A|\Gamma(t)|}{q((1-\theta)t)} \right] e^{-\alpha_1(\mathcal{J})\theta t} \|\varphi\|_{L^2(\gamma)},$$

from which it is an easy step to the existence of a $K(\theta, X) \in (0,\infty)$ for which (6.16) holds.

Concluding Remarks: The contents of these lectures is based on work in which I have been involved over a considerable period of time. The underlying principle on which that work is based is that GIBBS states become more interesting when they are viewed as the equilibrium state of a dynamical system and, in addition, the dynamics often provides a natural approach to the analysis of GIBBS state. This principle was taught to me by R. HOLLEY, and it was in joint work with him that many of the ideas in Lectures III and IV were developed. Furthermore, it was in [8]) that HOLLEY and I first applied logarithmic SOBOLEV inequalities to GIBBS states. At the time when we wrote [8], the only examples of GIBBS states for which a logarithmic SOBOLEV inequality was known to hold were the very special class of examples given in [2]. Indeed, this remained the only source of examples until the work of B. ZEGARLINSKI [14] and [15]; and the development given in Lectures V and VI here is a synthesis of three articles which he and I have written in the last year and a half (cf. [11], [12], and [13]). In fact, the only major point in those articles which I have not had time to explain here is that the DOBRUSHIN–SHLOSMAN mixing condition not only implies (6.15) but is *equivalent* to (6.15). The proof of this is not hard and can be accomplished by an argument very similar to the one which we have just given in our derivation of (6.16).

BIBLIOGRAPHY

1. D. Bakry and M. Emery, *Diffusions hypercontractives*, Séminaire de Probabilités XIX, Springer Lecture Notes in Mathematics **1123**, 1985, pp. 179–206.

2. E.A. Carlen and D.W. Stroock, *An application of the Bakry-Emery criterion to infinite dimensional diffusions*, Séminaire de Probabilités XX, Springer Lecture Notes in Mathematics **1204**, 1986, pp. 341–347.

3. R.L. Dobrushin and S.B. Shlosman, *Constructive criterion for the uniqueness of Gibbs field*, Statistical Physics and Dynamical Systems, Rigorous Results (Fritz, Jaffe, and Szasz, eds.), Birkhäuser, 1985, pp. 347–370.

4. ———, *Completely analytical Gibbs fields*, Statistical Physics and Dynamical Systems, Rigorous Results (Fritz, Jaffe, and Szasz, eds.), Birkhäuser, 1985, pp. 371–403.

5. J.-D. Deuschel and D. Stroock, *Large Deviations*, Academic Press, Pure & Appl. Math. Series # 137, Boston, 1989.

6. ———, *Hypercontractivity and spectral gap of symmetric diffusions with applications to the stochastic Ising Model*, J. Fnal. Anal. **92** (1990), 30–48.

7. L. Gross, *Logarithmic Sobolev inequalities*, Amer. J. Math. **97** (1976), 1061–1083.

8. R. Holley and D. Stroock, *Logarithmic Sobolev inequalities and stochastic Ising models*, J. Stat. Phys. **46** (1987), 1159–1194.

9. Korzeniowski, A. and Stroock, D., *An example in the theory of hypercontractive semigroups*, Proc. A.M.S. **94** #1 (1985), 87–90.

10. Revuz, Z., *Markov Chains*, North-Holland, Amsterdam & NY.

11. Stroock, D. and Zegarlinski, B., *The logarithmic Sobolev inequality for continuous spin systems on a lattice*, J. Fnal. Anal. **104** (1992), 299–326.

12. ———, *The equivalence of the logarithmic Sobolev inequality and the Dobrushin–Shlosman mixing condition*, Comm. Math. Phys. **144** (1992), 303–323.

13. ———, *The logarithmic Sobolev inequality for descrete spin systems on a lattice* (to appear) in Comm. Math. Phys..

14. Zegarlinski, B., *On log-Sobolev inequalities for infinite lattice systems*, Lett. Math. Phys. **20** (1990), 173–182.

15. ———, *Log-Sobolev inequalities for infinite one-dimensional lattice systems*, Comm. Math. Phys. **133** (1990), 147–162.

M.I.T., rm. 2-272, Cambridge, MA 02139, U.S.A; email: dws@math.mit.edu

C.I.M.E. Session on "Dirichlet Forms"

List of Participants

M. AMAR, SISSA, Via Beirut 2/4, 34014 Trieste

A. ANTONJUK, Dept. of Funct. Anal., Inst. of Math., ul.Repina 3, Kiev, Ukraine

I. BIRINDELLI, Dip.to di Mat., Univ. La Sapienza, P.le Aldo Moro 2, 00185 Roma

C. BONDIOLI, Dip.to di Mat., Strada Nuova 65, 27100 Pavia

Z. BRZEZNIAK, Univ. Bielefeld, Fak. f. Physik/BiBoS, 4-4800 Bielefeld 1

P. BUTTA', SISSA, Via Beirut 2-4, 34014 Trieste

X. CABRE, Courant Inst. of Math. Sci., 251 Mercer St., New York, NY 10012

I. CAPUZZO DOLCETTA, Dip.to di Mat., Univ. La Sapienza, P.le Aldo Moro 2, 00185 Roma

V. CHIADO' PIAT, Dip.to di Mat., Pol. di Torino, Corso Duca degli Abruzzi 24, 10129 Torino

M. CHICCO, Ist. Mat., Fac. Ing., P.le Kennedy, Pad. D, 16129 Genova

F. CIPRIANI, SISSA, Via Beirut 2-4, 34014 Trieste

A. CUTRI', Dip.to di Mat., Univ. La Sapienza, P.le Aldo Moro 2, 00185 Roma

G. DAL MASO, SISSA, Via Beirut 2-4, 34014 Trieste

V. DE CICCO, SISSA, Via Beirut 2-4, 34014 Trieste

A. DEFRANCESCHI, Dip.to di Mat., Univ. di Trento, 38050 Povo, Trento

L. DIOMEDA, Ist. di Mat. Fin., Fac. di Ec. e Comm., Via C. Rosalba, 70124 Bari

L. FONTANA, Dip.to di Mat., Pol. di Torino, Corso Duca degli Abruzzi 24, 10129 Torino

A. FRIGERIO, Dip.to di Sc. dell'Inf., Via Comelico 39-41, 20135 Milano

A. GARRONI, SISSA, Via Beirut 2-4, 34014 Trieste

G. GRILLO, Dip.to di Mat. e Inf., Via Zanon 6, 33100 Udine

H. HAMBLY, Stat. Lab., Univ. of Cambridge, 16 Mill Lane, Cambridge, CB2 1SB, U.K.

E.P. HSU, Max-Planck-Inst., Gottfried-Claren-Strasse 26, 5300 Bonn 3

P. IMKELLER, Math. Inst. der LMU, Theresienstr. 39, 8000 Munchen 2

S. JACQUOT, Dépt. de Math. et d'Inf., Univ. d'Orléans, BP 6759, 45067 Orléans Cedex 2

Y. KONDRATJEV, BIBOS, Dept. of Phys., Bielefeld Univ., 4800 Bielefeld 1

T. KUMAGAI, Statistical Lab., 16 Mill Lane, Cambridge CB2 1SB

B. LISENA, Dip.to di Mat., Univ., Via Orabona 5, 70125 Bari

Y. LU, Centro V. Volterra, Dip.to di Mat., Univ.

A. MALUSA, SISSA, Via Beirut 2-4, 34014 Trieste

P. MANSELLI, Ist. di Mat., Fac. di Arch., Piazza Brunelleschi 4, 50121 Firenze

L. NOTARANTONIO, SISSA, Via Beirut 2-4, 34014 Trieste

A. POSILICANO, Dept. of Math., Princeton Univ., Princeton, NJ 08544-1000

A. ROZKOSZ, Inst. Matematyki, Uniwersytet Mikolaja Kopernika, ul. Chopina 12/18, 87-100 Torun, Poland

E. SCACCIATELLI, Dip.to di Mat., Univ. La Sapienza, P.le Aldo Moro 2, 00185 Roma

S. SCARLATTI, Dip.to di Mat., Univ. de L'Aquila, Via Vetoio, 67010 Coppito, L'Aquila

B. SCHMULAND, Dept. of Stat. and Appl. Prob., Univ. of Alberta, Edmonton, Alberta, Canada T6G 2G1

R. SERAPIONI, Dip.to di Mat., Pol. di Milano, P.za L. da Vinci 32, 20133 Milano

K.-Th. STURM, Math. Inst., Bismarckstr. 1 1/2, D-W-8520 Erlangen

G. SWEERS, Dept. of Pure Math., P.O. Box 5021, 2600 GA Delft

G. TALENTI, Dip.to di Mat., Viale Morgagni 67/A, 50134 Firenze

A. TETA, Dip.to di Mat., Univ. La Sapienza, P.le Aldo Moro 2, 00185 Roma

L. TUBARO, Dip.to di Mat., Univ. di Trento, 38050 Povo, Trento

S. UGOLINI, Dip.to di Fisica "G.Galilei", Via F. Marzolo 8, 35131 Padova

M.A. VIVALDI, Dip.to di Metodi e Modelli Matematici, Univ. La Sapienza, Via A. Scarpa 10, 00161 Roma

T.-S. ZHANG, Dept. of Math., Univ. of Oslo, P.O.Box 1053, Blindern, N-0316 Oslo 3

FONDAZIONE C.I.M.E.
CENTRO INTERNAZIONALE MATEMATICO ESTIVO
INTERNATIONAL MATHEMATICAL SUMMER CENTER

"Integrable Systems and Quantum Groups"

is the subject of the First 1993 C.I.M.E. Session.

The Session, sponsored by the Consiglio Nazionale delle Ricerche and by the Ministero dell'Università e della Ricerca Scientifica e Tecnologica, will take place under the scientific direction of Professors Mauro FRANCAVIGLIA (Università di Torino), Silvio GRECO (Politecnico di Torino), Franco MAGRI (Università di Milano) at Villa "La Querceta", Montecatini Terme (Pistoia), **from June 14 to June 22, 1993.**

Courses

a) **Spectral covers, algebraically completely integrable Hamiltonian systems, and moduli of bundles.** (6 lectures in English)
Prof. Ron DONAGI (University of Pennsylvania)

Outline

Spectral covers allow a uniform treatment of a wide variety of algebraically completely integrable Hamiltonian systems, ranging from classical systems such as Jacobi's geodesic flow on an ellipsoid, to recent ones such as Hitchin's commuting flows on the cotangent bundle to the moduli space of stable vector bundles on a curve [H1], and Treibich-Verdier's theory of elliptic solitons [TV]. Our goal is to present an outline of this theory, together with some of the important special cases and applications. Topics to be discussed include:

. Construction of spectral covers, isotypic decomposition of their Picard varieties into generalized Pryms, the distinguished Prym [D2] and its modular interpretation via principal bundles with twisted endomorphisms (generalized 'Higgs bundles'), Kanev's Prym-Tyurin varieties [K] and the n-gonal constructions in Prym theory [D1], the structure of nilpotent cones [L] and its relation with fibers of the Springer resolution.

. Existence of symplectic and Poisson structures, considered both from the modular point of view (following [Ma], [Mu], [T]) and via their infinitesimal cubic invariant (compare [BG]).

. Existence of Lax structures linearizing a given system via spectral covers, and Griffiths' cohomological criterion for linearization [G].

. Examples and applications include:

- Jacobi's system and its generalizations by Beauville and by Adams, Harnad, Hurtubise and Previato.
- Elliptic (and abelian) solitons.
- Hitchin's system for an arbitrary reductive group, with its various applications to the structure of moduli spaces [H1] and to the projectively flat connection in Conformal Field Theory [H2].
- Some non-linear variants of Higgs bundles, living on Mukai spaces.

References

[BG] R. Bryant and P.A. Griffiths, Some observations on the infinitesimal period relations for regular threefolds with trivial canonical bundle, in: Arithmetic and Geometry II, Birkhäuser (1983), 77-102.

[D1] R. Donagi, The tetragonal construction, Bull. AMS 4 (1981), 181-185.

[D2] R. Donagi, Spectral covers, preprint, 1983.

[G] P.A. Griffiths, Linearizing flows and a cohomological interpretation of Lax equations, Amer. J. Math. 107 (1985), 1445-1484.

[H1] N. Hitchin, Stable bundles and integrable systems, Duke Math. J. 54 (1987), 91-114.

[H2] N. Hitchin, Flat connections and geometric quantization, Comm. Math. Phys. 131 (1990), 347-380.

[K] V. Kanev, Spectral curves, simple Lie algebras, and Prym-Tyurin varieties, Proc. Symp. Pure Math. 49 (1989, 627-645.
[L] G. Laumon, Un analogue global du cone nilpotent, Duke Math. J. 57 (1988), 647-671.
[Ma] E. Markman, Spectral curves and integrable systems, UPenn dissertation, 1992.
[Mu] S. Mukai Symplectic structure of the moduli space of sheaves on an abelian or K3 surface, Inv. Math. 77 (1984), 101-116.
[TV] A. Treibich and J.L. Verdier, Solitons elliptiques, The Grothendieck Festschrift, vol. 3, Birkhäuser (1990), 437-480.
[T] A. Tyurin, Symplectic structure on the varieties of moduli of vector bundles on an algebraic surface with p,g>0, Math. USSR Izv. 33 (1989), 139-117.

b) **Geometry of two-dimensional topological field theories.** (6 lectures in English).
Prof. Boris DUBROVIN (Moscow State University and SISSA, Trieste)

Lecture plan

1) Topological symmetric lagrangians and their quantization. Atiyah's axioms of a topological field theory (TFT). Intersection theory on moduli spaces as example of TFT. Topological conformal field theories (TCFT) as twisted N=2 susy theories. Topological deformations of a TCFT.

2) Equations of associativity of the primary chiral algebra as defining relations of a 2D TFT. Differential geometry of the small phase space of a TFT. Classification of massive TCFT by isomonodromy deformation method.

3) Integrable hierarchies associated with arbitrary 2D TFT, their hamiltonian formalism, solutions, and tau-functions. Coupling to topological gravity.

4) Ground state metric as a hermitian metric on the small phase space of a 2D TFT. Calculation of the ground state metric of a massive TCFT by isomonodromy deformations method. Relation to the theory of harmonic maps.

References

1. E. Witten, Surv. Diff. Geom. 1 (1991), 243.
2. R. Dijkgraaf, Intersection theory, integrable hierarchies, and topological field theory. Preprint IASSNS-HEP-91/91, to appear in the Proceedings of the Cargese Summer School on New Symmetry Principles in Quantum Field Theory (1991).
3. B. Dubrovin, Nucl. Phys. B379 (1992), 627.
4. B. Dubrovin, Integrable systems and classification of two-dimensional topological field theories, Preprint SISSA 162/92/FM, Semptember 1992, to appear in the J.-L. Verdier memorial volume, Integrable systems, 1992.
5. B. Dubrovin, Geometry and integrability of topological-antitopological fusion. Preprint INFN-8/92, April 1992, to appear in Comm. Math. Phys.

c) **Integrals of motion as cohomology classes.**. (6 lectures in English).
Prof.Edward FRENKEL (Harvard University)

Outline

Integrals of motion of Toda field theory can be interpreted as cohomology classes. For the classical theory they are cohomology classes of the nilpotent subalgebra of the corresponding finite-dimensional or affine Kac-Moody algebra. For the quantum theory they are cohomology classes of the quantized universal enveloping algebra of the nilpotent subalgebra. This definition makes possible to prove the existence of "big" algebras of integrals of motion in these theories, associated to finite-dimensional Lie algebras, these algebras are nothing but the W-algebras. For the affine Toda field theories, these algebras constitute infinite-dimensional abelian subalgebras of the W-algebras, and they are algebras of integrals of motion of certain deformations of conformal field theories.

References

- B. Feigin, E. Frenkel, Phys. Lett. B 276 (1992), 79-86.
- B. Feigin, E. Frenkel, Int. J. Mod. Phys. 7 (1992), Supplement 1A, 197-215.

d) **Integrable equations and moduli of curves and vector bundles.** (6 lectures in English)
Prof.Emma PREVIATO (Boston University)

Description

The theory of integrable systems/integrable equations of KdV type brought about profound interactions between physics and algebraic geometry over the past twenty years. This course will be an illustration of roughly three areas in the field and of open directions branching out of them: area one, the linearization of certain Hamiltonian flows over Jacobian varieties and generalizations to moduli spaces of vector bundles; two, moduli of special (elliptic) solutions; three, projective realizations of moduli spaces of vector bundles.
A list of topics follows.

Prerequisites

Classical Riemann-surface theory and rudiments of algebraic geometry.

Lecture I	:	Burchnall-Cauchy-Krichever map [ADCKP]
Lecture II	:	Generalization to vector bundles [KN]
Lecture II	:	The elliptic case [K], [M]
Lecture IV	:	The hyperelliptic case [VG]
Lecture V	:	The two-theta map [B]
Lecture VI	:	Verlinde formulas and Kummer varieties [vGP]

References

[ADCKP] E. Arbarello, C. De Concini, V.G. Kac and C. Procesi, Moduli spaces of curves and representation theory, Comm. Math. Phys. 117 (1988), 1-36.
[B] A. Beauville, Fibres de rang 2 sur une courbe, fibre determinant et fonctions theta, II, Bull. Soc. Math. France 119 (1991), 259-291.
[vG] B. van Geemen, Schottky-Jung relations and vector bundles hyperelliptic curves, Math. Ann. 281 (1988), 431-449.
[vGP] B. van Geemen and E. Previato, Prym varieties and the Verlinde formula, Math. Ann. (1933).
[K] I.M. Krichever, Elliptic solutions of the Kadomtsev-Petviashvili equation and integrable systems of particles, Functional Anal. Appl. 14 (1980), 282-290.
[KN] I.M. Krichever and S.P. Novikov, Holomorphic fiberings and nonlinear equations, Finite zone solutions of rank 2, Sov. Math. Dokl. 20 (1979), 650-654.
[M] O.I. Mokhov, Commuting differential operators of rank 3, and nonlinear differential equations, Math. USSR Izvestiya, 35 (1990), 629-655.

FONDAZIONE C.I.M.E.
CENTRO INTERNAZIONALE MATEMATICO ESTIVO
INTERNATIONAL MATHEMATICAL SUMMER CENTER

"Algebraic Cycles and Hodge Theories"

is the subject of the Second 1993 C.I.M.E. Session.

The Session, sponsored by the Consiglio Nazionale delle Ricerche and by the Ministero dell'Università e della Ricerca Scientifica e Tecnologica, will take place under the scientific direction of Prof. Fabio BARDELLI (Università di Pisa) at Villa Gualino, Torino, Italy, **from June 21 to June 29, 1993.**

C o u r s e s

a) **Infinitesimal methods in Hodge theory.** (8 lectures in English)
Prof. Mark GREEN (University of California, Los Angeles)

Lecture plan

1) The Hodge Theorem. Hodge decomposition and filtrations. The operators L, Λ and H, and the Hodge identities. Principle of two types. Degeneration of the Hodge-De Rham spectral sequence.

2) The Griffiths intermediate Jacobians, The Abel-Jacobi map. Infinitesimal Abel-Jacobi map and the extension class of the normal bundle sequence. Image of cycles algebraically equivalent to zero under the Abel-Jacobi map.

3) Variation of Hodge structure. The Hodge filtration varies analytically. The period map and its derivative. Infinitesimal period relations (Griffiths transversality). Griffiths computation of the infinitesimal period map as a cup product.

4) Hodge theory of hypersurfaces and complete intersections. Derivative of the period map for hypersurfaces. Infinitesimal Torelli for hypersurfaces and complete intersections. Examples of Hodge classes of cycles on hypersurfaces.

5) Mixed Hodge structures. Examples of extension classes.

6) Normal functions. Normal function associated to a primitive Deligne class. Analyticity and infinitesimal relation for normal functions. Infinitesimal invariant of normal function.

7) Koszul cohomology techniques in Hodge theory. Macaulay-Gotzmann theorem. Codimension of the Noether-Lefschetz locus for surfaces. Donagi's generic Torelli theorem for hypersurfaces. Vanishing of the infinitesimal invariant of normal functions for hypersurfaces of high degree.

8) Further applications of Koszul techniques. Nori's connectedness theorem. Abel-Jacobi map for general 3-fold of degree$\geq$ 6. Surjectivity of the general restriction map for rational Deligne cohomology, and the Poincaré-Lefschetz-Griffiths approach to the Hodge conjecture.

Suggested reading

- P. Griffiths and J. Harris, Principles of Algebraic Geometry, Chapters 0-1. This is a good source for the basic facts of Hodge theory, e.g. lecture 1.
- P. Griffiths, Topics in Transcendental Algebraic Geometry. The chapters (by various authors) include some useful surveys as well as more specialized research articles. Chapters I, III, XII, XIII, XIV, XVI, and XVII are probably the most helpful for this course.
- J. Carlson, M. Green, P. Griffiths, J. Harris, "Infinitesimal Variation of Hodge Structure", I-III. Compositio Math. 50 (1083), 109-

324. These contain a lot of information, including of course many interesting topics that won't be covered in these courses. Worth dipping into.
- M. Green, "Koszul cohomology and geometry", in "Lecture on Riemann Surfaces", Proceedings of the ICTP College on Riemann Surfaces, World Scientific 1989. This represents my best effort at an elementary exposition of the Hodge theory of hypersurfaces and Koszul-theoretic techniques. Sections 1,2 and 4 are relevant.
- M. Cornalba and P. Griffiths, "Some transcendental aspects of algebraic geometry" in "Algebraic Geometry - Arcata 1974", Proc. Symp. in Pure Math. 9, AMS (1975), 3-110. This has a relatively painless introduction to mixed Hodge structures. These lectures mostly deal with the differential-geometric aspects of the period map, a beautiful aspect of Hodge theory that we won't cover.
- J. Carlson and C. Peters have a new book on Hodge theory in the works. If it is available in time, it should be an outstanding introduction to many of the topics to be covered.

b) **Algebraic cycles and algebraic aspects of cohomology and K-theory.** (6 lectures in English).
Prof. Jacob MURRE (Universiteit Leiden)

The following subjects will be discussed:

1) Algebraic cycles.
Basic notions. Discussion of the most important equivalence relations. The Chow ring. The Griffiths group. Statement of the principal known facts in codimension greater than one. The definition of higher Chow groups of Bloch. Definition and main properties of Chern classes of vector bundles. The Grothendieck group of vector bundles and sheaves, and its relation to the Chow groups.

2) Deligne-Beilinson cohomology.
Definition and main properties. Examples. Construction of the cycle map; its relation to the classical cycle map and the Abel-Jacobi map.

3) Algebraic cycles and algebraic K-theory.
Introduction to algebraic K-theory. The functors K_0 (see also 1.), K_1 and K_2. The Bloch formula for the Chow groups. Discussion of the regulator map for $K_2(X)$ when X is an algebraic curve.

4) The Hodge Conjecture.
Statement of the (p,p)-conjecture. Survey of the typical known cases. Discussion of some examples. Statement of the generalized Hodge-conjecture as corrected by Grothendieck. Discussion of an example of Bardelli.

5) Some results in codimension 2.
a. Applications of the Merkurjev-Suslin theorem of algebraic K-theory.
b. Incidence equivalence and its relation to Abel-Jacobi equivalence.

6) Introduction to motives.
The standard conjectures and something about motives.

References

1. Bloch, S.: Lectures on algebraic cycles. Duke Univ. Math. Ser. IV, 1980.
2. Fulton, W.: Intersection theory. Erg. der Math., 3 Folge, Bd. 2. Springer Verlag, 1984.
3. Esnault, H. and Viehweg, E.: Deligne-Beilinson Cohomology. In: "Beilinson's conjectures and special values of L-function". Perspectives in Math., Vol. 4, Academic Press 1988.
4. Shioda, T.: What is known about the Hodge conjecture? In: Advances Studies in Pure Math., Vol. 1 Kinokuniya Comp. and North Holland, Tokyo 1983.
5. Murre, J.P.: Applications of algebraic K-theory to algebraic geometry. Proc. Conf. Alg. Geom. Sitges 1983, Springer LNM 1124.

c) **Transcendental methods in the study of algebraic cycles.** (8 lectures in English).
Prof. Claire VOISIN (Université de Orsay, Paris)

Outline of the lectures

1) Divisors.
Weil divisors. Cartiers divisors and line bundles; rational and linear equivalence; GAGA principle. The exponential exact sequence and its consequences:

- homological equivalence=algebraic equivalence for divisors
- the Lefschetz theorem on (1,1) classes; Neron-Severi group
- Hodge structure on H^1 and abelian varieties: the Picard variety
- the existence of Poincaré divisor

2) Topology and Hodge theory
Morse theory on affine varieties and the weak Lefschetz theorem. The Hodge index theorem. Consequences: The hard Lefschetz theorem and the Lefschetz decomposition. Applications:
- reduction to the primitive middle dimensional cohomology; degeneracy of Leray spectral sequences; semi-simplicity of the category of polarized Hodge structures.

3) Noether-Lefschetz locus
Deformations of Hodge classes. The Noether-Lefschetz loci; algebraicity of the components; local study (application of transversality of the period map to the codimension, infinitesimal description). Relation with the deformation theory of cycles; ghe semi-regularity property and Bloch-Kodaira theorem.

4) Monodromy
Nodal varieties. Lefschetz degenerations and Lefschetz pencils. Vanishing cycles and cones over them. The Picard-Lefschetz formula and applications of Noether-Lefschetz type. Discussion of the Hodge theory of the vanishing cycles on the central fibre.

5) O-cycles I
O-cycles and holomorphic forms on varieties; Mumford's theorem on the infinite dimensionality of the CH_0 group. Roitman's theorem: $CH^o{}_0$ finite dimensional $\Leftrightarrow$ $CH^o{}_0 \cong$ Alb. The Bloch's conjecture for surfaces and Bloch-Kas-Lieberman theorem.

6) O-cycles II
The proof of the Bloch conjecture for Godeaux type surfaces; Bloch-Srinivas theorem and consequences of "CH_0 small" on algebraic cycles and Hodge theory of a variety.

7) Griffiths group
One cycles on threefolds; Abel-Jacobi map on cycles algebraically equivalent to zero. Normal functions and their Hodge classes: The theorem of Griffiths. Statement of Clemens theorem and further examples.

8) Application of the NL locus to threefolds
M. Green's criterion for density of the Noether-Lefschetz locus. Applications to one-cycles on threefolds:
- parametrization of certain sub-Hodge structures by algebraic cycles
- infinitesimal proof of Clemens theorem
- generalization of Griffiths theorem to any Calabi-Yau threefold

References

- A. Weil: Variétés Kahleriennes, Actualités scientifiques et industrielles.
- J. Milnor: Morse theory, Annals of Math. Studies, Study 21, Princeton Univ. Press.
- Carlson, Green, Griffiths, Harris: Compositio Math. Vol. 50 (three articles).
- P. Deligne: Théorie de Hodge II, I.H.E.S. Publ. Math. 40, (1971), 5-58.
- P. Griffiths: On the periods of certain rational integrals I, II, Ann. of Math. 90 (1969), 460-541.
- P. Griffiths: Topics in transcendental algebraic geometry, Annals of Math. Studies, Study 106, Princeton Univ. Press.
- H. Clemens: Double solids, Advances in Math. Vol. 47 (1983).

FONDAZIONE C.I.M.E
CENTRO INTERNAZIONALE MATEMATICO ESTIVO
INTERNATIONAL MATHEMATICAL SUMMER CENTER

"Modelling and Analysis of Phase Transition and Hysteresis Phenomena"

is the subject of the Third 1993 C.I.M.E. Session.

The Session, sponsored by the Consiglio Nazionale delle Ricerche and by the Ministero dell'Università e della Ricerca Scientifica e Tecnologica, will take place under the scientific direction of Prof. Augusto VISINTIN (Università di Trento) at Villa "La Querceta", Montecatini Terme (Pistoia), **from July 13 to July 21, 1993.**

Courses

a) **Hysteresis operators.** (6 lectures in English)
Prof. Martin BROKATE (Universität Kaiserslautern)

Course outline

1) Scalar hysteresis operators.
Example of hysteresis models. Hysteresis operators. Continuity properties.
Memory properties. Applications.

2) Vector hysteresis operators

3) Hysteresis operators and differential equations.
Ordinary differential equations with hysteresis. Parabolic equations with hysteresis. Hyperbolic equations with hysteresis. Shape memory alloys.
Control problems with hysteresis.

References

1. Books:
- Brokate, M.: Optimal control of ordinary differential equations with nonlinearities of hysteresis type. Peter Lang Verlag, Frankfurt 1987. (In German; English translation in: Automation and Remote Control, 52 (1991) and 53 (1992)).
- Krasnoselskii, M.A., Pokrovskii, A.V.: Systems with hysteresis. Springer 1969.
- Mayergoyz, I.D.: Mathematical models of hysteresis. Springer 1991.
2. Survey:
- Visintin, A.: Mathematical models of hysteresis. In: Topics in nonsmooth mechanics (eds. J. J. Moreau, P.D. Panagiotopoulos, G. Strang), Birkhäuser 1988, 295-326.
3. Papers:
- Brokate, M., Visintin, A.: Properties of the Preisach model for hysteresis, J. Reine Angew. Math. 402 (1989), 1-40.
- Krejcí, P.: A Monotonicity method for solving hyperbolic problems with hysteresis. Apl. Mat. 33 (1988), 197-203.
- Krejcí, P.: Hysteresis memory preserving operators. Applications of Math. 36 (1991), 305-326.
- Krejcí, P.: Vector hysteresis models. European J. Appl. Math. 22 (1991), 281-292.
- Visintin, A.: A model for hysteresis of distributed systems. Ann. Mat. Pura Appl. 131 (1982), 203-231.
- Visintin, A.: Rheological models and hysteresis effects. Rend. Sem. Mat. Univ. Padova 77 (1987), 213-243.

b) **Systems of nonlinear PDEs arising from dynamical phase transition.** (6 lectures in English).
Prof. Nobuyuki KENMOCHI (Chiba University)

Outline of the contents

Systems of nonlinear PDEs are proposed as mathematical models for thermodynamical phase transition processes such as solidification and melting in solid-liquid systems. These are nonlinear parabolic PDEs and variational inequalities with obstacles and the unknowns are the absolute temperature and the order parameter representing the physical situation of the materials. We analyze these models from the following points (1)-(4) of view:

(1) Physical background of the problem
(2) Abstract treatment of the problem
(3) Existence and uniqueness results
(4) Asymptotic stability for the solutions

The basic literature references for the subjects

Nonlinear PDEs:
- D. Gilbarg and N. S. Trudinger, Elliptic Partial Differential Equations of Second Order, Springer-Verlag, Berlin, 1983.
- H. Brézis, Problémes unilatéraux, J. Math. pures appl., 51 (1972), 1-168.

Convex Analysis:
- J. L. Lions, Quelques méthodes de résolution des problémes aux limites non linéaires, Dunod, Gauthier-Villars, Paris, 1969.
- H. Brézis, Opérateurs maximaux monotones et semi-groupes de contractions dans les espaces de Hilbert, North-Holland, Amsterdam, 1973.

c) **Quasiplasticity and Pseudoelasticity in Shape Memory Alloys.** (6 lectures in English).
Prof. Ingo MÜLLER (Technical University Berlin)

Course outline

1. Phenomena.
The phenomena of quasiplasticity and pseudoelasticity in shape memory alloys are described and documented. They are due to a martensitic-austenitic phase transition and to the twinning of the martensitic phase, which is the lowtemperature phase.

2. Model.
A strctural model is introduced wihich is capable of simulating the observed phenomena. The model consists of lattice layers in a potential which has three potential wells, one metastable. Adjacent layers are coherent and their formation requires an extra energy, the coherency energy.

3. Statistical Mechanics.
Statistical Mechanics of the model provides a non-convex free energy and - consequently - a nonmonotone load deformation curve. This is appropriate for pseudoelasticiy. The proper description of quasiplasticity requires a kinetic theory of the model, akin to the theory of activated processes in chemistry.

4. Hysteresis.
Minimization of the free energy under constant deformation leads us to conclude that the observed hysteresis in the pseudo-elastic range is due to the coherency energy. The phase equilibria are unstable and this explains the occurence of internal yield and recovery in pseudoelasticity. A simple mathematical construct for the non-convex free energy permits the description of many observed phenomena inside the hysteresis loop.

5. Thermodynamics.
A systematic exploitation of the first and second law of thermodynamics allows us to predict the thermal and caloric side effects of pseudoelastic deformation.

6. Metastability.
The nature of the metastable states inside the hysteresis loop is as yet not well understood. But there are partial results. They concern observations of the number of interfaces during the phase transition and the role of a "fluctuation temperature" which activates the body to the extent that its entropy can approach its maximum value.

d) **Variational methods in the Stefan problem.** (6 lectures in English)
Prof. José Francisco RODRIGUES (CMAF/Universidade de Lisboa)

Outline of contents

The Stefan problem is one of the simplest possible macroscopic models for phase changes in a pure material when they occur either by heat conduction or diffusion. Its history provides a helpful example of the interplay between free boundary problems and the real world. This course intends to introduce this model problem and to develop an exposition of the variational methods applied to the study of weak solutions for multidimensional problems.

Plan:

1. Introduction to the mathematical-physics models
2. Analysis of the one-phase problem via variational inequalities I
3. Analysis of the two-phase problem via variational inequalities II
4. Study of the enthalpy formulation via Galerkin method
5. Analysis of more complex Stefan problems

Some basic literature

1. G. Duvaut & J. L. Lions, Les inéquations en mécanique et en physique, Dunod, Paris, 1972 (English transl. Springer, Berlin, 1976).
2. A. Friedman, Variational principles and free boundary value problems, Wiley, New York, 1982.
3. D. Kinderlehrer & G. Stampacchia, An introduction to variational inequalities and their application, Academic Press, New York, 1980.
4. J. L. Lions, Sur quelques questions d'analyse, de mécanique et de control optimal, Press Univ. Montréal, 1976.
5. A. M. Mermanov, The Stefan problem, W. De Gruyter, Berlin, 1992.
6. I. Pawlow, Analysis and control of evolution multiphase problems with free boundaries, Polska Akad. Nauk, Warszawa, 1987.
7. J. F. Rodrigues, Obstacle problems in mathematical physics, North-Holland, Amsterdam, 1987.
8. J. F. Rodrigues (Editor), Mathematical models for phase change problems, ISNM n. 88, Birkhäuser, Basel, 1989.
9. E. Zeidler, Nonlinear functional analysis and its application, Vol. II/B, Nonlinear monotone operators, Springer Verlag, New York, 1990.

e) **Numerical aspects of free boundary and hysteresis problems.** (6 lectures in English).
Prof. Claudio VERDI (Università di Pavia)

Summary

1. Time discretization of strongly nonlinear parabolic equations
   - 1.1 Nonlinear methods
   - 1.2 Linear methods
   - 1.3 Applications to problem with hysteresis

2. Full discretization
   - 2.1 Finite element spaces
   - 2.2 Nonlinear schemes
   - 2.3 Linear schemes
   - 2.4 Stability of fully discrete schemes
   - 2.5 Error estimates
   - 2.6 Approximation of free boundaries

3. Adaptive finite element methods for parabolic free boundary problems

Basic references

1. P.G. Ciarlet, The finite element method for elliptic problems, North-Holland, Amsterdam, 1978.
2. V. Thomee, Galerkin Finite Element Methods for Parabolic Problems, Lecture Notes in Mathematics 1054, Sprin ger Verlag, Berlin, 1984.
3. J. M. Ortega and W. C. Rheinboldt, Iterative Solution of Nonlinear Equations in Several Variables, Academic Press, New York, 1970.
4. R. H. Nochetto, Finite element methods for parabolic free boundary problems, in: Advances in Numerical Analysis, Vol. I: Nonlinear Partial Differential Equations and Dynamical Systems, Oxford Academic Press, 1991, 34-95.

5. R. H. Nochetto and C. Verdi, Approximation of degenerate parabolic problems using numerical integration, SIAM J. Numer. Anal., 25 (1988), 784-814.
6. R. H. Nochetto and C. Verdi, An efficient linear scheme to approximate parabolic free boundary problems: error estimates and implementation, Math. Comp., 51 (1988), 27-53.
7. R. H. Nochetto, M. Paolini and C. Verdi, Adaptive finite element method for the two-phase Stefan problem in two space dimension. Part I: Stability and error estimates, Math. Comp., 57 (1991), 73-108); Supplement, Math. Comp. 57 (1991), S1-S11.

LIST OF C.I.M.E. SEMINARS Publisher

1954 -	1. Analisi funzionale	C.I.M.E.
	2. Quadratura delle superficie e questioni connesse	"
	3. Equazioni differenziali non lineari	"
1955 -	4. Teorema di Riemann-Roch e questioni connesse	"
	5. Teoria dei numeri	"
	6. Topologia	"
	7. Teorie non linearizzate in elasticità, idrodinamica,aerodinamica	"
	8. Geometria proiettivo-differenziale	"
1956 -	9. Equazioni alle derivate parziali a caratteristiche reali	"
	10. Propagazione delle onde elettromagnetiche	"
	11. Teoria della funzioni di più variabili complesse e delle funzioni automorfe	"
1957 -	12. Geometria aritmetica e algebrica (2 vol.)	"
	13. Integrali singolari e questioni connesse	"
	14. Teoria della turbolenza (2 vol.)	"
1958 -	15. Vedute e problemi attuali in relatività generale	"
	16. Problemi di geometria differenziale in grande	"
	17. Il principio di minimo e le sue applicazioni alle equazioni funzionali	"
1959 -	18. Induzione e statistica	"
	19. Teoria algebrica dei meccanismi automatici (2 vol.)	"
	20. Gruppi, anelli di Lie e teoria della coomologia	"
1960 -	21. Sistemi dinamici e teoremi ergodici	"
	22. Forme differenziali e loro integrali	"
1961 -	23. Geometria del calcolo delle variazioni (2 vol.)	"
	24. Teoria delle distribuzioni	"
	25. Onde superficiali	"
1962 -	26. Topologia differenziale	"
	27. Autovalori e autosoluzioni	"
	28. Magnetofluidodinamica	"

1963 -	29. Equazioni differenziali astratte	"
	30. Funzioni e varietà complesse	"
	31. Proprietà di media e teoremi di confronto in Fisica Matematica	"
1964 -	32. Relatività generale	"
	33. Dinamica dei gas rarefatti	"
	34. Alcune questioni di analisi numerica	"
	35. Equazioni differenziali non lineari	"
1965 -	36. Non-linear continuum theories	"
	37. Some aspects of ring theory	"
	38. Mathematical optimization in economics	"
1966 -	39. Calculus of variations	Ed. Cremonese, Firenze
	40. Economia matematica	"
	41. Classi caratteristiche e questioni connesse	"
	42. Some aspects of diffusion theory	"
1967 -	43. Modern questions of celestial mechanics	"
	44. Numerical analysis of partial differential equations	"
	45. Geometry of homogeneous bounded domains	"
1968 -	46. Controllability and observability	"
	47. Pseudo-differential operators	"
	48. Aspects of mathematical logic	"
1969 -	49. Potential theory	"
	50. Non-linear continuum theories in mechanics and physics and their applications	"
	51. Questions of algebraic varieties	"
1970 -	52. Relativistic fluid dynamics	"
	53. Theory of group representations and Fourier analysis	"
	54. Functional equations and inequalities	"
	55. Problems in non-linear analysis	"
1971 -	56. Stereodynamics	"
	57. Constructive aspects of functional analysis (2 vol.)	"
	58. Categories and commutative algebra	"

1972 -	59. Non-linear mechanics		"
	60. Finite geometric structures and their applications		"
	61. Geometric measure theory and minimal surfaces		"
1973 -	62. Complex analysis		"
	63. New variational techniques in mathematical physics		"
	64. Spectral analysis		"
1974 -	65. Stability problems		"
	66. Singularities of analytic spaces		"
	67. Eigenvalues of non linear problems		"
1975 -	68. Theoretical computer sciences		"
	69. Model theory and applications		"
	70. Differential operators and manifolds		"
1976 -	71. Statistical Mechanics		Ed Liguori, Napoli
	72. Hyperbolicity		"
	73. Differential topology		"
1977 -	74. Materials with memory		"
	75. Pseudodifferential operators with applications		"
	76. Algebraic surfaces		"
1978 -	77. Stochastic differential equations		"
	78. Dynamical systems		Ed Liguori, Napoli and Birhäuser Verlag
1979 -	79. Recursion theory and computational complexity		"
	80. Mathematics of biology		"
1980 -	81. Wave propagation		"
	82. Harmonic analysis and group representations		"
	83. Matroid theory and its applications		"
1981 -	84. Kinetic Theories and the Boltzmann Equation	(LNM 1048)	Springer-Verlag
	85. Algebraic Threefolds	(LNM 947)	"
	86. Nonlinear Filtering and Stochastic Control	(LNM 972)	"
1982 -	87. Invariant Theory	(LNM 996)	"
	88. Thermodynamics and Constitutive Equations	(LN Physics 228)	"
	89. Fluid Dynamics	(LNM 1047)	"

1983 -	90.	Complete Intersections	(LNM 1092)	"
	91.	Bifurcation Theory and Applications	(LNM 1057)	"
	92.	Numerical Methods in Fluid Dynamics	(LNM 1127)	"
1984 -	93.	Harmonic Mappings and Minimal Immersions	(LNM 1161)	"
	94.	Schrödinger Operators	(LNM 1159)	"
	95.	Buildings and the Geometry of Diagrams	(LNM 1181)	"
1985 -	96.	Probability and Analysis	(LNM 1206)	"
	97.	Some Problems in Nonlinear Diffusion	(LNM 1224)	"
	98.	Theory of Moduli	(LNM 1337)	"
1986 -	99.	Inverse Problems	(LNM 1225)	"
	100.	Mathematical Economics	(LNM 1330)	"
	101.	Combinatorial Optimization	(LNM 1403)	"
1987 -	102.	Relativistic Fluid Dynamics	(LNM 1385)	"
	103.	Topics in Calculus of Variations	(LNM 1365)	"
1988 -	104.	Logic and Computer Science	(LNM 1429)	"
	105.	Global Geometry and Mathematical Physics	(LNM 1451)	"
1989 -	106.	Methods of nonconvex analysis	(LNM 1446)	"
	107.	Microlocal Analysis and Applications	(LNM 1495)	"
1990 -	108.	Geoemtric Topology: Recent Developments	(LNM 1504)	"
	109.	H_∞ Control Theory	(LNM 1496)	"
	110.	Mathematical Modelling of Industrical Processes	(LNM 1521)	"
1991 -	111.	Topological Methods for Ordinary Differential Equations	(LNM 1537)	"
	112.	Arithmetic Algebraic Geometry	(LNM 1553)	"
	113.	Transition to Chaos in Classical and Quantum Mechanics	to appear	"
1992 -	114.	Dirichlet Forms	(LNM 1563)	"
	115.	D-Modules and Representation Theory	to appear	"
	116.	Nonequilibrium Problems in Many-Particle Systems	(LNM 1551)	"

1993 -	117. Integrable Systems and Quantum Groups	to appear	"
	118. Algebraic Cycles and Hodge Theories	to appear	"
	119. Modelling and Analysis of Phase Transition and Hysteresis Phenomena	to appear	"

Printing: Weihert-Druck GmbH, Darmstadt
Binding: Buchbinderei Schäffer, Grünstadt

Vol. 1472: T. T. Nielsen, Bose Algebras: The Complex and Real Wave Representations. V, 132 pages. 1991.

Vol. 1473: Y. Hino, S. Murakami, T. Naito, Functional Differential Equations with Infinite Delay. X, 317 pages. 1991.

Vol. 1474: S. Jackowski, B. Oliver, K. Pawałowski (Eds.), Algebraic Topology, Poznań 1989. Proceedings. VIII, 397 pages. 1991.

Vol. 1475: S. Busenberg, M. Martelli (Eds.), Delay Differential Equations and Dynamical Systems. Proceedings, 1990. VIII, 249 pages. 1991.

Vol. 1476: M. Bekkali, Topics in Set Theory. VII, 120 pages. 1991.

Vol. 1477: R. Jajte, Strong Limit Theorems in Noncommutative L_2-Spaces. X, 113 pages. 1991.

Vol. 1478: M.-P. Malliavin (Ed.), Topics in Invariant Theory. Seminar 1989-1990. VI, 272 pages. 1991.

Vol. 1479: S. Bloch, I. Dolgachev, W. Fulton (Eds.), Algebraic Geometry. Proceedings, 1989. VII, 300 pages. 1991.

Vol. 1480: F. Dumortier, R. Roussarie, J. Sotomayor, H. Żoładek, Bifurcations of Planar Vector Fields: Nilpotent Singularities and Abelian Integrals. VIII, 226 pages. 1991.

Vol. 1481: D. Ferus, U. Pinkall, U. Simon, B. Wegner (Eds.), Global Differential Geometry and Global Analysis. Proceedings, 1991. VIII, 283 pages. 1991.

Vol. 1482: J. Chabrowski, The Dirichlet Problem with L^2-Boundary Data for Elliptic Linear Equations. VI, 173 pages. 1991.

Vol. 1483: E. Reithmeier, Periodic Solutions of Nonlinear Dynamical Systems. VI, 171 pages. 1991.

Vol. 1484: H. Delfs, Homology of Locally Semialgebraic Spaces. IX, 136 pages. 1991.

Vol. 1485: J. Azéma, P. A. Meyer, M. Yor (Eds.), Séminaire de Probabilités XXV. VIII, 440 pages. 1991.

Vol. 1486: L. Arnold, H. Crauel, J.-P. Eckmann (Eds.), Lyapunov Exponents. Proceedings, 1990. VIII, 365 pages. 1991.

Vol. 1487: E. Freitag, Singular Modular Forms and Theta Relations. VI, 172 pages. 1991.

Vol. 1488: A. Carboni, M. C. Pedicchio, G. Rosolini (Eds.), Category Theory. Proceedings, 1990. VII, 494 pages. 1991.

Vol. 1489: A. Mielke, Hamiltonian and Lagrangian Flows on Center Manifolds. X, 140 pages. 1991.

Vol. 1490: K. Metsch, Linear Spaces with Few Lines. XIII, 196 pages. 1991.

Vol. 1491: E. Lluis-Puebla, J.-L. Loday, H. Gillet, C. Soulé, V. Snaith, Higher Algebraic K-Theory: an overview. IX, 164 pages. 1992.

Vol. 1492: K. R. Wicks, Fractals and Hyperspaces. VIII, 168 pages. 1991.

Vol. 1493: E. Benoît (Ed.), Dynamic Bifurcations. Proceedings, Luminy 1990. VII, 219 pages. 1991.

Vol. 1494: M.-T. Cheng, X.-W. Zhou, D.-G. Deng (Eds.), Harmonic Analysis. Proceedings, 1988. IX, 226 pages. 1991.

Vol. 1495: J. M. Bony, G. Grubb, L. Hörmander, H. Komatsu, J. Sjöstrand, Microlocal Analysis and Applications. Montecatini Terme, 1989. Editors: L. Cattabriga, L. Rodino. VII, 349 pages. 1991.

Vol. 1496: C. Foias, B. Francis, J. W. Helton, H. Kwakernaak, J. B. Pearson, H_∞-Control Theory. Como, 1990. Editors: E. Mosca, L. Pandolfi. VII, 336 pages. 1991.

Vol. 1497: G. T. Herman, A. K. Louis, F. Natterer (Eds.), Mathematical Methods in Tomography. Proceedings 1990. X, 268 pages. 1991.

Vol. 1498: R. Lang, Spectral Theory of Random Schrödinger Operators. X, 125 pages. 1991.

Vol. 1499: K. Taira, Boundary Value Problems and Markov Processes. IX, 132 pages. 1991.

Vol. 1500: J.-P. Serre, Lie Algebras and Lie Groups. VII, 168 pages. 1992.

Vol. 1501: A. De Masi, E. Presutti, Mathematical Methods for Hydrodynamic Limits. IX, 196 pages. 1991.

Vol. 1502: C. Simpson, Asymptotic Behavior of Monodromy. V, 139 pages. 1991.

Vol. 1503: S. Shokranian, The Selberg-Arthur Trace Formula (Lectures by J. Arthur). VII, 97 pages. 1991.

Vol. 1504: J. Cheeger, M. Gromov, C. Okonek, P. Pansu, Geometric Topology: Recent Developments. Editors: P. de Bartolomeis, F. Tricerri. VII, 197 pages. 1991.

Vol. 1505: K. Kajitani, T. Nishitani, The Hyperbolic Cauchy Problem. VII, 168 pages. 1991.

Vol. 1506: A. Buium, Differential Algebraic Groups of Finite Dimension. XV, 145 pages. 1992.

Vol. 1507: K. Hulek, T. Peternell, M. Schneider, F.-O. Schreyer (Eds.), Complex Algebraic Varieties. Proceedings, 1990. VII, 179 pages. 1992.

Vol. 1508: M. Vuorinen (Ed.), Quasiconformal Space Mappings. A Collection of Surveys 1960-1990. IX, 148 pages. 1992.

Vol. 1509: J. Aguadé, M. Castellet, F. R. Cohen (Eds.), Algebraic Topology - Homotopy and Group Cohomology. Proceedings, 1990. X, 330 pages. 1992.

Vol. 1510: P. P. Kulish (Ed.), Quantum Groups. Proceedings, 1990. XII, 398 pages. 1992.

Vol. 1511: B. S. Yadav, D. Singh (Eds.), Functional Analysis and Operator Theory. Proceedings, 1990. VIII, 223 pages. 1992.

Vol. 1512: L. M. Adleman, M.-D. A. Huang, Primality Testing and Abelian Varieties Over Finite Fields. VII, 142 pages. 1992.

Vol. 1513: L. S. Block, W. A. Coppel, Dynamics in One Dimension. VIII, 249 pages. 1992.

Vol. 1514: U. Krengel, K. Richter, V. Warstat (Eds.), Ergodic Theory and Related Topics III, Proceedings, 1990. VIII, 236 pages. 1992.

Vol. 1515: E. Ballico, F. Catanese, C. Ciliberto (Eds.), Classification of Irregular Varieties. Proceedings, 1990. VII, 149 pages. 1992.

Vol. 1516: R. A. Lorentz, Multivariate Birkhoff Interpolation. IX, 192 pages. 1992.

Vol. 1517: K. Keimel, W. Roth, Ordered Cones and Approximation. VI, 134 pages. 1992.

Vol. 1518: H. Stichtenoth, M. A. Tsfasman (Eds.), Coding Theory and Algebraic Geometry. Proceedings, 1991. VIII, 223 pages. 1992.

Vol. 1519: M. W. Short, The Primitive Soluble Permutation Groups of Degree less than 256. IX, 145 pages. 1992.

Vol. 1520: Yu. G. Borisovich, Yu. E. Gliklikh (Eds.), Global Analysis – Studies and Applications V. VII, 284 pages. 1992.

Vol. 1521: S. Busenberg, B. Forte, H. K. Kuiken, Mathematical Modelling of Industrial Process. Bari, 1990. Editors: V. Capasso, A. Fasano. VII, 162 pages. 1992.

Vol. 1522: J.-M. Delort, F. B. I. Transformation. VII, 101 pages. 1992.

Vol. 1523: W. Xue, Rings with Morita Duality. X, 168 pages. 1992.

Vol. 1524: M. Coste, L. Mahé, M.-F. Roy (Eds.), Real Algebraic Geometry. Proceedings, 1991. VIII, 418 pages. 1992.

Vol. 1525: C. Casacuberta, M. Castellet (Eds.), Mathematical Research Today and Tomorrow. VII, 112 pages. 1992.

Vol. 1526: J. Azéma, P. A. Meyer, M. Yor (Eds.), Séminaire de Probabilités XXVI. X, 633 pages. 1992.

Vol. 1527: M. I. Freidlin, J.-F. Le Gall, Ecole d'Eté de Probabilités de Saint-Flour XX – 1990. Editor: P. L. Hennequin. VIII, 244 pages. 1992.

Vol. 1528: G. Isac, Complementarity Problems. VI, 297 pages. 1992.

Vol. 1529: J. van Neerven, The Adjoint of a Semigroup of Linear Operators. X, 195 pages. 1992.

Vol. 1530: J. G. Heywood, K. Masuda, R. Rautmann, S. A. Solonnikov (Eds.), The Navier-Stokes Equations II – Theory and Numerical Methods. IX, 322 pages. 1992.

Vol. 1531: M. Stoer, Design of Survivable Networks. IV, 206 pages. 1992.

Vol. 1532: J. F. Colombeau, Multiplication of Distributions. X, 184 pages. 1992.

Vol. 1533: P. Jipsen, H. Rose, Varieties of Lattices. X, 162 pages. 1992.

Vol. 1534: C. Greither, Cyclic Galois Extensions of Commutative Rings. X, 145 pages. 1992.

Vol. 1535: A. B. Evans, Orthomorphism Graphs of Groups. VIII, 114 pages. 1992.

Vol. 1536: M. K. Kwong, A. Zettl, Norm Inequalities for Derivatives and Differences. VII, 150 pages. 1992.

Vol. 1537: P. Fitzpatrick, M. Martelli, J. Mawhin, R. Nussbaum, Topological Methods for Ordinary Differential Equations. Montecatini Terme, 1991. Editors: M. Furi, P. Zecca. VII, 218 pages. 1993.

Vol. 1538: P.-A. Meyer, Quantum Probability for Probabilists. X, 287 pages. 1993.

Vol. 1539: M. Coornaert, A. Papadopoulos, Symbolic Dynamics and Hyperbolic Groups. VIII, 138 pages. 1993.

Vol. 1540: H. Komatsu (Ed.), Functional Analysis and Related Topics, 1991. Proceedings. XXI, 413 pages. 1993.

Vol. 1541: D. A. Dawson, B. Maisonneuve, J. Spencer, Ecole d' Eté de Probabilités de Saint-Flour XXI - 1991. Editor: P. L. Hennequin. VIII, 356 pages. 1993.

Vol. 1542: J.Fröhlich, Th.Kerler, Quantum Groups, Quantum Categories and Quantum Field Theory. VII, 431 pages. 1993.

Vol. 1543: A. L. Dontchev, T. Zolezzi, Well-Posed Optimization Problems. XII, 421 pages. 1993.

Vol. 1544: M.Schürmann, White Noise on Bialgebras. VII, 146 pages. 1993.

Vol. 1545: J. Morgan, K. O'Grady, Differential Topology of Complex Surfaces. VIII, 224 pages. 1993.

Vol. 1546: V. V. Kalashnikov, V. M. Zolotarev (Eds.), Stability Problems for Stochastic Models. Proceedings, 1991. VIII, 229 pages. 1993.

Vol. 1547: P. Harmand, D. Werner, W. Werner, M-ideals in Banach Spaces and Banach Algebras. VIII, 387 pages. 1993.

Vol. 1548: T. Urabe, Dynkin Graphs and Quadrilateral Singularities. VI, 233 pages. 1993.

Vol. 1549: G. Vainikko, Multidimensional Weakly Singular Integral Equations. XI, 159 pages. 1993.

Vol. 1550: A. A. Gonchar, E. B. Saff (Eds.), Methods of Approximation Theory in Complex Analysis and Mathematical Physics IV, 222 pages, 1993.

Vol. 1551: L. Arkeryd, P. L. Lions, P.A. Markowich, S.R. S. Varadhan. Nonequilibrium Problems in Many-Particle Systems. Montecatini, 1992. Editors: C. Cercignani, M. Pulvirenti. VII, 158 pages 1993.

Vol. 1552: J. Hilgert, K.-H. Neeb, Lie Semigroups and their Applications. XII, 315 pages. 1993.

Vol. 1553: J.-L- Colliot-Thélène, J. Kato, P. Vojta. Arithmetic Algebraic Geometry. Trento, 1991. Editor: E. Ballico. VII, 223 pages. 1993.

Vol. 1554: A. K. Lenstra, H. W. Lenstra, Jr. (Eds.), The Development of the Number Field Sieve. VIII, 131 pages. 1993.

Vol. 1555: O. Liess, Conical Refraction and Higher Microlocalization. X, 389 pages. 1993.

Vol. 1556: S. B. Kuksin, Nearly Integrable Infinite-Dimensional Hamiltonian Systems. XXVII, 101 pages. 1993.

Vol. 1557: J. Azéma, P. A. Meyer, M. Yor (Eds.), Séminaire de Probabilités XXVII. VI, 327 pages. 1993.

Vol. 1558: T. J. Bridges, J. E. Furter, Singularity Theory and Equivariant Symplectic Maps. VI, 226 pages. 1993.

Vol. 1559: V. G. Sprindžuk, Classical Diophantine Equations. XII, 228 pages. 1993.

Vol. 1560: T. Bartsch, Topological Methods for Variational Problems with Symmetries. X, 152 pages. 1993.

Vol. 1561: I. S. Molchanov, Limit Theorems for Unions of Random Closed Sets. X, 157 pages. 1993.

Vol. 1562: G. Harder, Eisensteinkohomologie und die Konstruktion gemischter Motive. XX, 184 pages. 1993.

Vol. 1563: E. Fabes, M. Fukushima, L. Gross, C. Kenig, M. Röckner, D. W. Stroock, Dirichlet Forms. Varenna, 1992. Editors: G. Dell'Antonio, U. Mosco. VII, 245 pages. 1993.